VOLUME NINETY SIX

ADVANCES IN PROTEIN CHEMISTRY AND STRUCTURAL BIOLOGY

Biomolecular Modelling and Simulations

VOLUME NINETY SIX

Advances in
PROTEIN CHEMISTRY AND STRUCTURAL BIOLOGY
Biomolecular Modelling and Simulations

Edited by

TATYANA KARABENCHEVA-CHRISTOVA

Department of Applied Sciences
Northumbria University at Newcastle
Newcastle-upon-Tyne, United Kingdom

ELSEVIER

AMSTERDAM • BOSTON • HEIDELBERG • LONDON
NEW YORK • OXFORD • PARIS • SAN DIEGO
SAN FRANCISCO • SINGAPORE • SYDNEY • TOKYO
Academic Press is an imprint of Elsevier

Academic Press is an imprint of Elsevier
32 Jamestown Road, London NW1 7BY, UK
525 B Street, Suite 1800, San Diego, CA 92101-4495, USA
225 Wyman Street, Waltham, MA 02451, USA
The Boulevard, Langford Lane, Kidlington, Oxford OX5 1GB, UK

First edition 2014

ISBN: 978-0-12-800013-7
ISSN: 1876-1623

CONTENTS

CONTRIBUTORS

Workalemahu M. Berhanu
Department of Chemistry & Biochemistry, University of Oklahoma, Norman, Oklahoma, USA

Dhruva K. Chakravorty
Department of Chemistry, University of New Orleans, New Orleans, Louisiana, USA

Xi Cheng
Department of Molecular Biosciences and Center for Bioinformatics, The University of Kansas, Lawrence, Kansas, USA

G. Andrés Cisneros
Department of Chemistry, Wayne State University, Detroit, Michigan, USA

Matteo Thomas Degiacomi*
Laboratory for Biomolecular Modeling, Institute of Bioengineering, School of Life Sciences, École Polytechnique Fédérale de Lausanne—EPFL, Lausanne, Switzerland

Angela A. Elias
Department of Chemistry, Wayne State University, Detroit, Michigan, USA

Michael Feig
Department of Biochemistry and Molecular Biology, and Department of Chemistry, Michigan State University, East Lansing, Michigan, USA

Wei Han
Beckman Institute and Center for Biophysics and Computational Biology, University of Illinois at Urbana–Champaign, Urbana, Illinois, USA

Ulrich H.E. Hansmann
Department of Chemistry & Biochemistry, University of Oklahoma, Norman, Oklahoma, USA

Lei Huang
Department of Biochemistry and Molecular Biology, Gordon Center for Integrative Science, University of Chicago, Chicago, Illinois, USA

Wonpil Im
Department of Molecular Biosciences and Center for Bioinformatics, The University of Kansas, Lawrence, Kansas, USA

Shahidul M. Islam
Department of Biochemistry and Molecular Biology, Gordon Center for Integrative Science, University of Chicago, Chicago, Illinois, USA

*Present address: Physical and Theoretical Chemistry Laboratory, South Parks Road, Oxford, United Kingdom

Sunhwan Jo
Department of Biochemistry and Molecular Biology, Gordon Center for Integrative Science, University of Chicago, Chicago, Illinois, USA

Parimal Kar
Department of Biochemistry and Molecular Biology, Michigan State University, East Lansing, Michigan, USA

Hui Sun Lee
Department of Molecular Biosciences and Center for Bioinformatics, The University of Kansas, Lawrence, Kansas, USA

Alessio Lodola
Dipartimento di Farmacia, Università degli Studi di Parma, Parma, Italy

Alexander D. MacKerell Jr.
Department of Pharmaceutical Sciences, School of Pharmacy, University of Maryland, Baltimore, Maryland, USA

Karina Martinez-Mayorga
Instituto de Química, Universidad Nacional Autónoma de México, Mexico City, Mexico

José L. Medina-Franco
Mayo Clinic, Scottsdale, Arizona, USA

Oscar Méndez-Lucio
Unilever Centre for Molecular Science Informatics, Department of Chemistry, University of Cambridge, Cambridge, United Kingdom

Kenneth M. Merz Jr.
Department of Chemistry, and Department of Biochemistry and Molecular Biology, Michigan State University, East Lansing, Michigan, USA

Marco Mor
Dipartimento di Farmacia, Università degli Studi di Parma, Parma, Italy

Matteo Dal Peraro
Laboratory for Biomolecular Modeling, Institute of Bioengineering, School of Life Sciences, École Polytechnique Fédérale de Lausanne—EPFL, and Swiss Institute of Bioinformatics—SIB, Lausanne, Switzerland

Yifei Qi
Department of Molecular Biosciences and Center for Bioinformatics, The University of Kansas, Lawrence, Kansas, USA

Silvia Rivara
Dipartimento di Farmacia, Università degli Studi di Parma, Parma, Italy

Benoît Roux
Department of Biochemistry and Molecular Biology, Gordon Center for Integrative Science, University of Chicago, Chicago, Illinois, USA

Huan Rui
Department of Biochemistry and Molecular Biology, Gordon Center for Integrative Science, University of Chicago, Chicago, Illinois, USA

Enrico Spiga
Laboratory for Biomolecular Modeling, Institute of Bioengineering, School of Life Sciences, École Polytechnique Fédérale de Lausanne—EPFL, Lausanne, Switzerland

Mustafa Tekpinar
Department of Physics, Yuzuncu Yil University, Kampus, Turkey

Kenno Vanommeslaeghe
Department of Pharmaceutical Sciences, School of Pharmacy, University of Maryland, Baltimore, Maryland, USA

Wenjun Zheng
Department of Physics, University at Buffalo, Buffalo, New York, USA

Allen Zhu
Department of Biochemistry and Molecular Biology, Gordon Center for Integrative Science, University of Chicago, Chicago, Illinois, USA

PREFACE

The field of Biomolecular Modelling has evolved dramatically over the past 10 years. The 2013 Nobel Prize in Chemistry was awarded to pioneers in the field M. Karplus, M. Levitt, and A. Warshel, recognizing the impact of the computational and molecular modeling methods in molecular and life sciences (http://www.nobelprize.org/nobel_prizes/chemistry/) (Estrada-Tejedor, Ros-Blanco, & Teixido Closa, 2014). The enormous growth in the computer power, construction of massive parallel supercomputers, and development of novel computational methods made possible successful simulation of increasingly large biomolecular systems. Currently, it is a routine to model reaction mechanisms of solvated enzyme–substrate complexes, the dynamic behavior of proteins, nucleic acids and their complexes in nanosecond and microsecond timescale, the insertion of membrane proteins into lipid bilayer, and the docking of ligands and inhibitors to their targets. The development of the biomolecular modeling unleashed new opportunities for combined, integrated, and synergistic studies with experimental methods. For example, the QM/MM modeling of enzyme reaction mechanism can be integrated with crystallographic, mutagenesis, and kinetic studies (Christov et al., 2013; Senn & Thiel, 2009; van der Kamp & Mulholland, 2013). The atomistic molecular dynamics simulations of fast protein motions can be complemented with NMR and circular dichroism studies of larger conformational changes (Karabencheva-Christova, Carlsson, Balali-Mood, Black, & Christov, 2013; Karplus & Kuriyan, 2005; Ortega, Pons, & Millet, 2013; Zandarashvili, Esadze, & Iwahara, 2013). In bioinorganic chemistry, combining QM, QM/MM, and MD methods with multiple spectroscopic studies of metal centers (UV-Vis, MCD, EPR, and XAS) provides unique and validated insight into the interaction between the metal and the protein which govern the enzyme reactivity (Neidig & Solomon, 2005). In addition, considerable progress was achieved in integrating computational methods at different levels of theory which made easer the practical implementation of the output from one modeling study as an input for another one(Sherwood, Brooks, & Sansom, 2008). The development and improvement of molecular visualizing software including computer animations made the preparation of the inputs for computational studies and the analysis of the results much more straightforward and user friendly than before. The mentioned progress made possible broader application of biomolecular modeling methods in biotechnology and drug design.

This thematic volume of *APCSB* is focused on some of the most recent top contributions in biomolecular modeling both in application and method developments such as the interplay between molecular modeling and chemoinformatics, computational studies of DNA polymerase, dynamic modeling of molecular assembly, stability of amyloid oligomers, coarse-grained modeling of proteins, allosteric regulations in metal sensor proteins, modeling the *N*-acylethanolamine acid amidase inhibition and action, CHARMM-GUI online server for protein nonstandard residues, and high-resolution modeling of protein structures.

Tatyana Karabencheva-Christova
Department of Applied Sciences,
Northumbria University at Newcastle,
Newcastle-upon-Tyne,
United Kingdom

REFERENCES

Christov, C. Z., Lodola, A., Karabencheva-Christova, T. G., Wan, S., Coveney, P. V., & Mulholland, A. J. (2013). Conformational effects on the pro-(S) hydrogen abstraction reaction in cyclooxygenase-1: An integrated QM/MM and MD study. *Biophysical Journal*, *104*, L01–L03.

Estrada-Tejedor, R., Ros-Blanco, L., & Teixido Closa, J. (2014). Multiscale modeling for complex chemical systems: Highlights about the Nobel Prize in Chemistry 2013. *Afinidad*, *71*(566), 89–94.

Karabencheva-Christova, T. G., Carlsson, U., Balali-Mood, K., Black, G. W., & Christov, C. Z. (2013). Conformational effects on the circular dichroism of human carbonic anhydrase II: A multilevel computational study. *PLoS One*, *8*(2), e56874.

Karplus, M., & Kuriyan, J. (2005). Molecular dynamics and protein function. *Proceedings of the National Academy of Sciences of the United States of America*, *102*(19), 6679–6685.

Neidig, M. L., & Solomon, E. I. (2005). Structure-function correlations in oxygen activating non-heme iron enzymes. *Chemical Communications*, *47*, 5843–5863.

Ortega, G., Pons, M., & Millet, O. (2013). Protein functional dynamics in multiple timescales as studied by NMR spectroscopy. *Advances in Protein Chemistry and Structural Biology*, *92*, 219–251.

Senn, H. M., & Thiel, W. (2009). QM/MM methods for biomolecular systems. *Angewandte Chemie International Edition in English*, *48*(7), 1198–1229.

Sherwood, P., Brooks, B. R., & Sansom, M. S. P. (2008). Multiscale methods for macromolecular simulations. *Current Opinion in Structural Biology*, *18*(5), 630–640.

van der Kamp, M. W., & Mulholland, A. J. (2013). Combined quantum mechanics/molecular mechanics (QM/MM) methods in computational enzymology. *Biochemistry*, *52*(16), 2708–2728.

Zandarashvili, L., Esadze, A., & Iwahara, J. (2013). NMR studies on the dynamics of hydrogen bonds and ion pairs involving lysine side chains of proteins. *Advances in Protein Chemistry and Structural Biology*, *93*, 37–80.

CHAPTER ONE

The Interplay Between Molecular Modeling and Chemoinformatics to Characterize Protein–Ligand and Protein–Protein Interactions Landscapes for Drug Discovery

José L. Medina-Franco*[,1], Oscar Méndez-Lucio[†], Karina Martinez-Mayorga[‡]

*Mayo Clinic, Scottsdale, Arizona, USA
[†]Unilever Centre for Molecular Science Informatics, Department of Chemistry, University of Cambridge, Cambridge, United Kingdom
[‡]Instituto de Química, Universidad Nacional Autónoma de México, Mexico City, Mexico
[1]Corresponding author: e-mail address: MedinaFranco.Jose@mayo.edu; jose.medina.franco@gmail.com

Contents

Advances in Protein Chemistry and Structural Biology, Volume 96
ISSN 1876-1623
http://dx.doi.org/10.1016/bs.apcsb.2014.06.001

Abstract

Protein–ligand and protein–protein interactions play a fundamental role in drug discovery. A number of computational approaches have been developed to characterize and use the knowledge of such interactions that can lead to drug candidates and eventually compounds in the clinic. With the increasing structural information of protein–ligand and protein–protein complexes, the combination of molecular modeling and chemoinformatics approaches are often required for the efficient analysis of a large number of such complexes. In this chapter, we review the progress on the developments of *in silico* approaches that are at the interface between molecular modeling and chemoinformatics. Although the list of methods and applications is not exhaustive, we aim to cover representative cases with a special emphasis on interaction fingerprints and their applications to identify "hot spots." We also elaborate on proteochemometric modeling and the emerging concept of activity landscape, structure-based interpretation of activity cliffs and structure–protein–ligand interaction relationships. Target–ligand relationships are discussed in the context of chemogenomics data sets.

1. INTRODUCTION

Understanding protein–ligand interactions (PLIs) and protein–protein interactions (PPIs) is at the core of molecular recognition and has a fundamental role in many scientific areas. PLIs and PPIs have a broad area of practical applications in drug discovery including but not limited to molecular docking (Bello, Martinez-Archundia, & Correa-Basurto, 2013), structure-based design, virtual screening of molecular fragments, small molecules, and other type of compounds, clustering of complexes, and structural interpretation of activity cliffs, to name a few. Over the years, the scientific community has made significant progress on the understanding of PLIs and PPIs that have led to the development of algorithms to predict the putative interaction of two molecules. For example, Chupakhin et al. recently used a machine learning approach to predict protein–ligand binding modes based on the two-dimensional (2D) structure of the ligand and a previous set of PLIs (Chupakhin, Marcou, Baskin, Varnek, & Rognan, 2013). One of the goals of improving the description of the protein–ligand binding process is, as recently discussed, to reach a point where a more detailed description of protein–ligand complexes can be associated with a more accurate prediction of binding affinity (Ballester, Schreyer, & Blundell, 2014). Indeed, Ballester et al. noted that a typical issue of current scoring functions used in docking is the "difficulty of explicitly modeling the various contributions of intermolecular interactions to binding affinity." Ballester et al. also commented that novel scoring functions based on machine learning regression models have shown superior performance over commonly used scoring functions. Finally, the authors of this elegant work concluded that "a more precise chemical description of the protein–ligand complex does not generally lead to a more accurate prediction of binding affinity" (Ballester et al., 2014).

In a broad sense, PLIs and PPIs have been characterized using either molecular modeling or chemoinformatic applications. While molecular modeling techniques such as molecular mechanics, quantum mechanics, molecular dynamics, pharmacophore modeling capture, manage, and represent PLIs and PPIs in a three-dimensional (3D) manner, chemoinformatic approaches typically transform those interactions in 2D or one-dimensional (1D) representations for the rapid and easy visualization, clustering, and mining of those interactions. Of course, there is a large overlap between both types of approaches. In-depth reviews of the progress and current status in each of the above mentioned methods have been published in an individual manner (Durrant & McCammon, 2011; Langer, 2010; Scior et al., 2012). In this chapter, our goal is to discuss recent advances and exemplary applications of the integration between molecular modeling and chemoinformatic methods to characterize PLIs and PPIs. We put a special emphasis on the development and application of protein–ligand interaction fingerprints (PLIFs). While the list of applications is not comprehensive, we want to focus on representative combined applications of current interest in drug discovery. The chapter is organized in seven sections. After this introduction, Section 2 discusses an overview and recent advances and selected applications of the characterization of PLIs using fingerprints. Section 3 is dedicated to the visual representation of PLIs with 2D graphs, representation of PLIFs using 3D pharmacophore models, and chemoinformatic approaches used for the visualization of chemical spaces. Section 4 presents studies that aim to explore structure–protein–ligand interaction relationships (SPLIRs). In this section, we put a particular emphasis on the application of the emerging concept of activity landscape. Advances in the characterization of structure-based activity cliffs, structure-based activity cliff generators, and 3D activity cliffs are discussed. Section 5 discusses examples of the analysis of target–ligand relationships in chemogenomics data sets. Section 6 addresses the characterization of PPIs. Section 7 presents summary conclusions.

2. CHARACTERIZING PLIS WITH FINGERPRINTS

PLIFs, also called "structural interaction fingerprints," are designed to "capture a 1D representation of the interactions between ligand and protein either in complexes of known structure or in docked poses" (Brewerton, 2008). PLIFs are a primary example of combining molecular modeling—that can characterize and describe in detail the interactions at the molecular level—with chemoinformatics that can process large amounts of protein–ligand and protein–protein complexes. PLIFs can also be derived from crystallographic information. As recently pointed out by Desaphy, Raimbaud, Ducrot, and Rognan (2013), fingerprints are a very convenient way to simplify the atomic coordinates of PLIs. Fingerprints are easy to generate, manipulate, and compare a vast number of protein–ligand complexes (Desaphy et al., 2013). For example, PLIFs enable the systematic analysis of large amounts of data and are suitable to evaluate if similar binding sites identically recognize similar ligands, if PLI patterns are conserved across target families, and if different ligand structures or substructures have the same interaction patterns with a single target (Desaphy et al., 2013).

There are two general approaches to generate PLIFs:

(A) Annotating ligand descriptors with interaction features (Tan, Batista, & Bajorath, 2010).

(B) Annotate protein descriptors, typically amino acids in the binding site, with ligand interaction features (Deng, Chuaqui, & Singh, 2004).

Both general approaches have been recently summarized by Desaphy et al. (2013). As exemplified below with some representative cases, interaction fingerprints (IFPs) have a number of applications including postprocessing docking results, virtual screening (Chupakhin et al., 2013), data mining and clustering protein–ligand complexes (Weisel, Bitter, Diederich, So, & Kondru, 2012), and library design (Deng, Chuaqui, & Singh, 2006). Representative applications of PLIFs are summarized in Table 1.

An example of a PLIF is illustrated in Fig. 1. The example corresponds to nonpeptidic vinylsulfone cruzain inhibitors (Bryant et al., 2009). Cruzain is a cysteine protease essential for the parasite survival. Peptides, peptidomimetics, and small molecules have been explored as cruzain inhibitors. Systematic modification of the P3, P2, P1, and P1′ side chains aided with structure-based design have rendered highly active cruzain inhibitors (in the picomolar range). This wide range of activities is desirable to conduct structure–activity relationship (SAR) studies. In this example, taking the reported crystallographic structure (PDB ID: 3HD3) of a parent compound as reference, a total of nine analogues were overlaid. Then, the PLIF analysis was generated with the program Molecular Operating Environment (MOE, 2013). Figure 1A shows the barcodes for the nine complexes along the *y*-axis, and amino acid residues for which fingerprints were generated. The barcode is a visual representation of the PLIFs for the cruzain inhibitors.

Table 1 Examples of applications of protein–ligand interaction fingerprints

Application	Example/representative study	Reference
Postprocessing docking results	Development of APIF (atom-pairs-based interaction fingerprint), an interaction fingerprint tuned for postprocessing protein–ligand docking results	Perez-Nueno, Rabal, Borrell, and Teixido (2009)
Data mining	Development of PROLIX (Protein–Ligand Interaction Explorer), a tool that employs fingerprint representations of protein–ligand interaction patterns for rapid data mining in large crystal structure databases	Weisel et al. (2012)
Clustering of protein–ligand complexes	Clustering of inhibitors of DNMT1 based on predicted docking poses	Medina-Franco and Yoo (2013)
Analysis of structure–protein–ligand interaction relationships (SPLIRs)	Relationship between the similarity of protein–ligand interactions with the ligand and/or protein binding similarities of 9877 high-resolution X-ray complexes stored in the sc-PDB data set	Desaphy et al. (2013)
Virtual screening	Novel approach to predict protein–ligand binding modes using neural networks trained on protein–ligand interaction fingerprints. The method was used on three molecular targets (CDK2, p38-α, HSP90-α)	Chupakhin et al. (2013)
Library design	Design of combinatorial libraries using structural information of a biding site. Approach exemplified with MAP kinase p38	Deng et al. (2006)

The rows represent the compounds following the order of the input file and the columns indicate the amino acid residues that make at least one contact with one of the compounds. A cell colored in black means that the compound makes an interaction with the corresponding intersecting residue

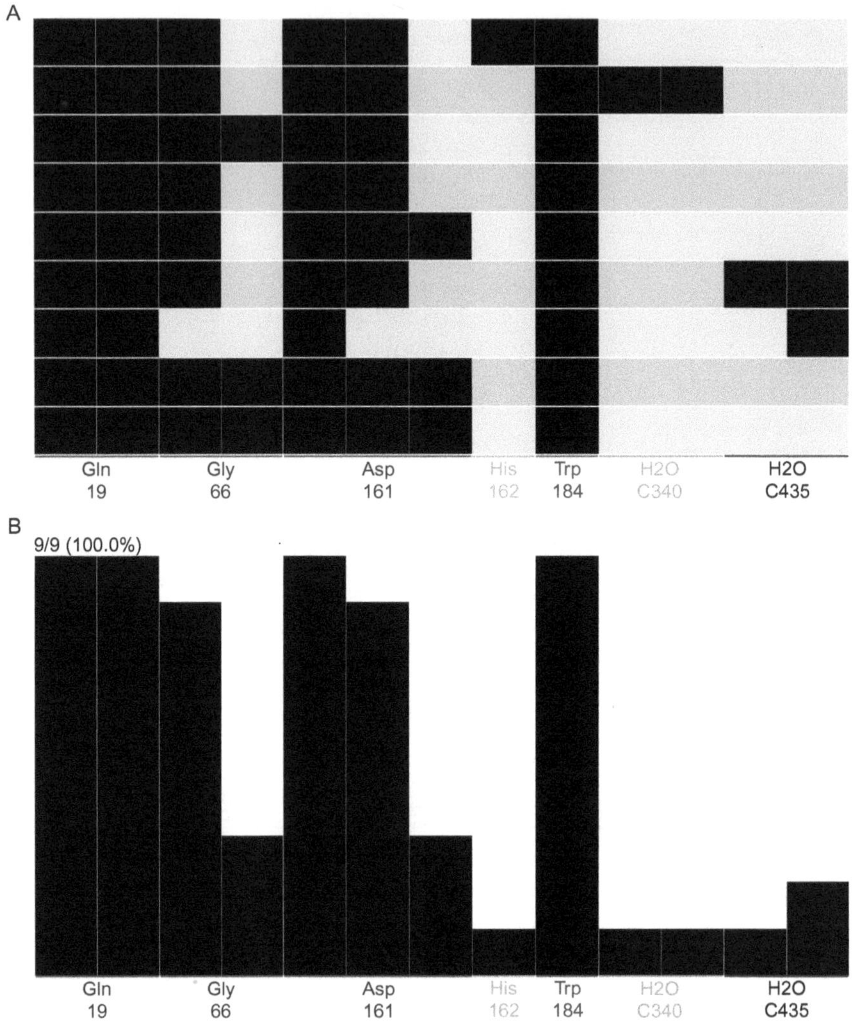

Figure 1 PLIF of cruzain inhibitors (PDB ID: 3HD3). (A) Barcode representation of binding interactions: rows correspond to each of the compounds studied (9 in this example). The interacting residues are shown at the bottom of each graph. Black indicates the presence of an interaction. (B) Population of each interaction. The types of interactions are listed on Table 2.

in the bar code. In contrast, the white cells denote that the compound does not make interactions with the corresponding residue. Therefore, the group of black and white cells for each compound in the barcode of Fig. 1A represents the PLIF for each molecule. This approach is reminiscent of the pioneering work of Deng et al. that developed the structural interaction fingerprint (SIFt) (Deng et al., 2004). The hallmark feature of SIFt is the representation of important target–ligand interactions as 1D binary bit strings. For example, molecule 1 (top to bottom) of Fig. 1A is the least active of the series, and molecule 3 (third row) is 10 times more active. Quick comparison of the PLIF shows that these two molecules share common interaction, as well as some differences that could be explored as responsible for the difference in biological activity, for example, molecule 1 makes contacts with His 162, in fact, none of the other molecules on the set show this contact. Molecule 3 makes contacts with Gly 66, an interaction that is only present in three cases. For a more detailed analysis, the types of interactions are reported on Table 2, providing the residue number and whether the interaction is as acceptor or donor and if it is to a side chain, backbone, or from the solvent. Note how the PLIF analysis is able to provide information in a compact manner easy to analyze. Lastly, Fig. 1B shows a histogram of the frequencies of the different interactions made by this set of cruzain inhibitors. Thus, analysis of Fig. 1A and B facilitates the comparison of interactions among the different molecules, the development of SAR, as well as the easy count of what interactions are the most common for this particular set.

Table 2 Types of interactions derived for cruzain inhibitors based on PLIF (cf. Fig. 1)

Number of interacting residue	Type of interaction
19	Acceptor from side chain
	Acceptor from side chain
66	Acceptor from backbone
	Acceptor from backbone
161	Donor to backbone
	Donor to backbone
	Surface contact
162	Acceptor from side chain
184	Acceptor from side chain
20,435	Acceptor from solvent
	Acceptor from solvent
20,435	Acceptor from solvent
	Surface contact

A detailed example of the use of PLIF to generate SAR can be found in the literature (Lopez-Vallejo & Martinez-Mayorga, 2012).

Yoo et al. recently reviewed the application of PLIFs for inhibitors of DNA methyltransferase 1 (DNMT1). DNMT1 is one of the family members of DNMTs which are promising epigenetic targets for the treatment of cancer and other diseases. Several computational studies have been conducted to analyze the activity of known inhibitors at the molecular level and to identify inhibitors with novel molecular scaffolds (Medina-Franco & Yoo, 2013). PLIFs were developed based on the results of docking studies with a modified crystal structure of DNMT. A total of 17 inhibitors of DNMT1 were docked into the catalytic site of the crystallographic structure of human DNMT1 modified to an active conformation (Yoo, Kim, Robertson, & Medina-Franco, 2012). As negative control, 19 compounds that previously have shown very weak or no enzymatic inhibitory activity were used as inactive/decoys (Kuck, Singh, Lyko, & Medina-Franco, 2010; Siedlecki et al., 2006; Yoo & Medina-Franco, 2012). In that analysis, molecules were classified as "active" or "inactive" based on the published experimental activity (Kuck et al., 2010; Siedlecki et al., 2006; Yoo & Medina-Franco, 2012). The lowest energy conformation of each ligand was selected from the docking results. Fingerprints were generated using PLIF tools implemented in MOE (2013). The raw interactions between ligands and receptors were calculated through the preparation step with Receptor + Solvent option. For that calculation, one protein which was the modified crystallographic structure of human DNMT1 without SAH was loaded and all of the ligands in the underlying database that have 3D coordinates relative to the active site of this protein were provided. Then, fingerprint bits were generated using the calculated raw PLI data with default parameters and the maximum number of bits (Medina-Franco & Yoo, 2013).

Kelly and Mancera developed an IFP method for analyzing the binding poses of ligands and structure-based approaches (Kelly & Mancera, 2004). In a more recent work, Perez-Nueno et al. developed the atom-pairs-based interaction fingerprint (APIF) for postprocessing protein–ligand docking results (Perez-Nueno et al., 2009). A distinctive feature of APIF over other fingerprints is that it considers the relative position of pairs of interacting atoms. In that work, the IFPs were used to derive a score that captures the similarity of the bit strings for each docked compound with the reference compound. Such score was compared with the score obtained from docking alone in virtual screening showing a superior performance as measured by enrichment plots. The IFPs were also used to analyze and compare binding modes of docked poses with the binding mode of the cocrystal ligand (Perez-Nueno et al., 2009).

Figure 2 illustrates the use of PLIFs to postprocess results of virtual screening based on molecular docking. In the example illustrated in this figure, we docked a database with 1200 approved drugs with a crystallographic structure of DNMT1 in complex with sinefungin (PDB ID: 3SWR). Docking was performed using Glide XP (2012). The docking protocol was validated by redocking the cocrystal ligand with an excellent root mean square deviation (RMSD) of 0.547 Å. This is part of a computer-guided drug repurposing strategy ongoing in our laboratory that previously identified olsalazine, an approved anti-inflammatory drug, as a novel hypomethylating agent (Méndez-Lucio, Tran, Medina-Franco, Meurice, & Muller, 2014). In order to analyze the results of the virtual screening, we generated the PLIFs of selected poses and compared the PLIF profile with PLIFs of the cocrystal ligand using the Tanimoto coefficient. Figure 2 shows

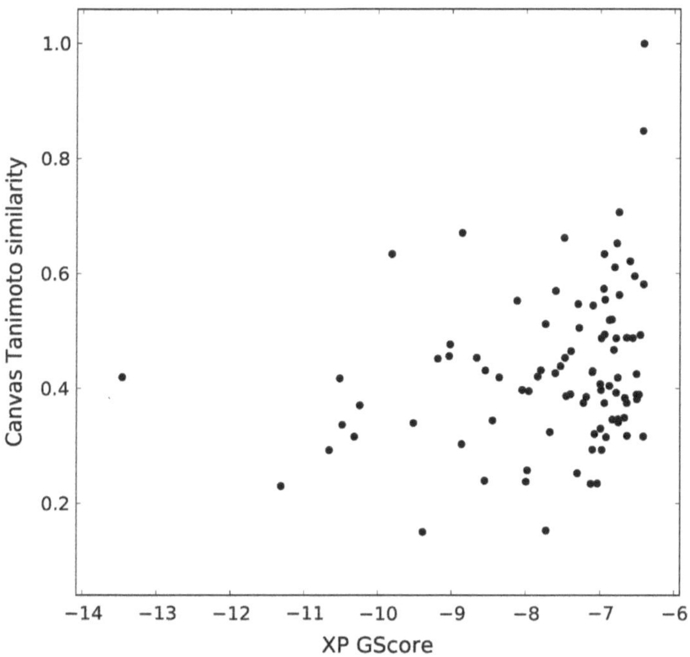

Figure 2 Relationship between protein–ligand contact similarity and docking scores of 92 compounds docked with DNMT1. The protein-contact similarity was measured using as a reference the binding pose of the cocrystal ligand sinefungin and the Tanimoto coefficient. This is an example of the postprocessing docking-based virtual screening with chemoinformatic methods.

a plot of the Glide XP score versus protein-contact similarity to cocrystal ligand of 91 representative hits. The position of sinefungin in the plot (XP Score = −6.42 kcal/mol and Tanimoto similarity = 1) is shown as reference. Almost all the 91 hits selected in this plot have a docking score better (more negative) than the docking score of sinefungin. However, the hits selected in this example showed a wide range of protein-contact similarities to the reference. In this case study, one may want to select compounds in the upper-left quadrant of the plot, i.e., molecules with favorable docking scores and protein–ligand contacts similar to the reference. Interestingly, most of the data points are located in the lower-right quadrant of the graph indicating that the predicted binding poses for these hits make different contacts with the protein as compared to sinefungin. Of course, in order to further select compounds for experimental validation, one needs to take into consideration that the most important contacts with the protein are preserved. The compound with the most favored docking score (−13.46 kcal/mol) showed a low Tanimoto similarity of 0.42 indicating that this molecule could bind in a different orientation (this could be highly influenced by the 2D similarity of the hit compound and the reference compound). In this regard, depending on the goals of the screening campaign, one may want to select molecules with similar 2D structures to the reference (for example, in lead optimization) or select molecules with different 2D structures (for instance, in scaffold hoping). Taken together the insights of this discussion, Fig. 2 illustrates a combination of molecular docking with PLIFs to analyze results of virtual screening. Brewerton has discussed in detail the role of the SIFt method in rescoring binding modes predicted with docking and virtual screening (Brewerton, 2008).

Other type of related fingerprints implemented to rescore solutions has been proposed by Balius et al. which developed a docking–rescoring method based on the comparison of per-residue van der Waals, electrostatic, or hydrogen bond energies, or the sum of docked ligands with the interaction signatures of a reference. Those signature interactions were termed "molecular footprints" that served as basis to compute the so-called footprint similarity score (Balius, Mukherjee, & Rizzo, 2011). Of note, the reference can be not only a known inhibitor but can be other biologically relevant entity such as know drugs, substrates, transition states, or even side chains that are involved in PPIs. Molecular footprints and the footprint similarity score were recently applied to antiviral inhibitors targeting HIVgp41 (Holden, Allen, Gochin, & Rizzo, 2014).

In a recent full paper, Desaphy et al. (2013) elegantly summarized progress on the development of PLIFs, highlighting advantages and disadvantages of different methods, and proposed a novel set of descriptors and approaches to make comparisons. The new method proposed by the authors enables the evaluation of PLIs regardless the size and sequence of the target binding site, allows the description of molecular interactions quantitatively using a specific frame-invariant descriptor, and provides an alternative 3D alignment of protein–ligand complexes to protein-based or ligand-based matches by only focusing on molecular interactions (Desaphy et al., 2013). Desaphy et al. pointed out that their method facilitates the interpretation of pairwise comparisons of protein–ligand complexes. This approach has several applications such as postprocessing docking results, identification of off-targets having common interactions patterns to a known ligand, and detection of bioisosteric fragments with a conserved interaction mode to a given molecular target (Desaphy et al., 2013).

Recently, Van Linden et al. conducted a comprehensive analysis of the PLIs of 1252 human kinase–ligand cocrystal structures present in the Protein Data Bank (PDB) (van Linden, Kooistra, Leurs, de Esch, & de Graaf, 2014). This study includes 190 different human kinases, which represent over 35% of the human kinome. The data were assembled in the so-called Kinase–Ligand Interaction Fingerprints and Structure database (KLIFS). This database has a consistent alignment of 85 kinase–ligand binding site residues. The different kinase–ligand interaction features were mapped by using PLIFs calculated with MOE resulting in seven types of interaction for each amino acid in the aligned sequence, i.e., seven binary bits per amino acid depending if it interacts with the ligand or not. The seven bits correspond to the following interactions: hydrophobic contact, face-to-face aromatic interactions, face-to-edge aromatic interactions, protein H-bond donor, protein H-bond acceptor, protein cationic interactions, and protein anionic interactions. A total 595 bits were obtained for each complex corresponding to 85 amino acids. KLIFS, which is freely available at http://www.vu-compmedchem.nl, enables the identification of family-specific interaction features and classification of ligands according to their binding modes. Additionally, the use of PLIFs facilitated the description of conserve hot spots and crucial interactions to gain selectivity. From the ligand point of view, PLIFs allowed the analysis of important chemical features that can be related to specific interactions with the kinase binding sites useful as a guide for design of new drugs.

3. VISUALIZATION OF PLIs AND PLIFs: THE PLIs SPACE

Data visualization plays a key role in science providing a key support to quantitative approaches. In addition to common 3D representation of protein–ligand complexes, graphical methods have been developed to generate 2D interaction plots of such complexes. Also, key interactions captured by PLIFs can be tracked back to 3D coordinates as pharmacophoric elements. Finally, PLIFs can be conveniently visualized using common approaches used in chemoinformatics. In Sections 3.1–3.3, we illustrate different graphical approaches to generate visual representations of the PLI landscapes for a given protein–ligand complex of a series of complexes.

3.1. 2D Schematic diagrams of PLIs

Visualization is usually the first approach to retrieve information from a protein–ligand complex (O'Donoghue et al., 2010). The analysis of 3D crystal structures or results from molecular modeling can be difficult due to the large number of atoms and interatomic distances involved in the PLI. Given the importance of visualization, software tools have been developed to rapidly generate 2D diagrams of protein–ligand contacts from 3D coordinates, highlighting essential information such as formation of hydrogen bonds and hydrophobic, π–π, and π–cation interactions. Such diagrams enable the easy interpretation of protein–ligand complexes that in many cases are difficult to interpret. Although such plots do no capture the details of 3D representations, they facilitate an initial assessment of structural information not only for experts but also for researchers not familiar with molecular modeling packages. In addition, 2D diagrams are excellent resources for communicating ideas between molecular modelers, chemoinformatians, and medicinal chemists, to name a few. In general, these tools use intuitive interfaces and generate the 2D plots with mouse-click operations. Some of these tools have been integrated with commercial software packages and others are freely available as Web-based servers or as stand-alone versions.

One of the first tools that became of common use was LIGPLOT (Wallace, Laskowski, & Thornton, 1995), recently superseded by LigPlot^{+} (Laskowski & Swindells, 2011). The full version of LigPlot^{+} can be downloaded from https://www.ebi.ac.uk/thornton-srv/software/LigPlus/. This tool runs from a Java interface which allows the user to edit the plots easily. Improvement of LigPlot^{+} over the previous version includes superposition of related diagrams either side by side or superposed and links to 3D viewers such as PyMol and RasMol.

PoseView is other example of a program free for academics that automatically generates 2D diagrams of complexes with known 3D structure according to chemical drawing conventions (Fricker, Gastreich, & Rarey, 2004; Stierand & Rarey, 2007). This tool is available as a Web-based service (http://poseview.zbh.uni-hamburg.de/), and it has been integrated in the PDB to the rapid and Web-based visualization of PLIs (Stierand & Rarey, 2010). Recently, PoseView was tested on a large-scale study to compute the 2D representations of nearly 210,000 protein–ligand complexes included in PDB succeeding in 85% of cases by generating plots (Stierand & Rarey, 2010). In this study, 90% of the computed diagrams contained less than 11 direct interactions between the ligand and the receptor and authors report a direct relationship between the number of interactions and the quality of the diagram.

Tools implemented in commercial software and frequently used are the Ligand Interactions application in MOE (Clark & Labute, 2007) and Ligand Interaction Diagram tool implemented in Maestro (2012).

A more detailed review and comparison of these methods can be found in the review carried out by Stierand & Rarey (2011).

Figure 3 shows examples of 2D interaction maps generated with tools implemented in MOE (2013), Maestro (2012), LigPlot^{+}, and PoseView. The figure shows the PLIs of the crystallographic structure of furosemide bound to *Ancylostoma ceylanicum* macrophage migration inhibitory factor (rAceMIF) (PDB ID: 3RF4). Furosemide is an approved drug for the treatment of hypertension and heart failure. rAceMIF is a molecular target for the treatment of infections by hookworms, blood-feeding intestinal nematode parasites. With the aim of identifying promising molecules for the therapeutic treatment of hookworm disease, furosemide was detected as an inhibitor of the rAceMIF tautomerase activity following a drug repositioning approach (Cho et al., 2011). To generate the 2D interaction maps in Fig. 3, the crystallographic structure was prepared using the Protein Preparation Wizard protocol implemented in Maestro (Schrödinger Suite 2012 Protein Preparation Wizard, 2012). In almost all diagrams in this figure, relevant protein–ligand contacts are represented with dashed lines and/or solid lines with different color codes and the ligands are visualized as structure diagrams. Tools such as MOE and Maestro display a proximity contour around the ligand and represent the ligand exposure to the solvent in the diagram. Not surprisingly, similar interactions are captured by all four

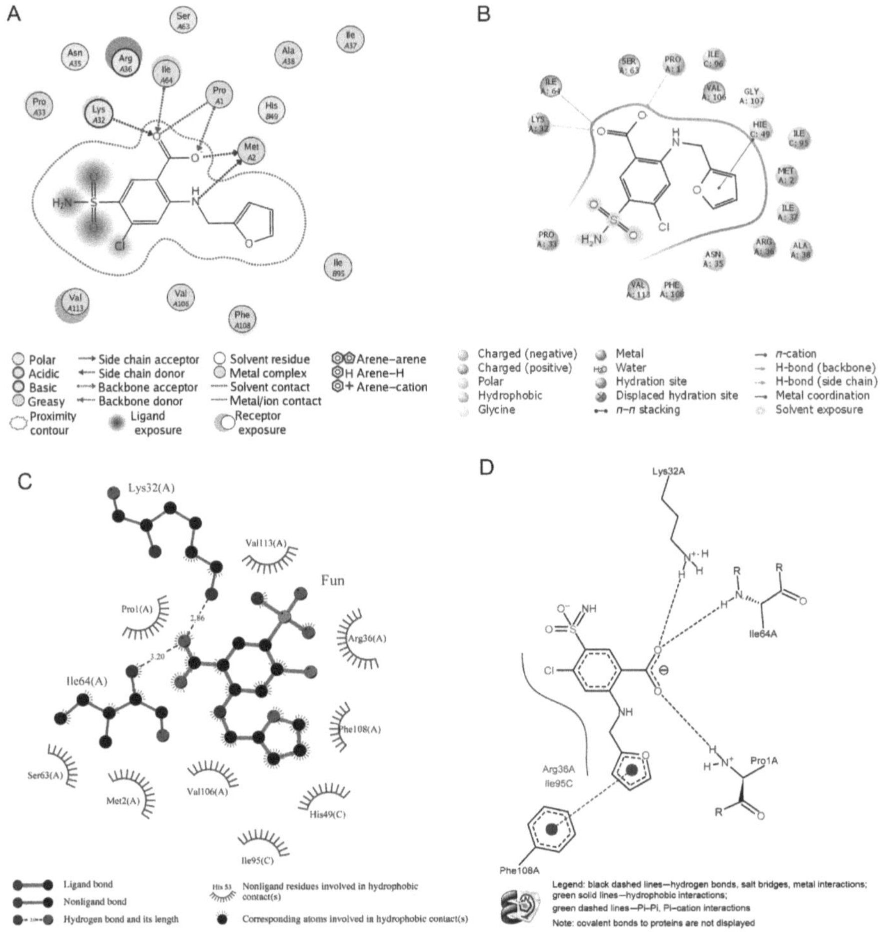

Figure 3 Example of 2D diagrams of protein–ligand interactions of furosemide with rAceMIF obtained with tools implemented in (A) MOE, (B) Maestro, (C) LigPlot^{+}, and (D) PoseView. The corresponding legend for each tool is displayed below each diagram.

programs, for example, hydrogen bond interactions with the side chain of Lys32 and the backbone of Ile64. In addition, the 2D maps generated with MOE, Maestro, and PoseView captured a hydrogen bond interaction with Pro 1. Figure 3A and B also clearly shows the exposure of the sulfonamide group to the solvent. However, some differences can be seen in the plots, for example, the total number of hydrogen bond interactions, which depends on the specific parameters of each program. Nonetheless, each 2D diagram in Fig. 3 clearly presents key interactions involved in the recognition process of furosemide with rAceMIF.

3.2. Representation and application of PLIFs as 3D pharmacophore models

Since PLIFs are derived from structural information, it is possible to track the information encoded into the PLIFs in 3D interactions. For this purpose, MOE has implemented the "Query Generator" tool that operates on the principle that "a modest selection of poses with a homogeneous set of interaction fingerprints will most likely also possess a homogeneous set of pharmacophoric feature points." The pharmacophoric feature points can be clustered and filtered according to which residues they interact with, and those with a sufficiently tight grouping can be converted into a "pharmacophore query feature."

As reviewed in detail below, Seebeck et al. introduced a novel approach to generate visual representation of structure-based activity cliffs (Seebeck, Wagener, & Rarey, 2011). In that work, authors presented a method to distinguish atoms of a protein frequently involved in activity cliff events. Using a quantitative measure and a visual approach, protein atoms frequently involved in activity cliffs were identified as "hot spots." Visualization of hot spots was useful to define pharmacophoric hypothesis that were further validated in structure-based virtual screening (Seebeck et al., 2011).

In this regard, Fingerprint for Ligands and Proteins (FLAP) is other well-known software that is at the interface between molecular modeling and chemoinformatics to characterize PLI landscapes (Baroni, Cruciani, Sciabola, Perruccio, & Mason, 2007). FLAP uses fingerprints obtained from GRID molecular interaction fields (MIFs) and GRID atom types are characterized as quadruplets of pharmacophoric characteristics. The GRID approach was designed to capture energetically favorable interaction sites in molecules with known structure using chemical probes which describes the shape, hydrogen bond acceptor, hydrogen bond donor, and hydrophobic interactions. As summarized by Poongavanam and Kongsted, FLAP creates a common reference framework in two stages: first, the MIFs of the molecules are calculated using the GRID force fields, and the resulting MIFs are summarized by deriving points (quadruplets or hotspot) representing the most favorable interactions. In a subsequent step, each quadruplet of these points is used to generate different superpositions of the test molecules onto a template molecule. The quadruplets of each molecule are stored as pharmacophoric fingerprints and used to evaluate their similarity (Poongavanam & Kongsted, 2013). FLAP has been recently used in a comparative study of virtual screening approaches to identify inhibitors of HIV-1 reverse transcriptase-associated ribonuclease H (RNase H) function

(Poongavanam & Kongsted, 2013) and to identify novel Fyn tyrosine kinase inhibitors (Poli et al., 2013). FLAP has also been recently used in virtual fragment screening to identify new fragment-like histamine H_3 receptor (H_3R) ligands that can be used as a starting point to design drugs targeting H_3R (Sirci et al., 2012).

3.3. Visualization of PLIFs using the concept of chemical space

PLIFs can be visualized using approaches employed in the visual representation of chemical space. There are several definitions of chemical space. For example, Virshup et al. define chemical space as "an *M*-dimensional Cartesian space in which compounds are located by a set of *M* physicochemical and/or chemoinformatic descriptors" (Virshup, Contreras-García, Wipf, Yang, & Beratan, 2013). The interested reader may refer to other works that review alternative definitions and conceptualizations of chemical space (Bohanec & Zupan, 1991; Pearlman & Smith, 1998; Virshup et al., 2013). One of the general applications of the concept of chemical space is library selection and design. Here, the chemical space is useful not only to visualize the distribution and relative position in space of entire compound libraries or subsets of libraries but also to make quantitative assessments of the degree of coverage and overlap of compound collections. A second no less important application is the clustering of bioactive molecules according to a "confinement criteria." That is, focused or confined chemical spaces can be divided in two major groups, namely (A) library design focused on a relevant therapeutic target or disease and (B) library design focused on the chemistry (e.g., peptides, macrocycles, and metal-based compounds) or a desired molecular function (e.g., PPI modulators). Further details are discussed elsewhere (Medina-Franco, Martinez-Mayorga, & Meurice, 2014).

Two methods frequently used to generate visual representations of the chemical space are principal component analysis and self-organizing maps (Digles & Ecker, 2011). Other multidimensional data mining tools are Principal Moments of Inertia plots (Sauer & Schwarz, 2003) and Multi-fusion Similarity maps (Medina-Franco, Maggiora, Giulianotti, Pinilla, & Houghten, 2007) which have been widely used (Akella & DeCaprio, 2010; Clemons et al., 2011; Medina-Franco, Martínez-Mayorga, Giulianotti, Houghten, & Pinilla, 2008). Additional approaches are multidimensional scaling, neural networks, support vector machines, genetic algorithms, decision trees, and hierarchical clustering. Recent advances in chemoinformatic methods to mine and generate visual representations of the chemical space are the generation of the Delimited Reference Chemical Subspaces, the Latent Trait Model for visualization of molecular fingerprints (Owen, Nabney, Medina-Franco, & López-Vallejo, 2011), and the development of a framework to navigate through a reference-independent Biologically Relevant Chemical Space (BRCS). Navigation through the BRCS is based on ligand–protein interactions and has found applications in key areas in drug discovery including SAR analysis of patents, comparison of compound libraries, and selection of reagents to design new chemical analogues (Rabal & Oyarzabal, 2012). These and other techniques are reviewed elsewhere (Akella & DeCaprio, 2010; Medina-Franco et al., 2008; Ritchie, Ertl, & Lewis, 2011; Wawer, Lounkine, Wassermann, & Bajorath, 2010). By analogy with chemical space, the PLIs as encoded in PLIFs space can be visually represented using techniques commonly used to generate chemical space representations.

4. EXPLORING SPLIRs

Desaphy et al. explored the relationship between the similarity of PLIs with the ligand and/or protein binding similarities of 9877 high-resolution X-ray complexes stored in the sc-PDB data set (Meslamani, Rognan, & Kellenberger, 2011). In that work, the pairwise similarity of protein–ligand complexes was measured using three metrics: (1) pairwise similarity of ligands using two fingerprint representations of different design, (2) the pairwise similarities of their binding sites, and (3) the pairwise similarities of their interaction patterns (Desaphy et al., 2013). Figure 4A and B shows the relationship between ligand similarity (as measured with MACCS keys and the extended connectivity fingerprints ECFP4) and PLI similarity showing a lack of linear correlation. Figure 4C shows the high linear correlation ($r=0.876$) between the pairwise binding site similarity and PLI similarity. Figure 4D illustrates the relationship between the three metrics. Desaphy et al. noted that there are few cases of similar interaction patterns between dissimilar ligands and dissimilar binding sites (several cases correspond to small ligands with common hydrophobic interactions). Authors concluded that the observations of this analysis (considering that there is still a limited ligand diversity in sc-PDB) suggest that "a single interaction mode to a single druggable cavity remains the rule because a few key interactions to a few key residues need to be fulfilled to achieve significant binding" (Desaphy et al., 2013).

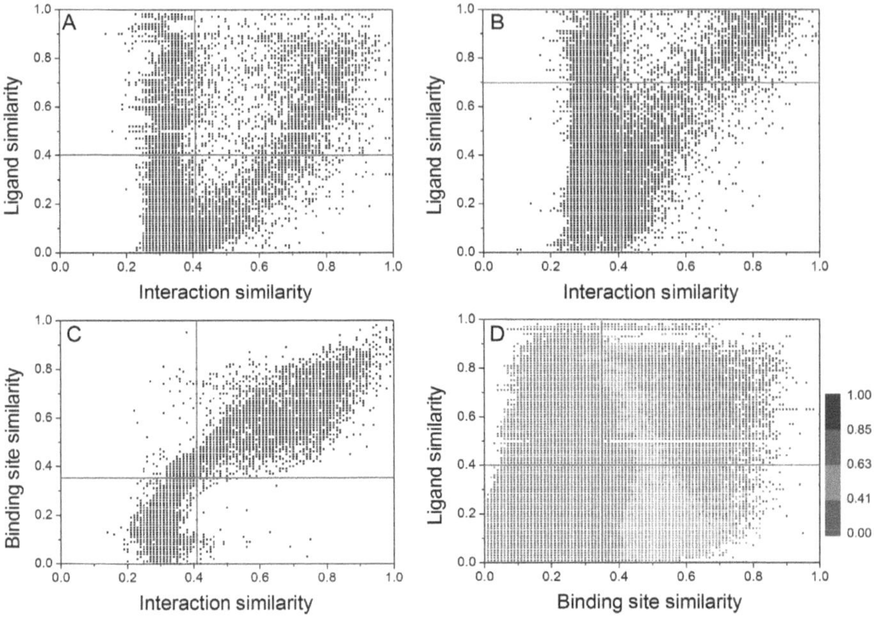

Figure 4 Relationships between ligand similarity, binding site similarity, and interaction pattern similarity for 9877 sc-PDB entries. (A) Ligand similarity (ECFP4/Tanimoto) versus interaction pattern similarity (IShape similarity score). (B) Ligand similarity (MACCS/Tanimoto) versus interaction pattern similarity (IShape similarity score). (C) Binding site similarity (Shaper similarity score) versus interaction pattern similarity (Ishape similarity score). (D) Ligand similarity (ECFP4/Tanimoto) versus binding site similarity (Shaper29 similarity score). Data are colored according to the interaction pattern similarity score (IShape similarity). *Reprinted with permission from Desaphy et al. (2013). Copyright 2013 American Chemical Society.*

4.1. Activity landscape: Activity cliffs and hot spots

The interaction between molecular modeling and chemoinformatics has encountered several applications in the analysis of SAR using the concept of activity landscape modeling. This concept is gaining relevance in the medicinal and computational chemistry communities (Guha, 2012; Stumpfe, Hu, Dimova, & Bajorath, 2014). It is well recognized that the identification of activity cliffs, defined as compounds with high structure similarity but unexpectedly large potency difference (Maggiora, 2006), has a high impact on lead optimization efforts. As such, activity cliffs have a "nice face" because they provide key structural information of specific and frequently subtle changes in the structure associated with large changes in activity. At the same time, activity cliffs have an "ugly face" representing the bottle neck of computational predictive models that often assume smooth regions of the SAR. The "duality" of the roles of activity cliffs in drug discovery has been recently commented (Cruz-Monteagudo et al., 2014). Also, it has been argued that activity cliffs may be artifacts of the molecular representation or artifacts due to, for example, errors in the measurement of potency (Medina-Franco, 2013).

In fact, one of the major issues in activity landscape modeling is the molecular representation. One approach to address this issue is consider using multiple representations and obtaining consensus conclusions (Medina-Franco et al., 2009). Other approach is using substructure relationships instead of computed similarity values. In this regard, Bajorath et al. have employed the concept of matched molecular pairs (MMPs) and define MMP-cliffs, which are extremely easy to interpret from a chemical perspective. However, as pointed out by Bajorath et al., substructure-based representation of activity cliffs has their own restrictions and complements whole-molecule similarity approaches (Hu, Hu, Vogt, Stumpfe, & Bajorath, 2012). In case of considering 3D structures, one approach is to derive consensus conclusions obtained from multiple conformations (Yongye et al., 2011).

Attempts to rationalize activity cliffs in terms of the PLIs have recently been proposed leading to the concepts of structure-based activity cliffs (Seebeck et al., 2011) and 3D activity cliffs (Hu, Furtmann, Gütschow, & Bajorath, 2012). These approaches give information concerning hot spots in the target protein, that is, key interactions between the ligand and the target protein that can lead to an activity cliff.

4.2. 3D Activity Cliffs

Hu et al. described an extensive study to systematically identify in public domain X-ray structures deposited in PDB, pairs of ligands with high 3D similarity (at least 80% of similarity), and potency difference of at least two orders of magnitude (Hu, Furtmann, et al., 2012). 3D similarity was measured using a property density function-based method that takes into account conformational, positional, and chemical differences. Authors of that work found in PDB 216 well-defined 3D activity cliffs distributed in 38 different targets. In a separate work, Hu and Bajorath compared 3D and 2D activity cliffs, finding a low degree of conservation between the two types of representations (Hu & Bajorath, 2012). That study confirmed the high dependence of activity landscape with molecular representation previously noted (Medina-Franco et al., 2009).

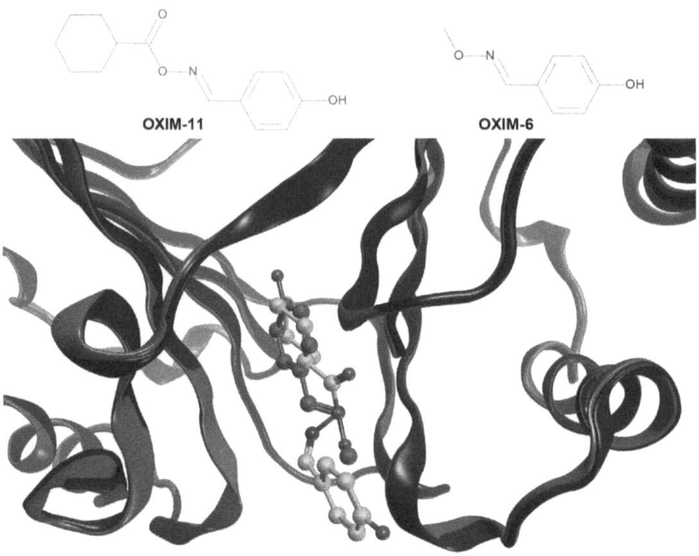

Figure 5 Example of a 3D activity cliff. OXIM-11 and OXIM-6 are carbonyloxime inhibitors of the macrophage migration inhibitory factor (MIF). The crystal structures of two highly similar compounds (PDB IDs: 2OOH and 2OOZ, respectively) revealed opposite orientations in the binding site.

An example of a 3D activity cliff is illustrated in Fig. 5. OXIM-6 and OXIM-11 are carbonyloxime-based compounds that inhibit the macrophage migration inhibitory factor (MIF), a proinflammatory cytokine critically involved in the pathogenesis of sepsis. Sepsis is still a lethal inflammatory disorder and a substantial health problem. Several small molecules have been identified as inhibitors of MIF using synthetic chemistry and virtual screening (Al-Abed et al., 2011; Cournia et al., 2009; McLean et al., 2010). As part of the efforts to validate that inhibition of the catalytic site could produce therapeutic benefits, the crystal structures of OXIM-6 and OXIM-11 (PDB IDs: 2OOZ and 2OOH, respectively) were obtained revealing two opposite and unexpected orientations in the binding site based on previous observations of other MIF inhibitors (Crichlow et al., 2007) (Fig. 5). The crystal structures of the two MIF inhibitory complexes provided valuable insights for later structure-based design efforts. Taken these insights together, 3D activity cliffs further illustrate the application of a chemoinformatics approach to advance the understanding of target–ligand interactions.

4.3. Structure-based activity cliffs and hot spots

Seebeck et al. introduced an approach for the identification of structure-based activity cliffs (ISAC) (Seebeck et al., 2011). This approach uses the valuable information of activity cliffs in a structure-based context by analyzing interaction energies of protein–ligand complexes. The authors of that work also presented a novel visualization of hot spots in the active site of a protein using the relative frequency at which a protein atom is involved in activity cliff events. ISAC is valuable to uncover the key interacting atoms of the binding site and facilitates the development of pharmacophore hypotheses that can be used as filters in virtual screening campaigns (Seebeck et al., 2011). As such, ISAC represents a comprehensive method that links activity cliff analysis, PLIs, and pharmacophore hypothesis. The ISAC approach uses ligand–receptor interactions of crystallized or docked complexes as descriptors for the similarity measure enabling the identification of activity cliffs at a structure-based level. In the method presented by Seebeck et al., a matrix of interaction scores is calculated per protein atom and interaction type. Each row in the matrix represents one ligand (one compound in the data set) and each column depicts the score for one specific protein atom of the active site and a certain interaction type (e.g., hydrogen bonds, ionic interactions, aromatic interactions, hydrophobic contacts). For each pair of compounds, the relationship between potency difference and protein–ligand contact similarity is assessed using the Structure–Activity Landscape Index (SALI) approach. SALI values are calculated with the expression (Guha & Van Drie, 2008a, 2008b):

$$\mathrm{SALI}_{i,j} = \frac{|A_i - A_j|}{1 - \mathrm{sim}(i,j)} \tag{1}$$

where A_i and A_j are the activities of the ith and jth molecules and sim(i,j) is the similarity coefficient between the two molecules. SALI was initially developed to compare compounds measuring molecular similarity using a fingerprint-based representation. However, as shown by Seebeck et al., the molecular similarity can be assessed using PLI information (protein–ligand contact similarity). Thus, compound pairs with high SALI values represent structure-based activity cliffs: pairs of compounds with very similar interaction patterns but very different activities. The authors state that "the use of protein–ligand interaction descriptors has the advantage of investigating activity cliffs completely independently from functional groups and the topology of the ligand. Thus, structurally different ligands with similar potencies, which can be explained by similar interaction profiles, are captured by the ISAC approach."

Note that, in the work of Seebeck et al. the matrix of protein–ligand energies (that is generic in terms of the scoring function) was transformed to binary bit vectors by using thresholds for each interaction score. However, the approach can be extended to accommodate similarities between protein–ligand contacts using basically any other schemes of PLIFs.

4.4. Activity cliff generators and structural interpretation

An activity cliff generator has been defined as a molecule with high probability to form activity cliffs with structurally similar molecules tested in the same biological assay (Mendez-Lucio, Perez-Villanueva, Castillo, & Medina-Franco, 2012). Mendez-Lucio et al. reported the identification of activity cliff generators based on Structure–Activity Similarity (SAS) maps and frequency counts (Mendez-Lucio et al., 2012). SAS maps are 2D plots of activity similarity (or potency difference) versus structural similarity. All possible pairs of compounds can be represented in a SAS map (Medina-Franco, 2012; Shanmugasundaram & Maggiora, 2001). Pairs of compounds that correspond to activity cliffs can be easily recognized in the quadrant that intersects pairs of molecules with high structure similarity but low activity similarity (or high potency difference). Activity cliff generators can be easily recognized as compounds with very high frequency (e.g., two standard deviations above average) in the "activity cliff quadrant" (or region) of the SAS map (Mendez-Lucio et al., 2012). In order to illustrate this approach, Mendez-Lucio et al. systematically identified and analyzed the activity cliff generators present in a data set of 168 compounds tested against three peroxisome-proliferator-activated receptor (PPAR) subtypes. PPARs are nuclear lipid-activated transcription factors that have been identified as major regulators in glucose and lipids metabolism; thereby they contribute significantly to some disorders such as diabetes, obesity, and cardiovascular complications (Nevin, Lloyd, & Fayne, 2011; Willson, Brown, Sternbach, & Henke, 2000). Results of that work gave rise to the identification of activity cliff generators for PPARα and δ, as well as dual-activity cliff generators for those receptors. Molecular docking calculations and a deeper analysis of PLIs of the activity cliff generators helped to uncover common structural features that have a great impact on activity providing a structure-based interpretation of the cliff-forming features of these compounds.

A word of caution of identifying activity cliff generators using SAS maps is the threshold used to define the activity cliff region of the landscape. Certainly, the thresholds to define quantitatively "high" (or "low") structural or activity similarity are tailored to the specific project needs (Medina-Franco,

2012; Stumpfe et al., 2014). An alternative approach to identify activity cliff generators is to identify the most frequent compounds among the pairs with the highest SALI values. A SALI value can be considered "high" relative to the distribution of the data set.

4.5. Interaction cliffs

In order to gain direct structural interpretation of activity cliffs, Mendez-Lucio et al. carried out a study integrating PLIs to a multitarget kinase activity landscape. In this study, the authors used three data sets, containing the crystallographic structure of the ligand bound to a kinase, extracted from KLIFS database (see above). Pairwise interaction similarity was assessed using PLIFs and the Tanimoto coefficient, whereas twelve 2D and 3D molecular descriptors were used to compute pairwise molecular similarity. Results show that pairwise structure similarity has no correlation with interaction similarity in none of the data sets, even the kinase ATP binding site is highly conserved. In average, only 33% of the molecular pairs categorized as highly similar showed similar interactions. This approach not only provided structural information of activity cliffs but also was useful to identify hot spots in the target protein associated with selectivity.

Figure 6A shows an example of SAS map generated using ComboScore as 3D molecular descriptor and with added interaction information. Colored points show those pairs of molecules that present similar interactions with the target kinase. As showed in this figure, not only those compounds with high molecular similarity show high interaction similarity, but also those molecular pairs with different chemical structures can present similar ligand–target interactions. Figure 6B shows an example of interaction cliff with the aurora kinase (AURKA) inhibitor Tozasertib (VX-680) bound to two different kinases. In these cases, the inhibitor forms the same hydrogen bond interactions with both, AURKA (APDB ID: 3E5A) and a mutant of cAMP-dependent protein kinase (PDB ID: 3AMB). Nevertheless, the π-staking interaction observed with the AURKA binding site increases the potency by two log units. Using the same strategy, authors also were able to identify pairs of compounds with different chemical structure, but presenting similar PLIs and hence similar potency. These pairs of compounds are the so-called *scaffold hops* as the one presented in Fig. 6C. In this example, both compounds form hydrogen bonds with amino acids in the same positions in both targets. Moreover, they present conserved hydrophobic interactions suggesting that the binding site of both targets have similar shape.

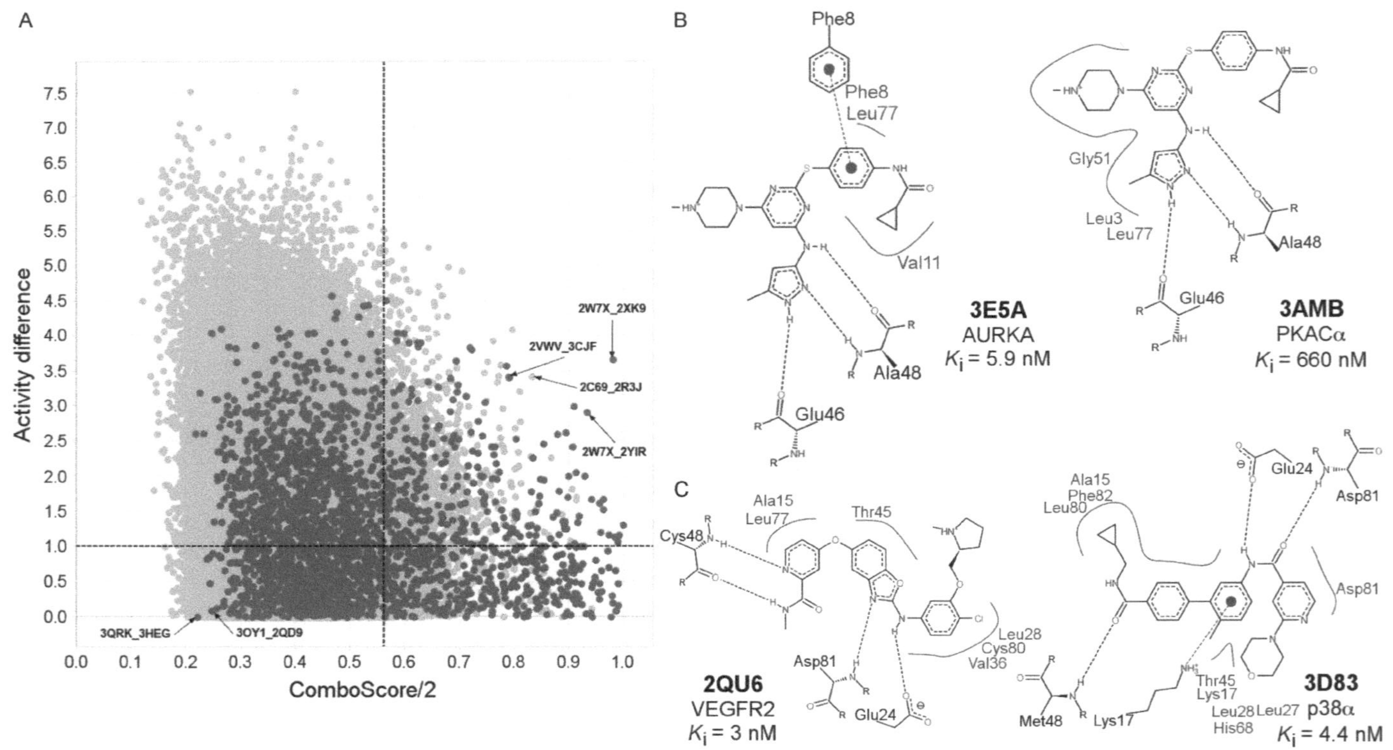

Figure 6 (A) Example of a Structure–Activity–Interaction Similarity (SAS) map containing 83,436 data points, resulting from the pairwise comparisons of 409 kinase crystal structures. Data points are color coded to highlight those molecular pairs with high interaction similarity, that is, two standard deviations above mean similarity for each data set. (B, C) An example of interaction cliff and scaffold hop, respectively, identified in the Kinase–Ligand Interaction Fingerprints and Structure database using a chemoinformatic approach.

The authors of this chapter showed that the added information given by the IFPs is very valuable to understand and rationalize activity cliffs from both the ligand and target point of view.

5. TARGET–LIGAND RELATIONSHIPS IN CHEMOGENOMICS DATA SETS

The augmented awareness of polypharmacology, i.e., that a drug may have its clinical effect through the interaction of multiple targets, is shifting the drug discovery paradigm from a single to a multitarget approach (Medina-Franco, Giulianotti, Welmaker, & Houghten, 2013). In line with the increasing importance of polypharmacology, there is an increase in chemogenomics data sets that capture the ligand–target relationships (Rognan, 2013). As such, experimental and computational approaches are emerging for the generation, storage, analysis, mining, and visualization of target–ligand interactions that define chemogenomic spaces (Bajorath, 2013; Medina-Franco & Aguayo-Ortiz, 2013).

Significant advances have been made to compile in public repositories activity data of compound data sets screened against one or multiple targets. Notable examples of large databases are PubChem, ChEMBL, and Binding Database (Nicola, Liu, & Gilson, 2012). Significant efforts are being made to develop chemoinformatic tools to efficiently mine and navigate through such large bioactive collections of chemical compounds (Kim, Bolton, & Bryant, 2013; Takada, Ohmori, & Okada, 2013).

Other example is the large microarray data published by Clemons et al. that contain the binding profile of more than 15,000 compounds including natural products, commercial compounds, and synthetic molecules from academic groups across 100 sequence-unrelated proteins (Clemons et al., 2010). Structure–multiple activity relationship studies have been conducted with this data set. For instance, Yongye and Medina-Franco developed a general approach for identifying structural changes that have a significant impact on the number of proteins to which a compound binds using the Structure–Promiscuity Index Difference (SPID) metric. SPID encodes the relationship between structure similarity and the number of different proteins to which each pair of compound binds (Yongye & Medina-Franco, 2012). In a subsequent study, Dimova et al. employed the concept of MMP to analyze the same data set to identify single-site substitutions that are associated with large magnitude differences in apparent compound promiscuity (Dimova, Hu, & Bajorath, 2012). The results of Dimova et al.

further confirmed the results of Yongye and Medina-Franco previously published in that promiscuity can be induced by small chemical substitutions.

5.1. Analyzing chemogenomic sets using target–ligand networks

Analyzing ligand–target relationships taking into account polypharmacologic interactions is not an easy task because of the different number of targets involved for each ligand. Recently, the use of network theory for the analysis of drug–target interactions has increased due to "the ability to capture complexity in a simple, compact and illustrative manner" (Vogt & Mestres, 2010). Ligand–target networks are mathematical models where nodes represent ligands and targets and the edge linking two nodes represents a cross-linking interaction, e.g., IC_{50} or affinity above a predefined threshold (Medina-Franco & Aguayo-Ortiz, 2013; Vogt & Mestres, 2010). In this regard, novel methods, such as sparse canonical correspondence analysis (Yamanishi, Pauwels, Saigo, & Stovent, 2011) and sparsity-induced binary classifiers (Tabei, Pauwels, Stoven, Takemoto, & Yamanishi, 2012), have been developed to study associations between chemical substructures and protein domains as a technique to extract more information from ligand–target interaction networks. These methods are applied to detect chemical substructures associated with selectivity and molecular scaffolds that increase activity against a protein family. A more detailed description of these methods and their applications has been reviewed previously by Yamanishi (2013).

One of the applications of PLI networks is the visualization and analysis of large chemogenomic data set. In this sense, Paolini et al. were able to perform a large-scale network analysis on 275,000 bioactive compounds and over 600,000 activity data points (Paolini, Shapland, van Hoorn, Mason, & Hopkins, 2006). With this analysis, the authors identified new and unexpected relationships between chemical structures and distant targets in a pharmacology interaction network. Interestingly, 700 proteins resulted to be connected by 12,119 interactions (i.e., the same compound active in two proteins). In the same analysis, it was possible to identify the most promiscuous targets which are G protein-coupled receptors (GPCRs), cytochrome P450s, and protein kinases (Paolini et al., 2006). The use of networks to analyze ligand–target interactions also has been extended to analyze natural products such as the components of Traditional Chinese Medicine (Li & Zhang, 2013; Zhao, Zhou, Ma, & Wei, 2013). In this context, Gu et al. generated a PLI network for the 676 molecules contained in the eleven Chinese herb medicines of Tangminling pills (Gu et al., 2011). The authors of this work identified the action mechanism of Tangminling pills as a treatment for diabetes mellitus 2 (DM2) using interaction networks. Moreover, they identified five novel compounds, whose relevance to DM2 was unknown.

The application of ligand–target interaction networks goes beyond visualization and analysis. They also have been applied for target prediction and drug repurposing (Cheng et al., 2012). One example is the work conducted by Cheng et al. where they used 12,483 FDA-approved and experimental drug–targets interactions and were able to predict and validate new targets for five drugs, namely montelukast, diclofenac, simvastatin, ketoconazole, and itraconazole (Cheng et al., 2012). In a separate work, the same authors integrated chemical and therapeutic spaces with side effects using interaction networks to predict pharmacological profiles (Cheng et al., 2013). The network was generated from 621 approved drugs and 856 targets and developed the drug side effect similarity inference method.

5.2. Proteochemometric modeling

An alternative to analyze PLIs from chemogenomic data is proteochemometric modeling (PCM). This technique has been developed as a bioactivity modeling method that combines the chemical (drug) and biological (protein target) space (van Westen, Wegner, Ijzerman, van Vlijmen, & Bender, 2011). The simultaneous extrapolation of both spaces allows the quantitative evaluation of target and ligand structural similarity across related ligands in order to find multitarget SAR. For this technique, compounds can be encoded by topological descriptors, physicochemical descriptors, molecular interaction fields, etc., whereas the target information is captured by sequential protein descriptors or 3D protein descriptors. In some cases, a cross-term, e.g., PLIFs, is added to model particular interaction between ligands and targets (van Westen, Wegner, Ijzerman, et al., 2011). All this information is modeled simultaneously using machine learning methods, for example, random forest, support vector machines, and neural networks, but other linear or nonlinear methods can be used. PCM has been nicely reviewed by van Westen, Wegner, Ijzerman, et al. (2011).

PCM has been used to study ligand–target interaction in several biological situations that include HIV reverse transcriptase mutants (van Westen et al., 2013; van Westen, Wegner, Geluykens, et al., 2011), to predict

CYP450 inhibitors (Lapins et al., 2013), to study dengue virus protease inhibitors (Prusis et al., 2008), protein kinases (Fernandez, Ahmad, & Sarai, 2010; Lapins & Wikberg, 2010), and GPCRs (van Westen et al., 2012), among others. One elegant example of PCM was performed on GPCRs by van Westen et al. (2012). In this work, the authors generated a PCM for four adenosine receptor subtypes using combined activity data of rats and humans. The final model was used to screen a library of >10,000 compounds, identifying six highly active ligands; in some cases, the potency was in the nanomolar range. PCM can also be applied to reveal new ligand–target interactions, e.g., in the deorphanization of drug targets (van Westen, Wegner, Ijzerman, et al., 2011).

6. PROTEIN–PROTEIN INTERACTIONS

PPIs are part of the so-called interactome, i.e., the complete set of interactions in a living organism (Garcia-Garcia et al., 2012). The regulation of PPIs is an attractive strategy in drug discovery. This is because many cellular functions are regulated by multiprotein complexes that are controlled by PPIs between protein subunits. It is well known that human diseases can be caused by abnormal PPIs. Therefore, PPI modulators, either inhibitors or stabilizing agents, are attractive in drug discovery (Zinzalla & Thurston, 2009). For example, tirofiban and maraviroc are drugs that target PPIs and are approved for clinical use. Tirofiban is an antiplatelet drug and maraviroc is an antiretroviral drug used in the treatment of HIV infection.

The interaction between proteins can be analyzed experimentally and computationally at different levels of detail, from a high structural level (e.g., specific molecular interactions at the protein–protein interface, conformational changes that occur during the interaction) to lower levels such as the coexpression and colocalization. In an excellent review, Garcia-Garcia et al. discuss experimental and computational approaches used to characterize PPIs at different degrees of resolution, including goals and challenges of each method (Garcia-Garcia et al., 2012).

Protein–protein binding interfaces are characterized by the presence of "hot spots," that is, residues that provide a large fraction of the binding free energy. Using experimental approaches such as alanine scanning is known that residues frequently found in hot spots are tryptophan, arginine, and tyrosine. Tyrosine, phenylalanine, tryptofan, and leucine are considered as typical "anchor residues," that is, residues with large buried area whose presence should reveal druggable pockets (small-molecule binding pockets) at the interface of protein–protein complex (Falchi, Caporuscio, & Recanatini, 2014).

General approaches to design and develop PPIs modulators include biophysical methods such as NMR and X-ray crystallography, fragment-based approaches, high-throughput screening, and computational or *in silico* approaches. Bienstock has recently reviewed recent advances in computational approaches for computational protein–protein docking, *in silico* methods to identify protein interface hot spots, databases to classify protein–protein interfaces into categories and to generalize modes of protein interaction, and successful design antagonists and small-molecule inhibitors for PPIs (Bienstock, 2012).

A recent successful example of virtual screening for PPI inhibitors is exemplified by the work of the group of Meurice and coworkers that discovered a small molecule that disrupts the interaction between TWEAK and Fn14. TWEAK is a multifunctional cytokine controlling a number of cellular activities and exerts its effect by binding to Fn14, a member of the TNFR superfamily. Dysregulation of TWEAK–Fn14 signaling is observed in cancer and several other disease states. Protein–protein docking followed by data-driven prioritization suggested two promising TWEAK–Fn14 binding hypotheses. Mutagenesis analysis confirmed one hypothesis, providing a novel structural basis for target-based identification of small-molecule inhibitors of the TWEAK–FN14 interaction. A focused compound data set was built using high-throughput docking and pharmacophore-based virtual screening. Experimental iterative screening of the targeted library led to identification of molecules producing up to 37% inhibition of TWEAK–Fn14 binding and acting on mechanism (Dhruv et al., 2013). Several other successful applications of virtual screening protocols based on pharmacophore modeling, docking, and prediction of hot spots and druggable pockets have been extensively reviewed by Falchi et al. (2014).

Similar to drug discovery efforts focused on the interactions between small molecules and proteins, the increasing information related to identify modulators of PPIs demands the integration of chemoinformatic tools with classical molecular modeling for the efficient use of PPIs for drug discovery.

Using chemoinformatics and machine learning methods, Neugebauer et al. constructed and validated a decision tree to differentiate a set of 25 inhibitors of PPIs structurally diverse from 1137 approved drugs and small molecules stored in the ZINC database (Neugebauer, Hartmann, & Klein, 2007). The decision tree contains three descriptors of each; the authors identified that one constitutional descriptor related to shape and size is of major importance (Neugebauer et al., 2007).

More recently, Hamon et al. developed a machine learning tool termed 2P2I HUNTER for filtering putative orthosteric PPI modulators. Using

2P2I HUNTER, the authors design a PPI-focused library with 143,218 small molecules from chemical providers. To design this library, the machine learning tool was applied to 8.3 million compounds from commercial sources (Hamon et al., 2013). A subset of 51,476 compounds was further prioritized based on chemical scaffolds considered as privileged scaffolds in medicinal chemistry. Further selection based on structural diversity and structural complexity leads to design a focused set of 1683 compounds as potential PPI modulators (Hamon et al., 2013).

Cao et al. developed the platform termed PyDPI (drug–protein interaction with Python). PyDPI is a phyton toolkit for calculating structural and physicochemical features of proteins and peptides from amino acid sequences, molecular descriptors of drug molecules from their topology, and PPI and PLI descriptors. This toolkit, freely available at https://sourceforge.net/projects/pydpicao/, is a "good example of the integration between chemoinformatics and bioinformatics into a chemogenomics platform for drug discovery." PyDPI computes six major types of descriptors for proteins and peptides that have previously used for predicting protein and peptide-related problems. For small molecules, the phyton package computes 12 groups of molecular descriptors (Cao et al., 2013).

Uchikoga and Hirokawa implemented IFPs to process results of protein–protein docking, with special emphasis on clustering solutions of docking flexible proteins. IFPs are based on binary states of interacting amino acid residues and were used as a means for measuring unique similarities between the complex structures. IFPs offer an alternative to comparing solutions of protein–protein docking based on RMSD (Uchikoga & Hirokawa, 2010). Uchikoga and Hirokawa commented that IFP allows examination of the properties of PPIs simply by comparing the docking structures in terms of their interaction patterns using a metric commonly used for small molecules such as the Tanimoto coefficient. Thus, by using IFP, one could select the near-native structures at the contact residue string level, rather than obtaining the exact complex structure at the Cartesian coordinate level.

7. CONCLUSIONS

The increasing availability of 3D structures of molecular targets and corresponding applications of structure-based design have boosted the need to handle, interpret, and visualize PLI and PPI in an intuitive manner. Moreover, several current drug discovery projects involve the analysis of large data sets of protein–ligand and protein–protein complexes. A notable example is the increasing development of chemogenomics data sets. The systematic analysis, mining, and visual representations of such data sets often require computational approaches. While typical molecular modeling methods are developed to analyze and quantify in great detail the interactions involved between protein–ligands and proteins–proteins, the management of large quantities of data requires the integration of chemoinformatic methods. The development and application of PLIFs is a primary example of the synergy between molecular modeling and chemoinformatics approaches to navigate through PLI landscapes. PLIFs have many applications such as the classification and selection of representative complex structures, analysis of docking results, filtering criteria in virtual screening, and starting points to generate pharmacophoric queries. PLIFs and 2D interaction diagrams are applications of computer-generated representations that are used not only by experts in molecular modeling and chemoinformatics but also by other research areas such as medicinal chemistry. A second major area with a significant overlap between molecular modeling and chemoinformatics is modeling activity landscapes. Modeling activity landscapes is an emerging concept to systematically analyze SAR. In particular, activity landscape methods are tuned for the rapid identification of activity cliffs. The identification and structure-based interpretation of compounds frequently involved in activity cliffs, i.e., activity cliff generators, has a significant impact on lead optimization and virtual screening. In addition, the structure-based interpretation of activity cliffs may lead to the identification of "hot spots" in the protein of the interacting partner. Computational methods commonly used to characterize PLIs can be adapted to analyze PPIs, for example, through the development of PPI fingerprints.

ACKNOWLEDGMENTS

O. M.-L. acknowledges CONACyT (No. 217442/312933) and the Cambridge Overseas Trust for funding. K. M.-M. thanks DGAPA-UNAM (PAPIIT IA200513-2). We thank Dr. Didier Rognan for providing Fig. 4 in high resolution and Dr. Roman A. Laskowski for providing an academic license of LigPlot$^+$.

REFERENCES

Akella, L. B., & DeCaprio, D. (2010). Cheminformatics approaches to analyze diversity in compound screening libraries. *Current Opinion in Chemical Biology*, *14*, 325–330.

Al-Abed, Y., Metz, C. N., Cheng, K. F., Aljabari, B., VanPatten, S., Blau, S., et al. (2011). Thyroxine is a potential endogenous antagonist of macrophage migration inhibitory factor (MIF) activity. *Proceedings of the National Academy of Sciences of the United States of America*, *108*, 8224–8227.

Bajorath, J. (2013). A perspective on computational chemogenomics. *Molecular Informatics, 32*, 1025–1028.
Balius, T. E., Mukherjee, S., & Rizzo, R. C. (2011). Implementation and evaluation of a docking-rescoring method using molecular footprint comparisons. *Journal of Computational Chemistry, 32*, 2273–2289.
Ballester, P. J., Schreyer, A., & Blundell, T. L. (2014). Does a more precise chemical description of protein–ligand complexes lead to more accurate prediction of binding affinity? *Journal of Chemical Information and Modeling, 54*, 944–955.
Baroni, M., Cruciani, G., Sciabola, S., Perruccio, F., & Mason, J. S. (2007). A common reference framework for analyzing/comparing proteins and ligands. Fingerprints for ligands and proteins (FLAP): Theory and application. *Journal of Chemical Information and Modeling, 47*, 279–294.
Bello, M., Martinez-Archundia, M., & Correa-Basurto, J. (2013). Automated docking for novel drug discovery. *Expert Opinion on Drug Discovery, 8*, 821–834.
Bienstock, R. J. (2012). Computational drug design targeting protein-protein interactions. *Current Pharmaceutical Design, 18*, 1240–1254.
Bohanec, S., & Zupan, J. (1991). Structure generation of constitutional isomers from structural fragments. *Journal of Chemical Information and Computer Sciences, 31*, 531–540.
Brewerton, S. C. (2008). The use of protein-ligand interaction fingerprints in docking. *Current Opinion in Drug Discovery & Development, 11*, 356–364.
Bryant, C., Kerr, I. D., Debnath, M., Ang, K. K. H., Ratnam, J., Ferreira, R. S., et al. (2009). Novel non-peptidic vinylsulfones targeting the s2 and s3 subsites of parasite cysteine proteases. *Bioorganic & Medicinal Chemistry Letters, 19*, 6218–6221.
Cao, D.-S., Liang, Y.-Z., Yan, J., Tan, G.-S., Xu, Q.-S., & Liu, S. (2013). PyDPI: Freely available python package for chemoinformatics, bioinformatics, and chemogenomics studies. *Journal of Chemical Information and Modeling, 53*, 3086–3096.
Cheng, F., Li, W., Wu, Z., Wang, X., Zhang, C., Li, J., et al. (2013). Prediction of polypharmacological profiles of drugs by the integration of chemical, side effect, and therapeutic space. *Journal of Chemical Information and Modeling, 53*, 753–762.
Cheng, F. X., Liu, C., Jiang, J., Lu, W. Q., Li, W. H., Liu, G. X., et al. (2012). Prediction of drug-target interactions and drug repositioning via network-based inference. *PLoS Computational Biology, 8*, e1002503.
Cho, Y., Vermeire, J. J., Merkel, J. S., Leng, L., Du, X., Bucala, R., et al. (2011). Drug repositioning and pharmacophore identification in the discovery of hookworm MIF inhibitors. *Chemistry & Biology, 18*, 1089–1101.
Chupakhin, V., Marcou, G., Baskin, I., Varnek, A., & Rognan, D. (2013). Predicting ligand binding modes from neural networks trained on protein–ligand interaction fingerprints. *Journal of Chemical Information and Modeling, 53*, 763–772.
Clark, A. M., & Labute, P. (2007). 2D depiction of protein–ligand complexes. *Journal of Chemical Information and Modeling, 47*, 1933–1944.
Clemons, P. A., Bodycombe, N. E., Carrinski, H. A., Wilson, J. A., Shamji, A. F., Wagner, B. K., et al. (2010). Small molecules of different origins have distinct distributions of structural complexity that correlate with protein-binding profiles. *Proceedings of the National Academy of Sciences of the United States of America, 107*, 18787–18792.
Clemons, P. A., Wilson, J. A., Dancik, V., Muller, S., Carrinski, H. A., Wagner, B. K., et al. (2011). Quantifying structure and performance diversity for sets of small molecules comprising small-molecule screening collections. *Proceedings of the National Academy of Sciences of the United States of America, 108*, 6817–6822.
Cournia, Z., Leng, L., Gandavadi, S., Du, X., Bucala, R., & Jorgensen, W. L. (2009). Discovery of human macrophage migration inhibitory factor (MIF)-CD74 antagonists via virtual screening. *Journal of Medicinal Chemistry, 52*, 416–424.

Crichlow, G. V., Cheng, K. F., Dabideen, D., Ochani, M., Aljabari, B., Pavlov, V. A., et al. (2007). Alternative chemical modifications reverse the binding orientation of a pharmacophore scaffold in the active site of macrophage migration inhibitory factor. *The Journal of Biological Chemistry*, *282*, 23089–23095.

Cruz-Monteagudo, M., Medina-Franco, J. L., Pérez-Castillo, Y., Nicolotti, O., Cordeiro, M. N. D. S., & Borges, F. (2014). Activity cliffs in drug discovery: Dr. Jekyll or Mr. Hyde? *Drug Discovery Today*. http://dx.doi.org/10.1016/j.drudis.2014.02.003.

Deng, Z., Chuaqui, C., & Singh, J. (2004). Structural interaction fingerprint (SIFt): A novel method for analyzing three-dimensional protein-ligand binding interactions. *Journal of Medicinal Chemistry*, *47*, 337–344.

Deng, Z., Chuaqui, C., & Singh, J. (2006). Knowledge-based design of target-focused libraries using protein–ligand interaction constraints. *Journal of Medicinal Chemistry*, *49*, 490–500.

Desaphy, J., Raimbaud, E., Ducrot, P., & Rognan, D. (2013). Encoding protein–ligand interaction patterns in fingerprints and graphs. *Journal of Chemical Information and Modeling*, *53*, 623–637.

Dhruv, H., Loftus, J. C., Narang, P., Petit, J. L., Fameree, M., Burton, J., et al. (2013). Structural basis and targeting of the interaction between fibroblast growth factor-inducible 14 and tumor necrosis factor-like weak inducer of apoptosis. *The Journal of Biological Chemistry*, *288*, 32261–32276.

Digles, D., & Ecker, G. F. (2011). Self-organizing maps for in silico screening and data visualization. *Molecular Informatics*, *30*, 838–846.

Dimova, D., Hu, Y., & Bajorath, J. (2012). Matched molecular pair analysis of small molecule microarray data identifies promiscuity cliffs and reveals molecular origins of extreme compound promiscuity. *Journal of Medicinal Chemistry*, *55*, 10220–10228.

Durrant, J., & McCammon, J. A. (2011). Molecular dynamics simulations and drug discovery. *BMC Biology*, *9*, 71.

Falchi, F., Caporuscio, F., & Recanatini, M. (2014). Structure-based design of small-molecule protein–protein interaction modulators: The story so far. *Future Medicinal Chemistry*, *6*, 343–357.

Fernandez, M., Ahmad, S., & Sarai, A. (2010). Proteochemometric recognition of stable kinase inhibition complexes using topological autocorrelation and support vector machines. *Journal of Chemical Information and Modeling*, *50*, 1179–1188.

Fricker, P. C., Gastreich, M., & Rarey, M. (2004). Automated drawing of structural molecular formulas under constraints. *Journal of Chemical Information and Computer Sciences*, *44*, 1065–1078.

Garcia-Garcia, J., Bonet, J., Guney, E., Fornes, O., Planas, J., & Oliva, B. (2012). Networks of protein-protein interactions: From uncertainty to molecular details. *Molecular Informatics*, *31*, 342–362.

Glide, v. (2012). *Glide*. New York: Schrödinger, LLC.

Gu, J. Y., Zhang, H., Chen, L. R., Xu, S., Yuan, G., & Xu, X. J. (2011). Drug-target network and polypharmacology studies of a traditional Chinese medicine for type II diabetes mellitus. *Computational Biology and Chemistry*, *35*, 293–297.

Guha, R. (2012). Exploring structure–activity data using the landscape paradigm. *Wiley Interdisciplinary Reviews: Computational Molecular Science*, *2*, 829–841.

Guha, R., & Van Drie, J. H. (2008a). Assessing how well a modeling protocol captures a structure-activity landscape. *Journal of Chemical Information and Modeling*, *48*, 1716–1728.

Guha, R., & Van Drie, J. H. (2008b). Structure-activity landscape index: Identifying and quantifying activity cliffs. *Journal of Chemical Information and Modeling*, *48*, 646–658.

Hamon, V., Brunel, J. M., Combes, S., Basse, M. J., Roche, P., & Morelli, X. (2013). 2P2Ichem: Focused chemical libraries dedicated to orthosteric modulation of protein-protein interactions. *Medicinal Chemistry Communications*, *4*, 797–809.

Holden, P. M., Allen, W. J., Gochin, M., & Rizzo, R. C. (2014). Strategies for lead discovery: Application of footprint similarity targeting HIVgp41. *Bioorganic & Medicinal Chemistry*, *22*, 651–661.

Hu, Y., & Bajorath, J. (2012). Exploration of 3D activity cliffs on the basis of compound binding modes and comparison of 2D and 3D cliffs. *Journal of Chemical Information and Modeling*, *52*, 670–677.

Hu, Y., Furtmann, N., Gütschow, M., & Bajorath, J. (2012). Systematic identification and classification of three-dimensional activity cliffs. *Journal of Chemical Information and Modeling*, *52*, 1490–1498.

Hu, X., Hu, Y., Vogt, M., Stumpfe, D., & Bajorath, J. (2012). MMP-cliffs: Systematic identification of activity cliffs on the basis of matched molecular pairs. *Journal of Chemical Information and Modeling*, *52*, 1138–1145.

Kelly, M. D., & Mancera, R. L. (2004). Expanded interaction fingerprint method for analyzing ligand binding modes in docking and structure-based drug design. *Journal of Chemical Information and Computer Sciences*, *44*, 1942–1951.

Kim, S., Bolton, E. E., & Bryant, S. H. (2013). PubChem3D: Conformer ensemble accuracy. *Journal of Cheminformatics*, *5*, 1.

Kuck, D., Singh, N., Lyko, F., & Medina-Franco, J. L. (2010). Novel and selective DNA methyltransferase inhibitors: Docking-based virtual screening and experimental evaluation. *Bioorganic & Medicinal Chemistry*, *18*, 822–829.

Langer, T. (2010). Pharmacophores in drug research. *Molecular Informatics*, *29*, 470–475.

Lapins, M., & Wikberg, J. E. S. (2010). Kinome-wide interaction modelling using alignment-based and alignment-independent approaches for kinase description and linear and non-linear data analysis techniques. *BMC Bioinformatics*, *11*, 339.

Lapins, M., Worachartcheewan, A., Spjuth, O., Georgiev, V., Prachayasittikul, V., Nantasenamat, C., et al. (2013). A unified proteochemometric model for prediction of inhibition of cytochrome P450 isoforms. *PLoS One*, *8*, e66566.

Laskowski, R. A., & Swindells, M. B. (2011). LigPlot^{+}: Multiple ligand–protein interaction diagrams for drug discovery. *Journal of Chemical Information and Modeling*, *51*, 2778–2786.

Li, S., & Zhang, B. (2013). Traditional Chinese medicine network pharmacology: Theory, methodology and application. *Chinese Journal of Natural Medicines*, *11*, 110–120.

Lopez-Vallejo, F., & Martinez-Mayorga, K. (2012). Furin inhibitors: Importance of the positive formal charge and beyond. *Bioorganic & Medicinal Chemistry*, *20*, 4462–4471.

Maestro, v. (2012). *Maestro*. New York: Schrödinger, LLC.

Maggiora, G. M. (2006). On outliers and activity cliffs—Why QSAR often disappoints. *Journal of Chemical Information and Modeling*, *46*, 1535.

McLean, L. R., Zhang, Y., Li, H., Choi, Y. M., Han, Z. N., Vaz, R. J., et al. (2010). Fragment screening of inhibitors for MIF tautomerase reveals a cryptic surface binding site. *Bioorganic & Medicinal Chemistry Letters*, *20*, 1821–1824.

Medina-Franco, J. L. (2012). Scanning structure–activity relationships with structure–activity similarity and related maps: From consensus activity cliffs to selectivity switches. *Journal of Chemical Information and Modeling*, *52*, 2485–2493.

Medina-Franco, J. L. (2013). Activity cliffs: Facts or artifacts? *Chemical Biology & Drug Design*, *81*, 553–556.

Medina-Franco, J. L., & Aguayo-Ortiz, R. (2013). Progress in the visualization and mining of chemical and target spaces. *Molecular Informatics*, *32*, 942–953.

Medina-Franco, J. L., Giulianotti, M. A., Welmaker, G. S., & Houghten, R. A. (2013). Shifting from the single to the multitarget paradigm in drug discovery. *Drug Discovery Today*, *18*, 495–501.

Medina-Franco, J. L., Maggiora, G. M., Giulianotti, M. A., Pinilla, C., & Houghten, R. A. (2007). A similarity-based data-fusion approach to the visual characterization and comparison of compound databases. *Chemical Biology & Drug Design, 70,* 393–412.

Medina-Franco, J. L., Martínez-Mayorga, K., Bender, A., Marín, R. M., Giulianotti, M. A., Pinilla, C., et al. (2009). Characterization of activity landscapes using 2D and 3D similarity methods: Consensus activity cliffs. *Journal of Chemical Information and Modeling, 49,* 477–491.

Medina-Franco, J. L., Martínez-Mayorga, K., Giulianotti, M. A., Houghten, R. A., & Pinilla, C. (2008). Visualization of the chemical space in drug discovery. *Current Computer-Aided Drug Design, 4,* 322–333.

Medina-Franco, J. L., Martinez-Mayorga, K., & Meurice, N. (2014). Balancing novelty with confined chemical space in modern drug discovery. *Expert Opinion on Drug Discovery, 9,* 151–165.

Medina-Franco, J. L., & Yoo, J. (2013). Molecular modeling and virtual screening of DNA methyltransferase inhibitors. *Current Pharmaceutical Design, 19,* 2138–2147.

Mendez-Lucio, O., Perez-Villanueva, J., Castillo, R., & Medina-Franco, J. L. (2012). Identifying activity cliff generators of PPAR ligands using SAS maps. *Molecular Informatics, 31,* 837–846.

Méndez-Lucio, O., Tran, J., Medina-Franco, J. L., Meurice, N., & Muller, M. (2014). Towards drug repurposing in epigenetics: Olsalazine as a novel hypomethylating compound active in a cellular context. *ChemMedChem, 9,* 560–565.

Meslamani, J., Rognan, D., & Kellenberger, E. (2011). Sc-PDB: A database for identifying variations and multiplicity of 'druggable' binding sites in proteins. *Bioinformatics, 27,* 1324–1326.

Molecular Operating Environment (MOE), version 2013.08. (2013). Montreal, Quebec, Canada: Chemical Computing Group Inc. http://www.chemcomp.com.

Neugebauer, A., Hartmann, R. W., & Klein, C. D. (2007). Prediction of protein–protein interaction inhibitors by chemoinformatics and machine learning methods. *Journal of Medicinal Chemistry, 50,* 4665–4668.

Nevin, D. K., Lloyd, D. G., & Fayne, D. (2011). Rational targeting of peroxisome proliferating activated receptor subtypes. *Current Medicinal Chemistry, 18,* 5598–5623.

Nicola, G., Liu, T., & Gilson, M. K. (2012). Public domain databases for medicinal chemistry. *Journal of Medicinal Chemistry, 55,* 6987–7002.

O'Donoghue, S. I., Goodsell, D. S., Frangakis, A. S., Jossinet, F., Laskowski, R. A., Nilges, M., et al. (2010). Visualization of macromolecular structures. *Nature Methods, 7,* S42–S55.

Owen, J. R., Nabney, I. T., Medina-Franco, J. L., & López-Vallejo, F. (2011). Visualization of molecular fingerprints. *Journal of Chemical Information and Modeling, 51,* 1552–1563.

Paolini, G. V., Shapland, R. H. B., van Hoorn, W. P., Mason, J. S., & Hopkins, A. L. (2006). Global mapping of pharmacological space. *Nature Biotechnology, 24,* 805–815.

Pearlman, R. S., & Smith, K. M. (1998). Novel software tools for chemical diversity. *Perspectives in Drug Discovery and Design, 9–11,* 339–353.

Perez-Nueno, V. I., Rabal, O., Borrell, J. I., & Teixido, J. (2009). APIF: A new interaction fingerprint based on atom pairs and its application to virtual screening. *Journal of Chemical Information and Modeling, 49,* 1245–1260.

Poli, G., Tuccinardi, T., Rizzolio, F., Caligiuri, I., Botta, L., Granchi, C., et al. (2013). Identification of new Fyn kinase inhibitors using a FLAP-based approach. *Journal of Chemical Information and Modeling, 53,* 2538–2547.

Poongavanam, V., & Kongsted, J. (2013). Virtual screening models for prediction of HIV-1 RT associated RNase H inhibition. *PLoS One, 8,* e73478.

Prusis, P., Lapins, M., Yahorava, S., Petrovska, R., Niyomrattanakit, P., Katzenmeier, G., et al. (2008). Proteochemometrics analysis of substrate interactions with dengue virus NS3 proteases. *Bioorganic & Medicinal Chemistry, 16,* 9369–9377.

Rabal, O., & Oyarzabal, J. (2012). Biologically relevant chemical space navigator: From patent and structure–activity relationship analysis to library acquisition and design. *Journal of Chemical Information and Modeling*, *52*, 3123–3137.
Ritchie, T. J., Ertl, P., & Lewis, R. (2011). The graphical representation of ADME-related molecule properties for medicinal chemists. *Drug Discovery Today*, *16*, 65–72.
Rognan, D. (2013). Towards the next generation of computational chemogenomics tools. *Molecular Informatics*, *32*, 1029–1034.
Sauer, W. H. B., & Schwarz, M. K. (2003). Molecular shape diversity of combinatorial libraries: A prerequisite for broad bioactivity. *Journal of Chemical Information and Computer Sciences*, *43*, 987–1003.
Schrödinger Suite 2012 Protein Preparation Wizard. Epik version 2.3. (2012). New York: Schrödinger; Impact version 5.8. (2005). New York: Schrödinger, LLC; Prime version 3.1. (2012). New York: Schrödinger, LLC.
Scior, T., Bender, A., Tresadern, G., Medina-Franco, J. L., Martínez-Mayorga, K., Langer, T., et al. (2012). Recognizing pitfalls in virtual screening: A critical review. *Journal of Chemical Information and Modeling*, *52*, 867–881.
Seebeck, B., Wagener, M., & Rarey, M. (2011). From activity cliffs to target-specific scoring models and pharmacophore hypotheses. *ChemMedChem*, *6*, 1630–1639.
Shanmugasundaram, V., & Maggiora, G. M. (2001). Characterizing property and activity landscapes using an information-theoretic approach. In *CINF-032 222nd ACS National Meeting, Chicago, IL*, Washington, DC: American Chemical Society.
Siedlecki, P., Boy, R. G., Musch, T., Brueckner, B., Suhai, S., Lyko, F., et al. (2006). Discovery of two novel, small-molecule inhibitors of DNA methylation. *Journal of Medicinal Chemistry*, *49*, 678–683.
Sirci, F., Istyastono, E. P., Vischer, H. F., Kooistra, A. J., Nijmeijer, S., Kuijer, M., et al. (2012). Virtual fragment screening: Discovery of histamine H3 receptor ligands using ligand-based and protein-based molecular fingerprints. *Journal of Chemical Information and Modeling*, *52*, 3308–3324.
Stierand, K., & Rarey, M. (2007). From modeling to medicinal chemistry: Automatic generation of two-dimensional complex diagrams. *ChemMedChem*, *2*, 853–860.
Stierand, K., & Rarey, M. (2010). Drawing the PDB: Protein–ligand complexes in two dimensions. *ACS Medicinal Chemistry Letters*, *1*, 540–545.
Stierand, K., & Rarey, M. (2011). Flat and easy: 2D depiction of protein-ligand complexes. *Molecular Informatics*, *30*, 12–19.
Stumpfe, D., Hu, Y., Dimova, D., & Bajorath, J. (2014). Recent progress in understanding activity cliffs and their utility in medicinal chemistry. *Journal of Medicinal Chemistry*, *57*, 18–28.
Tabei, Y., Pauwels, E., Stoven, V., Takemoto, K., & Yamanishi, Y. (2012). Identification of chemogenomic features from drug–target interaction networks using interpretable classifiers. *Bioinformatics*, *28*, i487–i494.
Takada, N., Ohmori, N., & Okada, T. (2013). Mining basic active structures from a large-scale database. *Journal of Cheminformatics*, *5*, 15.
Tan, L., Batista, J., & Bajorath, J. (2010). Computational methodologies for compound database searching that utilize experimental protein-ligand interaction information. *Chemical Biology & Drug Design*, *76*, 191–200.
Uchikoga, N., & Hirokawa, T. (2010). Analysis of protein-protein docking decoys using interaction fingerprints: Application to the reconstruction of CaM-ligand complexes. *BMC Bioinformatics*, *11*, 236.
van Linden, O. P. J., Kooistra, A. J., Leurs, R., de Esch, L. J. P., & de Graaf, C. (2014). KLIFS: A knowledge-based structural database to navigate kinase-ligand interaction space. *Journal of Medicinal Chemistry*, *57*, 249–277.

van Westen, G. J. P., Hendriks, A., Wegner, J. K., Ijzerman, A. P., van Vlijmen, H. W. T., & Bender, A. (2013). Significantly improved HIV inhibitor efficacy prediction employing proteochemometric models generated from antivirogram data. *PLoS Computational Biology*, *9*, e1002899.

van Westen, G. J. P., van den Hoven, O. O., van der Pijl, R., Mulder-Krieger, T., de Vries, H., Wegner, J. K., et al. (2012). Identifying novel adenosine receptor ligands by simultaneous proteochemometric modeling of rat and human bioactivity data. *Journal of Medicinal Chemistry*, *55*, 7010–7020.

van Westen, G. J. P., Wegner, J. K., Geluykens, P., Kwanten, L., Vereycken, I., Peeters, A., et al. (2011). Which compound to select in lead optimization? Prospectively validated proteochemometric models guide preclinical development. *PLoS One*, *6*, e27518.

van Westen, G. J. P., Wegner, J. K., Ijzerman, A. P., van Vlijmen, H. W. T., & Bender, A. (2011). Proteochemometric modeling as a tool to design selective compounds and for extrapolating to novel targets. *Medicinal Chemistry Communications*, *2*, 16–30.

Virshup, A. M., Contreras-García, J., Wipf, P., Yang, W., & Beratan, D. N. (2013). Stochastic voyages into uncharted chemical space produce a representative library of all possible drug-like compounds. *Journal of the American Chemical Society*, *135*, 7296–7303.

Vogt, I., & Mestres, J. (2010). Drug-target networks. *Molecular Informatics*, *29*, 10–14.

Wallace, A. C., Laskowski, R. A., & Thornton, J. M. (1995). Ligplot: A program to generate schematic diagrams of protein-ligand interactions. *Protein Engineering*, *8*, 127–134.

Wawer, M., Lounkine, E., Wassermann, A. M., & Bajorath, J. (2010). Data structures and computational tools for the extraction of SAR information from large compound sets. *Drug Discovery Today*, *15*, 630–639.

Weisel, M., Bitter, H.-M., Diederich, F., So, W. V., & Kondru, R. (2012). Prolix: Rapid mining of protein–ligand interactions in large crystal structure databases. *Journal of Chemical Information and Modeling*, *52*, 1450–1461.

Willson, T. M., Brown, P. J., Sternbach, D. D., & Henke, B. R. (2000). The PPARs: From orphan receptors to drug discovery. *Journal of Medicinal Chemistry*, *43*, 527–550.

Yamanishi, Y. (2013). Inferring chemogenomic features from drug-target interaction networks. *Molecular Informatics*, *32*, 991–999.

Yamanishi, Y., Pauwels, E., Saigo, H., & Stovent, V. (2011). Extracting sets of chemical substructures and protein domains governing drug-target interactions. *Journal of Chemical Information and Modeling*, *51*, 1183–1194.

Yongye, A., Byler, K., Santos, R., Martínez-Mayorga, K., Maggiora, G. M., & Medina-Franco, J. L. (2011). Consensus models of activity landscapes with multiple chemical, conformer and property representations. *Journal of Chemical Information and Modeling*, *51*, 1259–1270.

Yongye, A. B., & Medina-Franco, J. L. (2012). Data mining of protein-binding profiling data identifies structural modifications that distinguish selective and promiscuous compounds. *Journal of Chemical Information and Modeling*, *52*, 2454–2461.

Yoo, J., Kim, J. H., Robertson, K. D., & Medina-Franco, J. L. (2012). Molecular modeling of inhibitors of human DNA methyltransferase with a crystal structure: Discovery of a novel DNMT1 inhibitor. *Advances in Protein Chemistry and Structural Biology*, *87*, 219–247.

Yoo, J., & Medina-Franco, J. L. (2012). Trimethylaurintricarboxylic acid inhibits human DNA methyltransferase 1: Insights from enzymatic and molecular modeling studies. *Journal of Molecular Modeling*, *18*, 1583–1589.

Zhao, M. Z., Zhou, Q., Ma, W. H., & Wei, D. Q. (2013). Exploring the ligand-protein networks in traditional Chinese medicine: Current databases, methods, and applications. *Evidence-Based Complementary and Alternative Medicine*, *2013*, article ID 806072, 15 pages.

Zinzalla, G., & Thurston, D. E. (2009). Targeting protein-protein interactions for therapeutic intervention: A challenge for the future. *Future Medicinal Chemistry*, *1*, 65–93.

CHAPTER TWO

Computational Study of Putative Residues Involved in DNA Synthesis Fidelity Checking in *Thermus aquaticus* DNA Polymerase I

Angela A. Elias, G. Andrés Cisneros[1]
Department of Chemistry, Wayne State University, Detroit, Michigan, USA
[1]Corresponding author: e-mail address: andres@chem.wayne.edu

Contents

Abstract

A fidelity-checking site for DNA polymerase I has been proposed based on recent single-molecule Förster resonance energy transfer studies. The checking site is believed to ensure proper base pairing of the newly inserted nucleotide. Computational studies have been utilized to predict residues involved in this putative checking site on the Klenow and *Bacillus* fragments. Here, we employ energy decomposition analysis, electrostatic free energy response, and noncovalent interaction plots to identify the residues involved in the hypothesized checking site in the homologous Klenow fragment from *Thermus aquaticus* (Klentaq). Our results indicate multiple protein residues that show altered interactions for three mispairs compared to the correctly paired DNA dimer. Many of these residues are also conserved along A family polymerases.

Advances in Protein Chemistry and Structural Biology, Volume 96
ISSN 1876-1623
http://dx.doi.org/10.1016/bs.apcsb.2014.06.003

1. INTRODUCTION

The replication of DNA is required for the survival of every living organism. Errors in this process can cause disease and, in more severe cases, cell death (Kunkel, 2003). In *Escherichia coli*, the replication error rate is approximately 10^{-10} (Schaaper, 1993). Arguably, one of the most important enzyme families in the fidelity of DNA replication is the DNA polymerase family. DNA polymerases are divided into seven families depending on their structure and particular function, these include A, B, C, D, X, Y, and reverse transcriptase (RT). DNA polymerase I (Pol I) is a member of the A family of polymerases. In a recent detailed analysis of mutation rates, Fijalkowska *et al.* found that the contribution to the fidelity of DNA polymerase I is specific to filling the gaps in Okazaki fragments of the lagging strand (Fijalkowska, Schaaper, & Jonczyk, 2012). Pol I reaches high-fidelity rates due to its three enzymatic activities: a $5' \rightarrow 3'$ polymerase, a $5' \rightarrow 3'$ exonuclease, and a $3' \rightarrow 5'$ proofreading exonuclease (Joyce & Grindley, 1984).

Studies on the fidelity of DNA polymerases have been conducted on many of the families. Polymerases within the same family are highly related in both structure and function. In the A and B families, the polymerases are responsible for DNA repair and replication. It has been known that many DNA polymerases undergo a conformational change during the base insertion process. Studies on polymerase ß of the X family have shown that DNA polymerase ß remains in the open conformation longer and forms the product more slowly when an improper base has been inserted (Arora, Beard, Wilson, & Schlick, 2005; Radhakrishnan & Schlick, 2005). Along with polymerase ß, studies have also been conducted on polymerases in the A family. One study showed that the T7 polymerase contains three different structural states, and the mispair recognition is due to a rapid release of the misincorporated base (Tsai & Johnson, 2006).

The Klenow fragment (KF) of *E. coli* is a well-characterized polymerase that is made up of the polymerase and $3'$-$5'$ exonuclease domains of *E. coli* DNA polymerase I. Bebenek et al. have shown that DNA in KF is synthesized with medium fidelity and contains an average base substitution rate of approximately 2×10^{-5} (Bebenek, Joyce, Fitzgerald, & Kunkel, 1990). Kutcha et al. found that the lower binding affinity of KF for the incorrect nucleotide and a slower rate for misinsertion enable discrimination against nucleotide misinsertion (Kuchta, Benkovic, & Benkovic, 1988). Dzantiev and Romano found that insertion of the incorrect base leads to inhibition of the closed ternary complex in KF (Dzantiev & Romano, 1999).

The *Bacillus* fragment (BF) of *Bacillus stearothermophilus* and the Klentaq fragment of *Thermus aquaticus* are also a part of the A family of DNA polymerases, and therefore, can also be used to study the fidelity of DNA polymerase I. The fidelity of both enzymes, the BF and Klentaq, has been studied, and the structures have been characterized (Kiefer, Mao, Braman, & Beese, 1998; Kiefer et al., 1997). Klentaq has been more widely used in polymerase chain reaction experiments due to its ability to polymerize DNA at high temperatures (Lohff & Cease, 1992; Tindall & Kunkel, 1988).

Warshel et al. have established that electrostatic effects in enzyme catalysis are significant through computational studies of DNA polymerase fidelity (Florian, Goodman, & Warshel, 2002, 2003, 2005; Ishikita & Warshel, 2008; Ram Prasad & Warshel, 2011; Rucker, Oelschlaeger, & Warshel, 2010; Warshel et al., 2006; Xiang, Goodman, Beard, Wilson, & Warshel, 2008). In addition, Warshel et al. have developed the linear response approximation (LRA), which has been applied to biosimulations in order to investigate group contributions to ligand binding and enzyme catalysis (Muegge, Tao, & Warshel, 1997). LRA and PDLD/LRA-S use electrostatic and nonelectrostatic contributions to estimate ligand-binding free energies and group contributions to ligand binding (Muegge et al., 1997).

Christian and coworkers recently conducted single-molecule Forster resonance energy transfer studies and found a formerly unknown step in DNA polymerization of DNA polymerase I (KF) (Christian, Romano, & Rueda, 2009). Subsequent to the addition of each nucleotide insertion, KF translocates along the DNA strand two bases downstream ($n-1$) from the active site along the template, before returning the proper location, one base downstream (n), to begin the next base incorporation. The results suggest that the translocation is done in an open conformation of the KF. These results have also resulted in the proposal of a putative fidelity-checking site, located two bases downstream of the site for the incoming nucleotide, ensuring proper base incorporation to the template base (Fig. 1) (Christian et al., 2009).

Some of us recently used computational methods on KF and BF to investigate the location of potential residues that could play a role in the putative checking site (Graham, Syeda, & Cisneros, 2012). In order to determine the location of residues involved in the putative postinsertion fidelity-checking hypothesis, Graham et al. performed molecular dynamics (MD) on a total of 16 systems (eight from KF and eight from BF). Of the eight systems of each fragment, only two contained the reference pair (T:A), and the rest contained mispairs (C:A, G:A, and A:A). For each mispair and the reference

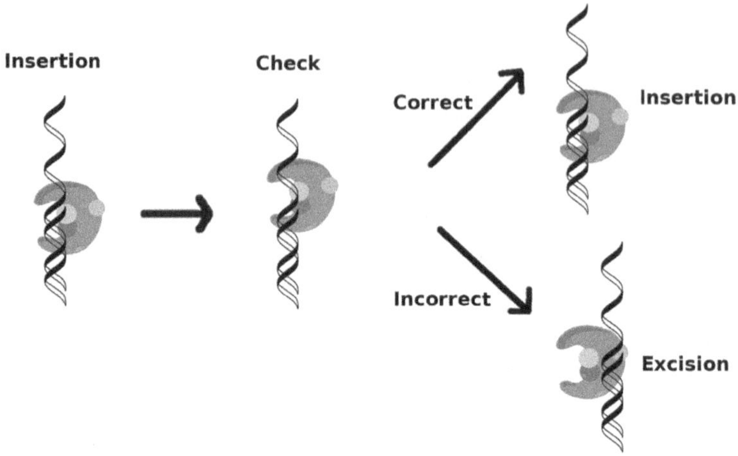

Figure 1 Schematic diagram of the proposed postinsertion checking mechanism. After insertion, the DNA is translocated to the checking site, if the newly inserted base is correctly paired, the DNA polymerase backtracks and inserts a new nucleotide. If there is a mismatch, the DNA is transferred to the exonuclease active site for excision of the incorrectly paired base (Christian et al., 2009).

pair, the DNA was positioned such that the last DNA nucleotides (blunt-end strand) were located either in the checking site ($n-1$) or the preinsertion site (n). Energy decomposition analysis (EDA), electrostatic free energy response (EFER) (Florian et al., 2003, 2005), and noncovalent interaction (NCI) analysis (Contreras-García et al., 2011) were used to analyze the MD results. Graham et al. found 11 residues that were conserved among A family DNA polymerases that could be involved in the putative checking mechanism.

In the present contribution we have employed, the same analysis techniques (EDA, EFER, and NCI) to determine the possible residues in the *Thermus aquaticus* DNA polymerase I (Klentaq fragment) that may be involved in the checking site for the postinsertion fidelity-checking hypothesis.

2. METHODS

2.1. Preparation of crystal structures

The *Thermus aquaticus* polymerase I crystal structure (3PY8) was obtained from the protein data bank (PDB) (Obeid et al., 2011). The crystal structure was originally missing the phosphate group in the DNA double helix between residues 112 and 113. In order to resolve this issue, residue 113 was removed from the PDB. The original PDB structure also contained a two base overhang. The base overhang was brought in so that it aligned with the rest of the DNA for the checking site, but for the preinsertion site, the overhang was left unmodified. The nucleotides located at the preinsertion site (primer base and complementary template) corresponded to cytosine and guanine, respectively. For the purpose of consistency with the previous study, the bases were changed to thymine and adenine, respectively. The incorrect pairs are comprised of the same structure as the reference pair, with the exception of the residue 112 mutation to A, C, and G, making the incorrect pairs A:A, C:A, and G:A. COOT was used to create the checking site by superposing the DNA one base pair downstream for all four pairs, creating eight structures total.

2.2. MD simulations

The modified crystal structures were used to run the MD simulations. AMBER was used to conduct MD simulations for DNA fragments in Klentaq. The MD simulations consisted of the Klentaq fragments in complex with DNA dimers having either a T:A pair (correctly paired) and three mispairs (C:A, A:A, and G:A) in the preinsertion or checking sites. The simulations were carried out with DNA in the preinsertion (*n*) and putative fidelity-checking (*n* − *1*) sites. The ff99SB force field from the AMBER11 suite was used for the purposes of the experiment (Case et al., 2005). AMBER refers to a set of molecular mechanical force fields and molecular simulation programs (Carroll, Cowart, & Benkovic, 1991). Each structure was subjected to minimization and single-run MD with the PMEMD program from the AMBER11 suite (Case et al., 2005). The initial minimization consisted of 1000 steps of conjugate gradient minimization, solvation, and NTP simulations to set the density to 1 g cm^{-3}. The Klentaq fragment was solvated in 103 Å × 127 Å × 108 Å box of TIP3P (Jorgensen, Chandrasekhar, Madura, Impey, & Klein, 1983). Once the system reached 300 K, the *NVT* ensemble was used to remove all constraints gradually over 100 ps until no constraints were left. Long-range electrostatics was calculated using the smooth particle mesh Ewald method (Darden, York, & Pedersen, 1993; Essmann et al., 1995). The total run time for each of the simulations was 20 ns. Snapshots for or every 1 ps were recorded. The RMSD (root-mean square deviation) for each system was calculated.

From each of the eight single-run MD simulations, snapshots corresponding to 10 ns were used for energy decomposition and EFER analyses. For both EDA and EFER, 10,000 structures for each system were employed. To obtain energy differences for the decomposition analysis, the nonbonded intermolecular interaction energies between each residue in the polymerase and the two nucleotides in the preinsertion or checking site were determined. We have selected a cutoff of ±kcal/mol for the interactions; thus, only residues having interactions of greater than ±1 kcal/mol have been considered as having a significant contribution.

2.3. NCI calculations

NCI analysis (Johnson et al., 2010) is a tool used for visualization. NCI analysis allows for identification of NCIs between molecules. It uses molecular densities and its derivatives for visualization. NCI gives a way to consider the intra- and intermolecular interactions. As Johnson et al. previously showed, NCI provides a way to analyze interactions between a ligand and a protein. The analysis provides isosurfaces, which are color coded to denote the strength and characteristics of the interactions. For example, green surfaces denote weak interactions, e.g., van der Waals, blue surfaces strong attractive interactions, e.g., hydrogen bonds, and red surfaces strong repulsive interactions.

In this study, NCI was used to analyze possible interactions between the protein and the DNA bases in the checking or preinsertion site. Three snapshots were subjected to NCI analysis for each of the eight systems. The DNA bases in each of the sites were considered as ligands interacting with a spherical surface with a 10 Å radius around the binding site. The calculations were obtained with a 0.2 step size for the cube and a cutoff of 5 Å for the calculation of the interactions between the nucleotides and the active site.

2.4. Energy decomposition analysis

Quantum mechanical/molecular mechanical studies have used EDA to successfully infer catalytic roles of residues surrounding enzymatic active sites (Cisneros, Liu, Zhang, & Yang, 2003; Cisneros et al., 2008, 2009; Cisneros, Wang, Silinski, Fitzgerald, & Yang, 2004; Cui, Karplus, & Valerie, 2003; Li & Cui, 2003; Marti et al., 2003; Senn, O'Hagan, & Thiel, 2005; Szefczyk, Claeyssens, Mulholland, & Sokalski, 2007). Further experiments of some of these predictions have confirmed the catalytic activity (Bebenek, Garcia-Diaz, Zhou, Povirk, & Kunkel, 2010; Metanis, Brik, Dawson, & Keinan, 2004). Following the method used previously to investigate what residues might be involved in the putative checking site

(Graham et al., 2012), the change in nonbonded intermolecular interactions between every protein residue and the two nucleic acids located in either the preinsertion or checking sites was calculated. Two sets of energy differences were determined, the first set consists of the energy difference between the change in energy between the preinsertion site and the checking site of the reference pair (T:A):

$$\Delta E_{\text{ref}} = \left\langle \Delta E_{i,\text{correct}}^{\text{nonbond}} \right\rangle_{\text{preinsertion}} - \left\langle \Delta E_{i,\text{correct}}^{\text{nonbond}} \right\rangle_{\text{check}} \tag{1}$$

where i represents an individual residue, $\langle \Delta E_{i,\text{correct}}^{\text{nonbond}} \rangle$ represents the nonbonded interaction (electrostatic or van der Waals) between residue i and the two correctly paired terminal bases (template and newly added correct nucleotide, T:A), and the broken brackets represent the averages over the conformational space ensemble sampled from MD simulations. The second set of energy differences is the difference between the change in energy between each protein residue and the preinsertion site and the change in energy between each protein residue and checking site of the incorrect pair (C:A, G:A, A:A):

$$\Delta E = \left\langle \Delta E_{i,\text{incorrect}}^{\text{nonbond}} \right\rangle_{\text{preinsertion}} - \left\langle \Delta E_{i,\text{incorrect}}^{\text{nonbond}} \right\rangle_{\text{check}} \tag{2}$$

The above calculations were used to determine the residues with the largest change in interaction between the residues in the protein and the correctly and incorrectly paired bases.

$$\Delta\Delta E = \Delta E_{\text{ref}} - \Delta E \tag{3}$$

Following the calculations of the ΔE for each pair (mispair and reference), $\Delta\Delta E$ was calculated by subtracting ΔE of the mispair from ΔE of the reference pair as in Eq. 3.

2.5. EFER analysis

In addition to EDA, the EFER method (Florian et al., 2003, 2005) was used to calculate the average free energy between the reference pair and each of the mispairs. The EFER free energy difference expression was used to calculate the free energy coming from correctly and incorrectly paired nucleotides in each site (the checking site and the preinsertion site) from the MD ensemble averages of electrostatic interaction energies between the DNA pair (correct or incorrect) and the rest of the system. The free energy change is calculated by

$$\Delta G = \frac{1}{2}\left(\left\langle V_{i,\mathrm{correct}}^{\mathrm{preinsertion}} - V_{i,\mathrm{correct}}^{\mathrm{check}}\right\rangle + \left\langle V_{i,\mathrm{incorrect}}^{\mathrm{preinsertion}} - V_{i,\mathrm{incorrect}}^{\mathrm{check}}\right\rangle\right) \quad (4)$$

where the broken brackets denote the ensemble average from the MD simulations, i represents an individual residue, and V_i is the electrostatic interaction energy between residue i and the DNA bases in the respective site. The T:A pair was considered the correct pair, while the C:A, G:A, and A:A were considered the mispairs.

3. RESULTS AND DISCUSSION

A total of eight structures from the Klentaq fragment have been considered in this study. The RMSD was calculated for each of the eight systems for the full protein as well as for each domain (palm, thumb, and fingers) and for combined domains (palm–thumb, palm–fingers, and thumb–fingers). In all cases, it was observed that the systems were stable throughout the simulation length.

For the reference pair (T:A), the RMSD was no higher than approximately 4 Å. As shown in Fig. 2, the RMSD between the thumb and fingers was the highest when compared to the rest of the parts of the polymerase. The larger range of movement between the thumb and fingers may be due to the flexibility needed for the open to closed transition during base insertion although we did not observe the change from the open to the closed conformation in any of our simulations. The RMSD graphs for the mispairs show interesting differences. In the C:A checking site, the RMSD reaches nearly 6 Å, while in the preinsertion, the RMSD peaks at approximately 4 Å (Fig. 3). The larger amount of movement in the checking site implies that the polymerase may be recognizing the mispair, resulting in larger movement of the protein structure, specifically, the thumb and fingers.

A similar pattern is shown in the graphs of the RMSDs of G:A in the preinsertion site and the checking site (Fig. 4). While the RMSDs of the checking site is more consistent across time, it shows a maximum value of about 5 Å. The RMSDs of the preinsertion site also contains spikes in which the RMSD increases and then decreases by approximately 0.5 Å at approximately the same time period as the spikes in the C:A checking site RMSD occurred.

The RMSDs of the A:A mispair in both the checking and preinsertion sites remained below 4 Å (Fig. 5). Also, the overall RMSD's for both sites

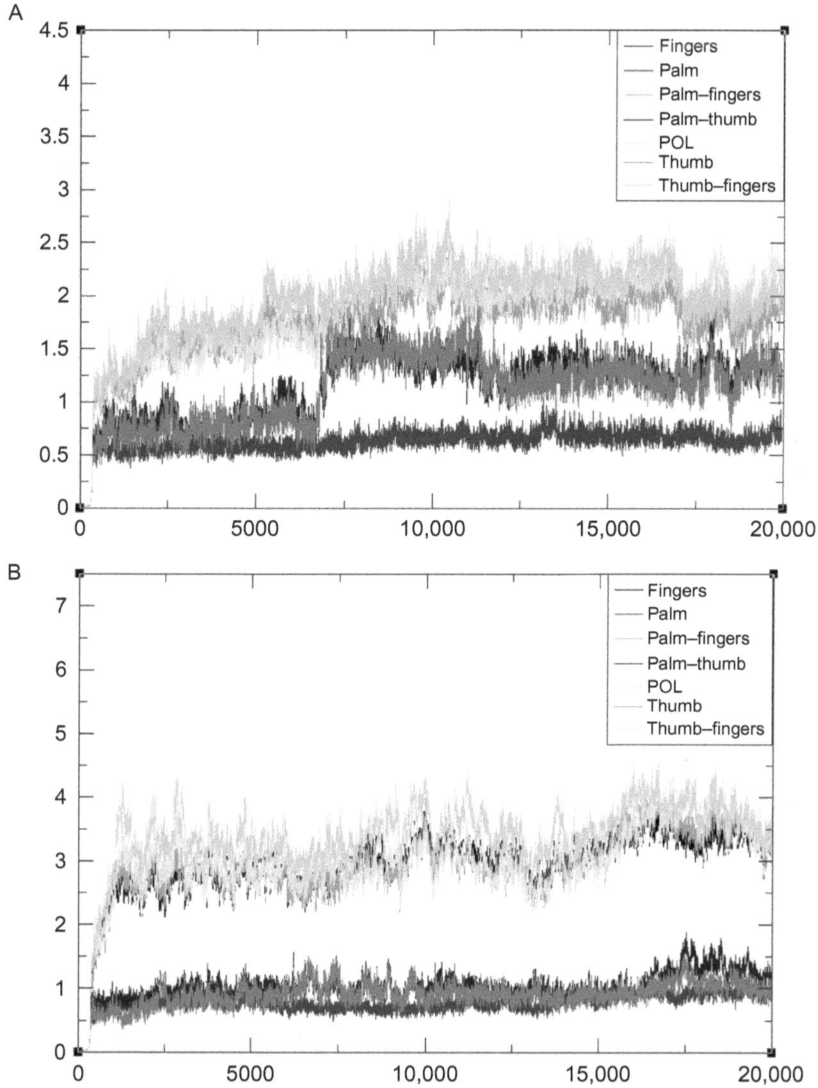

Figure 2 RMSD for Klentaq (full and by domains) with the correct pair, T:A, in the pre-insertion (A) and checking (B) sites.

with the A:A mispair are lower than the RMSD's for the C:A and G:A mispairs. The lower RMSD values in the A:A sites could be the consequence of the pairing of two bulky nucleic acids, thus not allowing significant movement.

The number of residues that showed large interaction energy varies between each mispair, suggesting distinctive selectivity dependent on the

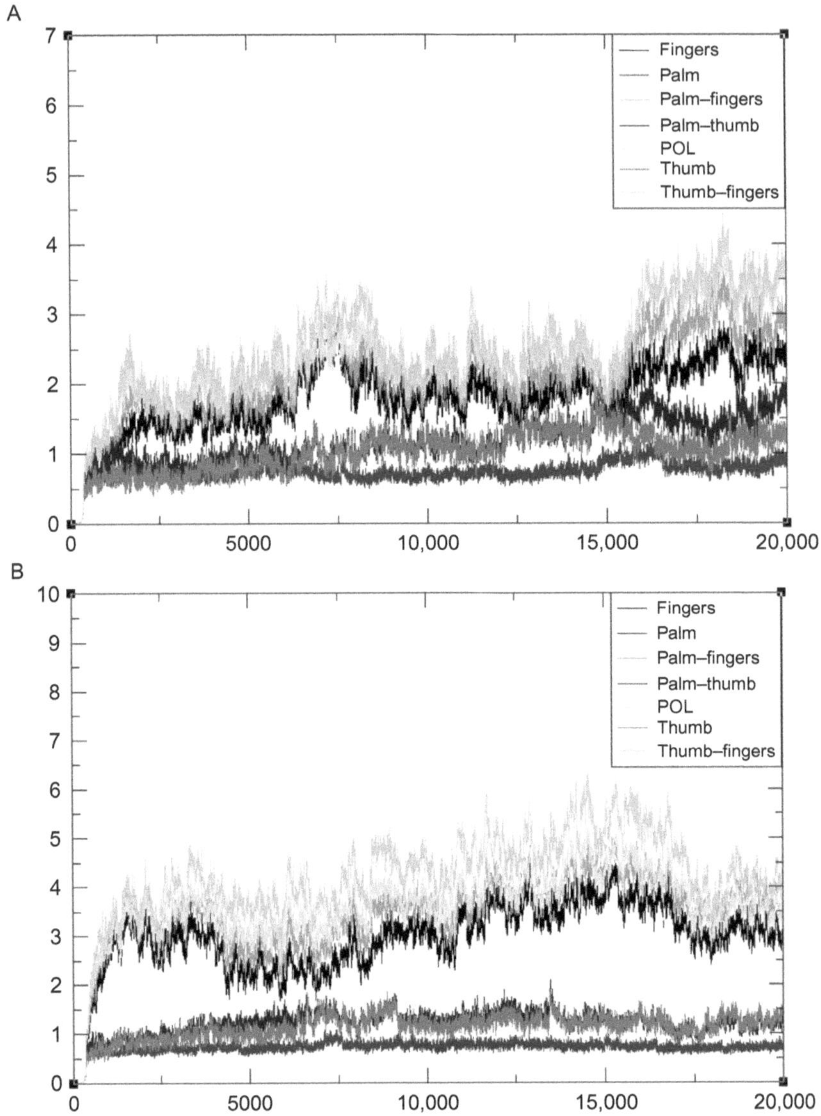

Figure 3 RMSD for klentaq (full and by domains) with the G:A mispair, in the pre-insertion (A) and checking (B) sites.

mismatch. This selectivity agrees with previous experimental results (Carroll et al., 1991). For the C:A, A:A, and G:A mispairs in the Klentaq fragment, there are 112, 56, and 66, residues that show a change in interaction energy, respectively. When compared to our previous results for KF, which was

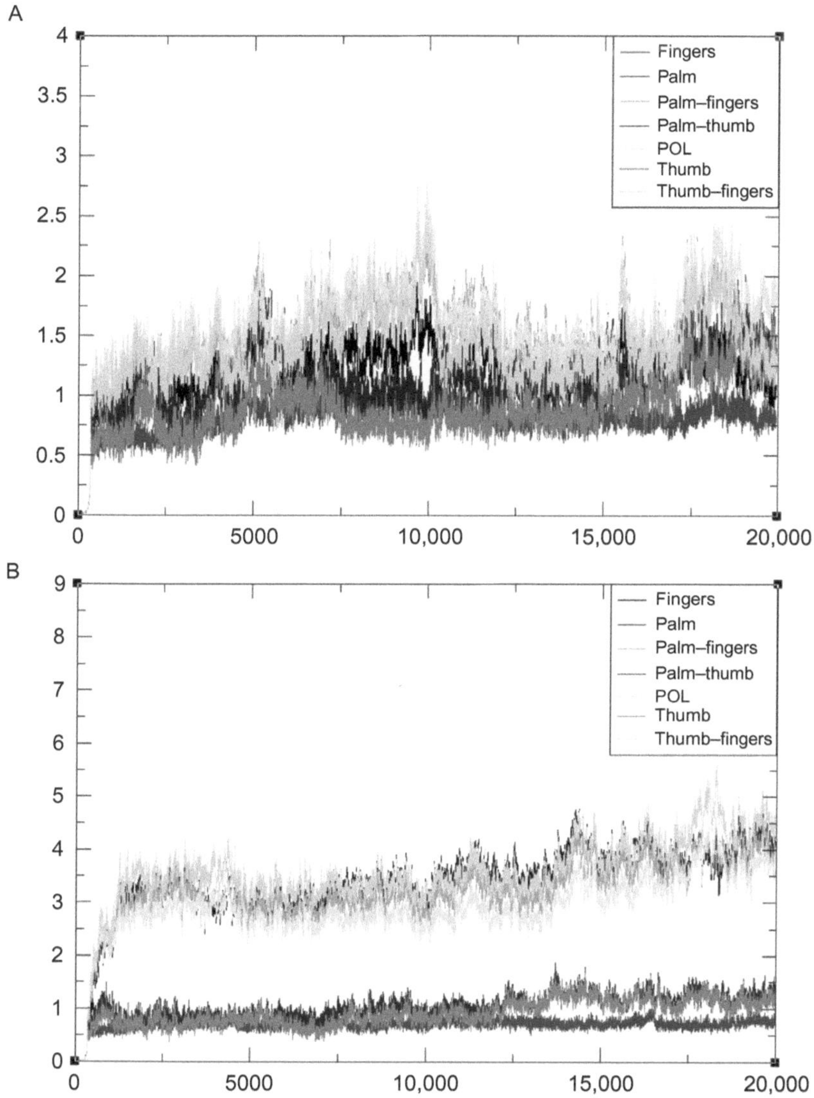

Figure 4 RMSD for klentaq (full and by domains) with the C:A mispair, in the pre-insertion (A) and checking (B) sites.

reported to have 157, 168, and 137 residues that show a change in interaction for the C:A, A:A, and G:A mispairs, respectively, the Klentaq fragment has less residues with a change in interaction (Graham et al., 2012). Conversely, the number of residues that showed a change in interaction for

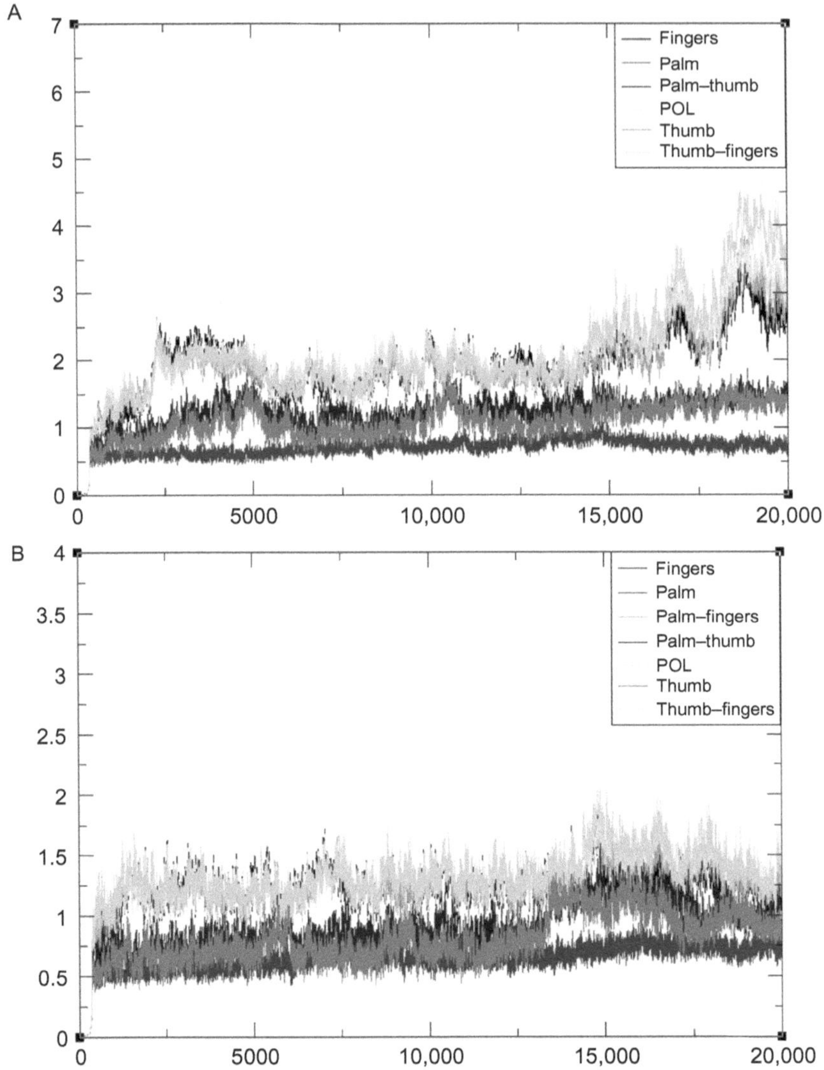

Figure 5 RMSD for klentaq (full and by domains) with the A:A mispair, in the pre-insertion (A) and checking (B) sites.

BF was reported to be 72, 76, and 70 residues for the C:A, A:A, and G:A mispairs, respectively, which is much closer to the Klentaq fragment.

EDA showed approximately 30 residues that had altered interactions for all three mispairs in Klentaq (Fig. 6 and Tables 1 and 2). The majority of the residues with altered interactions are located in the polymerase. Among the 30 residues, five of them also show large interactions in the

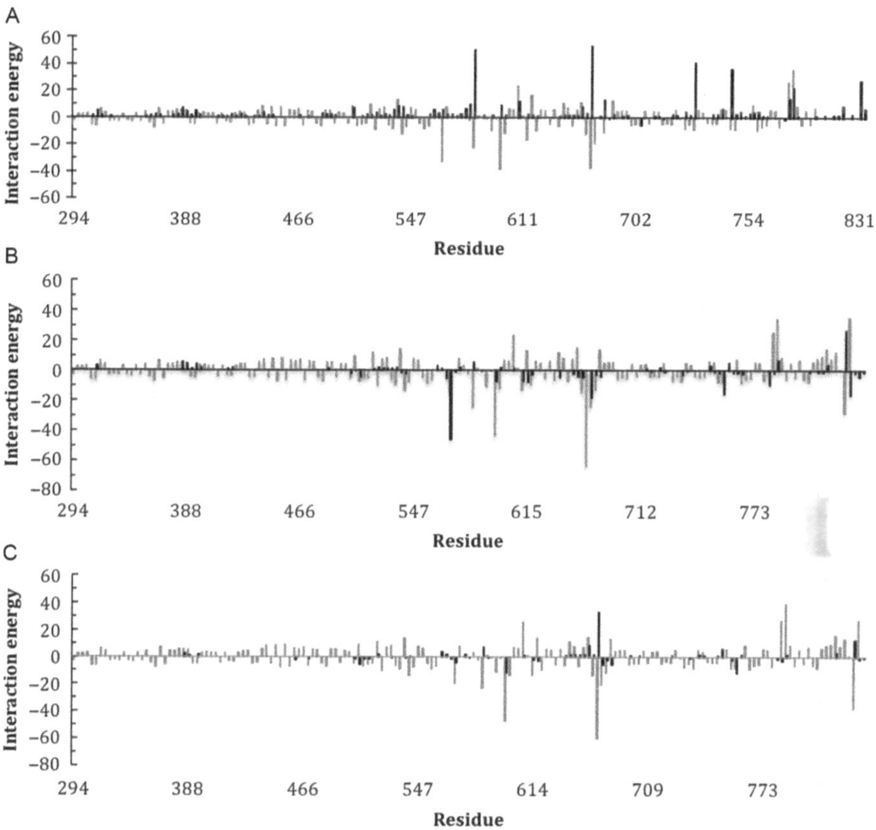

Figure 6 EDA and EFER results for C:A (A), G:A (B), and A:A (C) mispairs. EDA results are shown in black, and EFER results are shown in gray.

EFER analysis (see below) and are conserved among all A family polymerases. These five residues are R573(R668), R587(R682), D610(D705), E615 (E710), and R659(R754) for Klentaq(KF). Residues R573 and R587 are within the closest proximity of the active site base pairs (Fig. 7).

Three of the five conserved residues that show a large change in interaction are arginines. Arginine is a proton donor, implying that the interaction with the mispairs may be a result of a possible change in the hydrogen bonding when mispairs are present compared to the correct pair in the checking site. Although R587 shows a large interaction with all three mispairs, the only interaction greater than 10 kcal/mol is with the C:A mispair. The preceding residue, V586, also contains a significantly high interaction with the C:A mispair ($\geq$10 kcal/mol). While all of the altered interactions conserved among all three polymerases are Coulomb interactions, P550

Table 1 Results for the change in Coulomb interaction from EDA for residues with $|\Delta\Delta E| \geq 1$ (in kcal/mol)

Res.	$\Delta\Delta E_{GA}$	$\Delta\Delta E_{CA}$	$\Delta\Delta E_{AA}$
313	–	1.47	–
314	2.82	4.97	–
315	–	1.72	–
328	–	1.2	–
347	–	1.18	–
349	–	1.51	–
354	–	2.66	–
355	–	1.65	–
371	–	1.55	–
372	–	2.58	–
381	5.29	7.01	2.09
388	4.06	4.13	1.1
392	1.33	1.84	–
393	4.07	4.84	1.59
397	1.83	1.56	–
405	–	1.28	–
410	–	1.18	–
411	1.03	1.22	–
419	1.99	2.11	–
421	–	1.33	–
423	–	2.29	–
424	–	1.11	–
431	–	1.11	–
432	–	1.14	–
434	–	3.79	–
435	–	1.53	–
445	–	1.99	–
462	–	–	−2.23

Table 1 Results for the change in Coulomb interaction from EDA for residues with $|\Delta\Delta E| \geq 1$ (in kcal/mol)—cont'd

Res.	$\Delta\Delta E_{GA}$	$\Delta\Delta E_{CA}$	$\Delta\Delta E_{AA}$
466	–	1.61	–
469	–	–	−1.33
471	–	1.27	–
487	–	3.18	–
488	1.51	2.19	–
491	–	2.46	–
492	–	1.51	–
505	−2.14	–	−1.54
507	−0.74	7.25	−5.96
508	−4.58	–	−1.38
511	–	1.28	−1.38
512	–	2.28	–
520	0.17	2.04	2
523	1.95	4.06	–
524	1.47	1.16	–
530	1.5	2.12	−0.42
531	1.19	1.51	–
536	1.91	5.82	–
537	−1.8	8.41	−0.81
540	−2.3	7.86	0.44
542	–	1.55	–
547	–	1.02	–
563	–	2.49	–
569	3.36	5.95	3.8
570	1.38	3.08	1.98
571	−5.42	–	−1.35
573	−5.42	5.89	−4.49
574	−5.42	–	–

Continued

Table 1 Results for the change in Coulomb interaction from EDA for residues with $|\Delta\Delta E| \geq 1$ (in kcal/mol)—cont'd

Res.	$\Delta\Delta E_{GA}$	$\Delta\Delta E_{CA}$	$\Delta\Delta E_{AA}$
578	−5.42	2.67	–
582	–	2.29	1.9
583	–	1.11	−1.07
584	–	2.23	–
585	–	6.57	–
586	–	10.17	–
587	5.72	50.68	6.95
588	–	–	−1.29
591	–	1.4	–
593	–	1.58	–
595	−7.11	9.04	−11.61
596	1.61	1.75	–
610	1.03	11.78	1.01
614	−7.26	2.55	−2.25
615	−7.99	1.79	−2.98
617	−3.06	1.06	–
636	–	1.3	1.14
637	−4.9	1.47	1.48
641	–	1.72	1.7
651	–	1.78	1.72
652	–	1.96	1.56
655	−2.45	7.57	8.27
659	−4.02	2.86	1.95
660	−4.54	53.23	33.19
663	−17.92	–	−5.75
670	–	1.24	–
677	−2.82	13.33	−2.33
680	–	−1	–

Table 1 Results for the change in Coulomb interaction from EDA for residues with $|\Delta\Delta E| \geq 1$ (in kcal/mol)—cont'd

Res.	$\Delta\Delta E_{GA}$	$\Delta\Delta E_{CA}$	$\Delta\Delta E_{AA}$
681	−4.25	2.37	−5.5
688	–	1.09	–
694	–	1.46	–
695	–	–	1.06
704	–	−5.65	1.28
712	–	1.1	–
713	1.17	1.02	–
715	−1.17	−1.02	–
717	1.94	2.23	1.28
721	−1.02	1.66	–
726	–	4.77	–
727	–	1.52	−1.05
728	−3.05	41.26	–
732	−0.64	–	1.46
734	–	–	1.03
738	–	2.2	–
741	–	2.39	1.32
742	3.19	6.62	6.06
745	−1.18	–	–
746	−2.06	36.29	−2.33
747	−15.93	2.23	−11.53
750	4.83	4.34	1.8
751	–	1.21	–
754	−1.54	3.22	–
759	−1.75	3.93	–
762	−2.49	4.29	–
767	–	1.36	–
771	–	1.03	–

Continued

Table 1 Results for the change in Coulomb interaction from EDA for residues with $|\Delta\Delta E| \geq 1$ (in kcal/mol)—cont'd

Res.	$\Delta\Delta E_{GA}$	$\Delta\Delta E_{CA}$	$\Delta\Delta E_{AA}$
784	−9.75	−1.37	−1.45
785	−1.93	14.17	−2.92
786	6.91	22.34	1.57
790	–	1.11	–
797	–	1.26	–
804	−1.35	1.89	–
808	−1.11	1.71	–
818	−1.01	1.8	–
820	3.48	8.59	2.76
826	–	2.1	–
831	26.61	27.54	12.3
832	−16.26	6.41	−1.35

Table 2 EDA van der Waals results for residues with $|\Delta\Delta E| \geq 1$ (in kcal/mol)

Residue	$\Delta\Delta G_{GA}$	$\Delta\Delta G_{CA}$	$\Delta\Delta G_{AA}$
278.00	–	1.65	–
290.00	–	–	−1.01
378.00	–	−1.12	−4.79
490.00	1.63	1.60	–
550.00	–	28.80	–

shows a significantly large van der Waals' interaction for the C:A mispair, and not for the other two mispairs. This indicates that P550 is possibly responsible for identifying C:A mispairs. Additionally, multiple residues only show altered interactions for two of the three mispairs. These residues are E507, E537, K540, S578, and R728. Lastly, similar to our EDA and EFER analysis for KF and BF, numerous residues in the Klentaq fragment have a $|\Delta\Delta E| \geq 10$ kcal/mol for at least one of the mismatches.

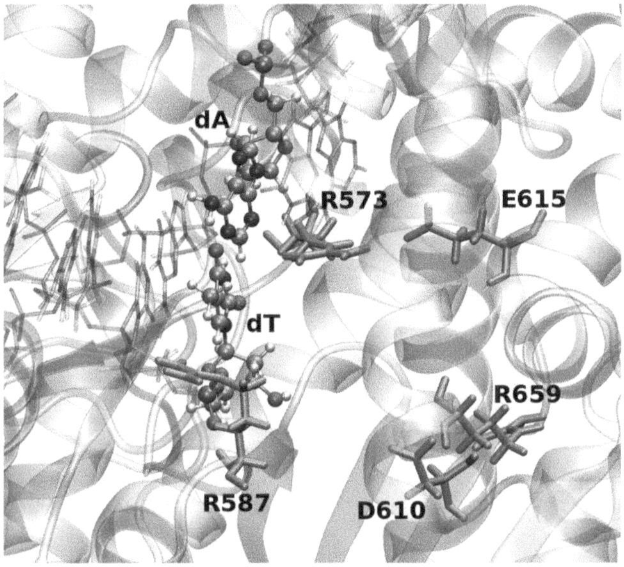

Figure 7 Positions of the residues with large interactions for all mispairs with the newest formed base pair in the putative checking site.

EFER shows a total of over 150 residues with large interactions (Fig. 6 and Table 3). Among these residues, 23 show significantly large interactions ($\geq$10 kcal/mol). Residue R660 (corresponding to R755 in KF) contains the highest free energy between all three mispairs. R660 also contains electrostatic interactions greater than 10 kcal/mol for the C:A and A:A mispairs. The corresponding residue in KF (R755) also shows electrostatic interactions greater than 10 kcal/mol in C:A and A:A. The results show that the interactions of the G:A and A:A in R755 (of KF) are less energetically favorable than the interaction of C:A. This could be due to strain caused by the bulkier structures of the purines compared to pyrimidines. In general, EFER analysis showed many more significant interactions than EDA analysis (Fig. 6). The C:A mispair shows more repulsive interactions than G:A and A:A (Fig. 6).

The NCI results show a total of 28 residues with noncovalent interactions between the protein and the base pairs in the active site in at least one of the eight systems (Table 4). Of these residues, R573, V586, K747, Q754, V783, and H784 contain NCIs with all eight systems. R573 shows significant interactions in all three methods of analysis; EDA, EFER, and NCI plot results.

Table 3 EFER results for residues with $|\Delta\Delta G| \geq 1$ (in kcal/mol)

Res.	$\Delta\Delta G_{CA}$	$\Delta\Delta G_{GA}$	$\Delta\Delta G_{AA}$
294	−2.54	−2.83	−2.73
295	2.23	2.51	2.42
296	2.26	2.56	2.65
303	2.83	3.01	3.06
313	−5.91	−6.07	−6.2
314	−6.56	−6.36	−6.24
315	6.21	6.41	6.39
320	3.99	4.25	4.27
328	−3.01	−3.03	−3.01
331	−2.13	−2.34	−2.38
334	−2.79	−2.93	−3.01
337	2.67	2.91	3.02
340	−2.52	−2.81	−2.86
343	−3.23	−3.33	−3.54
344	2.85	3.07	3.19
346	−3.09	−3.2	−3.28
347	4.23	4.26	4.4
349	−4.01	−3.98	−4.1
354	−7.88	−7.43	−7.88
355	6.51	6.49	6.77
362	−4.94	−5.28	−5.43
363	4.25	4.53	4.62
371	4.49	4.49	4.64
372	5.82	5.55	5.78
381	6.03	5.53	5.76
388	5.11	5.02	5.01
392	−4.52	−4.3	−4.7
393	−5.3	−4.97	−5.22

Table 3 EFER results for residues with $|\Delta\Delta G| \geq 1$ (in kcal/mol)—cont'd

Res.	$\Delta\Delta G_{CA}$	$\Delta\Delta G_{GA}$	$\Delta\Delta G_{AA}$
397	3.31	3.58	3.37
400	3.69	3.91	3.77
401	3.1	3.29	3.31
404	3	3.22	3.14
405	−4.26	−4.41	−4.42
410	3.11	3.2	3.21
411	−3.08	−3.02	−3.26
419	−3.88	−3.76	−3.93
421	3.25	3.17	3.35
423	4.58	4.29	4.53
424	4.04	4.05	4.08
425	−5.51	−5.04	−5.32
431	−4.4	−4.23	−4.4
432	5.38	5.23	5.47
434	8.52	7.57	8.15
435	−4.79	−4.75	−4.95
445	8.03	7.88	8.19
450	−6.67	−7.2	−7.41
452	7.87	8.65	8.96
457	−6.29	−6.83	−7.01
462	5.62	6.42	6.27
465	5.37	6.06	5.98
466	6.47	7.27	7.19
469	−6	−6.84	−6.3
471	5.13	5.86	5.74
473	4.68	5.33	5.06
476	−4.3	−4.94	−4.71
487	−6.22	−7.24	−6.9

Continued

Table 3 EFER results for residues with $|\Delta\Delta G| \geq 1$ (in kcal/mol)—cont'd

Res.	$\Delta\Delta G_{CA}$	$\Delta\Delta G_{GA}$	$\Delta\Delta G_{AA}$
488	4.95	5.59	5.74
491	4.94	5.74	5.43
492	−4.16	−4.91	−4.54
496	3.76	4.48	4.27
497	3.28	4.04	3.92
505	−4.64	−5.49	−5.23
507	8.45	9.47	8.94
508	−6.51	−7	−6.75
511	−4.95	−5.52	−5.44
512	−4.45	−5.33	−5.1
520	9.97	12.13	11.43
523	−9.09	−10.56	−10.23
524	6.42	7.29	7.08
530	7.79	8.81	8.18
531	−6.01	−6.59	−6.37
536	−8.14	−9.45	−8.84
537	13.16	14.55	13.92
540	−12.79	−13.52	−13.9
542	−7.01	−7.68	−7.49
547	7.01	7.69	7.51
551	5.99	6.43	6.48
556	−4.5	−4.85	−4.96
559	−8.31	−8.66	−8.95
563	−6.73	−6.75	−7.03
573	−33.09	−13.62	−19.46
578	7.77	7.73	7.99
584	2.58	2.85	2.85
586	−2.12	−2.12	−2.18

Table 3 EFER results for residues with $|\Delta\Delta G| \geq 1$ (in kcal/mol)—cont'd

Res.	$\Delta\Delta G_{CA}$	$\Delta\Delta G_{GA}$	$\Delta\Delta G_{AA}$
587	−22.52	−24.87	−22.86
588	0.91	1.07	1.17
592	−1.04	−1.45	−1.18
593	−9.81	−10.68	−10.74
594	−0.62	−0.64	−0.67
595	−38.26	−43.89	−46.92
596	−11.92	−12.19	−13.44
601	6.65	6.79	7.09
602	5.64	5.81	6.37
610	23.66	23.5	26.13
611	1	1.01	1.06
614	−16.88	−13.47	−13.93
615	16.32	13.59	13.85
617	−9.81	−9.49	−9.57
625	6.04	5.94	5.9
626	4.79	4.87	4.75
630	−5.17	−5.47	−5.35
634	5.64	6	5.91
636	−6.33	−6.85	−6.6
637	10.57	12.08	10.83
641	7.13	8.14	7.69
651	−6.63	−7.37	−7.12
652	6.04	7.16	7.13
655	11.4	15	14.2
659	−12.42	−14.69	−13.63
660	−37.67	−63.46	−59.8
663	−19.6	−24.25	−20.44
677	−11.39	−12.77	−11.52

Continued

Table 3 EFER results for residues with $|\Delta\Delta G| \geq 1$ (in kcal/mol)—cont'd

Res.	$\Delta\Delta G_{CA}$	$\Delta\Delta G_{GA}$	$\Delta\Delta G_{AA}$
681	12.87	13.81	12.79
687	4.63	4.89	4.72
688	5.32	5.8	5.4
694	5.06	5.3	5.32
695	−5.87	−6.18	−5.63
702	−4.93	−4.77	−4.76
704	−5.66	−5.24	−5.15
708	4.85	4.46	4.51
709	−4.69	−4.44	−4.41
712	4.14	3.76	3.9
713	4.68	4.34	4.43
715	−4.96	−4.42	−4.7
716	−4.11	−3.6	−3.78
717	−4.74	−4.28	−4.46
721	5.89	5.1	5.34
722	−0.58	−0.36	−0.43
726	−8.74	−7.02	−7.75
727	−5.21	−4.48	−4.86
728	−9.4	−7.17	−8.04
732	5.71	4.95	5.36
734	4.02	3.71	3.9
736	−3.52	−3.48	−3.52
738	−4.63	−4.75	−4.58
741	−4.4	−4.45	−4.38
742	6.31	6.26	6.25
745	6.05	5.75	5.67
746	−8.43	−8.04	−8.07
747	−9.18	−7.77	−8.51

Table 3 EFER results for residues with $|\Delta\Delta G| \geq 1$ (in kcal/mol)—cont'd

Res.	$\Delta\Delta G_{CA}$	$\Delta\Delta G_{GA}$	$\Delta\Delta G_{AA}$
759	9.4	7.57	8.26
762	−10.13	−8.14	−9.01
767	−6.94	−6.11	−6.53
771	−5.82	−5.43	−5.77
773	5.61	5.25	5.81
774	5.11	4.96	5.05
778	−6.86	−6.77	−7.01
784	−1.06	−3.06	−1.02
785	26.33	25.23	27.27
786	35.71	34.44	39.09
790	8.4	8.29	8.68
793	−6.05	−5.89	−6.33
794	4.86	4.77	4.96
795	−5.13	−5.02	−5.22
797	6.85	6.7	7.07
801	−6.1	−5.77	−6.22
804	−8.58	−8.1	−8.51
805	6.21	5.75	6.12
808	8.45	7.7	8.16
818	9.84	8.86	9.27
820	15.7	14.88	16.07
825	7.48	7.52	7.87
826	11.86	12.15	12.95
827	0.89	0.83	1.06
831	−29.94	−28.61	−37.7
832	26.4	35.46	27.12

Table 4 NCI results for all base pairs

Residue	A:A	C:A	G:A	T:A
E507	+(+)			+
T569	(+)			+(+)
A570	(+)			+(+)
T571	(+)	+(+)	(+)	+(+)
R573	+(+)	+(+)	+(+)	+(+)
Q582	+(+)	+	(+)	(+)
N583	(+)		(+)	+(+)
I584	(+)		(+)	(+)
P585	(+)		(+)	+(+)
V586	+(+)	+(+)	+(+)	+(+)
R587	(+)		(+)	+(+)
R595	+	+(+)	+(+)	+(+)
E614			+	
E615		+(+)	(+)	
R660	+(+)			+
K663	+		+	+
T664			+	+
F667				+
Y671	+	+	+	+
R728	+(+)	+(+)	(+)	+(+)
R746	+	+(+)	+	+
K747	+(+)	+(+)	+(+)	+(+)
N750	(+)	+(+)	+(+)	+(+)
Q754	+(+)	+(+)	+(+)	+(+)
V783	+(+)	+(+)	+(+)	+(+)
H784	+(+)	+(+)	+(+)	+(+)
D785	+	+(+)	+(+)	+
E786	+(+)	+(+)	+(+)	+
K831	+			

Residues with a mark in parentheses denote interaction in the checking site, whereas no parentheses denotes preinsertion site

Residue E786 exhibits noncovalent interactions in every system except within the T:A checking site, implying that this residue does not interact with the checking site when the correct base is present. T571 shows interactions in all systems except for when the purine mispairs, A:A and G:A, are located in the preinsertion site. K663 shows interactions with all base pairs in the preinsertion site except C:A. Residue F667 only contains an interaction for T:A in the preinsertion site. It is possible that this interaction indicates the correct base has been inserted. Y671 only contains interactions in the preinsertion sites of all four base pairs. The higher interaction in the preinsertion site is expected, as T671 is located above the DNA and the DNA in the checking site is superposed downstream one base pair. Three residues N583, I584, and P585 are shown to have interactions with two mispairs when the nucleotides are located only in the checking site.

Figures 8–11 show NCI plots for selected snapshots for each of the eight systems. In all cases, the interaction surfaces are larger when the nucleotide pair is located in the checking site compared to when they are located in the preinsertion site. This suggests that there are more close contacts between the protein and the base pairs in the $n-1$ position. Moreover, it can be seen that the noncovalent interactions between the nucleotide pair and the protein are different for each mispair and for the correctly paired nucleotides (Figs. 8–11). For example, the C:A mispair shows significantly less interactions than the G:A and A:A mispairs. In addition, the G:A mispair shows the largest interactions between the nucleotides and the protein (Fig. 10).

Yoshida et al. reported that the R660S mutation showed a reduction in T→C transitions (Yoshida et al., 2001). Thompson et al. measured the fidelity of 26 mutants with amino acid substitutions in the KF, and found that DNA mismatches are recognized not only by the difference in free energy for the partitioning of the DNA primer terminus between the polymerase and exonuclease sites for several mispairs but also between the residues near the active site and the mismatched pairs (Thompson, Bailey, van der Schans, Joyce, & Millar, 2002). Thompson and coworkers concluded that residues N845 and R668 are required for recognition of correct mispairs (Thompson et al., 2002). Our studies indicate highly altered interaction energy for R573 in Klentaq (corresponding to R668 in KF). Some studies have shown that the mutant R668A favors errors involving the insertion of dGTP, our results of the KF, BF, and Klentaq all show large interactions for this residue (Graham et al., 2012). In agreement with the negligible energetic change of the interaction of Q849 in KF

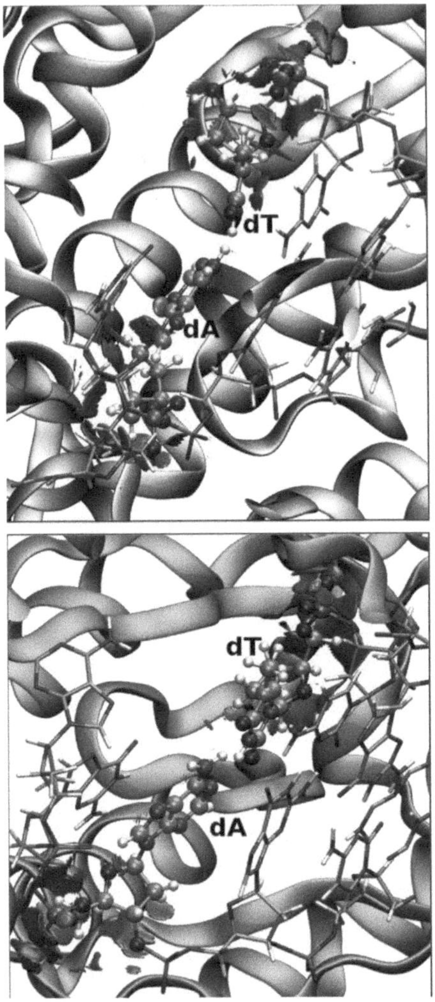

Figure 8 NCI analysis for a selected snapshot for T:A in the preinsertion (top) and the checking (bottom) sites. Colored surfaces denote noncovalent interactions: green (gray in the print version) surfaces denote weak interactions, e.g., van der Waals, red (black in the print version) surfaces denote repulsive interactions, and blue (black in the print version) surfaces denote attractive interactions, e.g., hydrogen bonds.

(Graham et al., 2012), the corresponding residue in Klentaq (Q754) also shows negligible energetic change (Fig. 12).

The mutations of R668, R682, E710, and N845 have been shown to cause decreased fidelity in the KF by other experimental studies. These residues correspond to residues R573, R587, E615, and N750 in Klentaq. In our study, all of those residues showed significant interactions with the

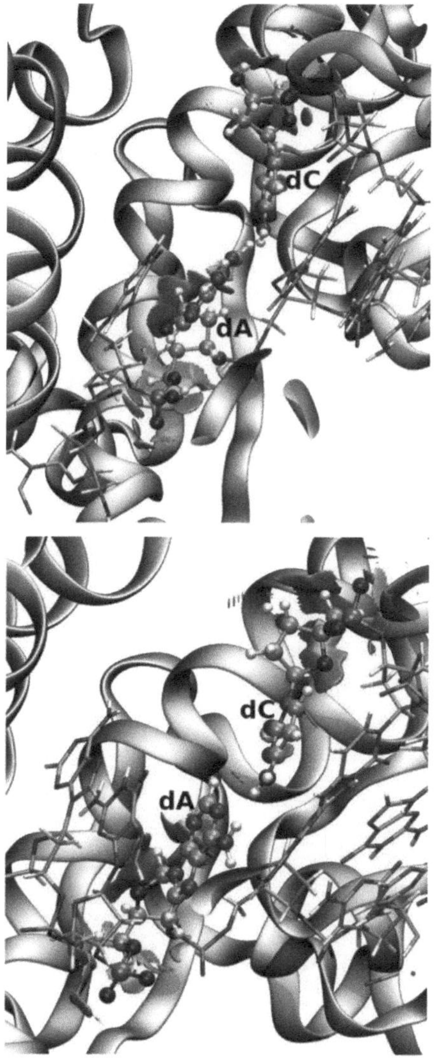

Figure 9 NCI plot for C:A in the preinsertion (top) and the checking (bottom) sites. Colored surfaces denote noncovalent interactions: green (gray in the print version) surfaces denote weak interactions, e.g., van der Waals, red (black in the print version) surfaces denote repulsive interactions, and blue (black in the print version) surfaces denote attractive interactions, e.g., hydrogen bonds.

mispairs, and four of them showed altered interactions from EFER, and EDA analysis across the KF, the BF, and the Klentaq fragment. Minnick et al. showed that the recognition of specific mispairs is a function of the mispair (Minnick et al., 1999). The E710A mutant has shown an increase

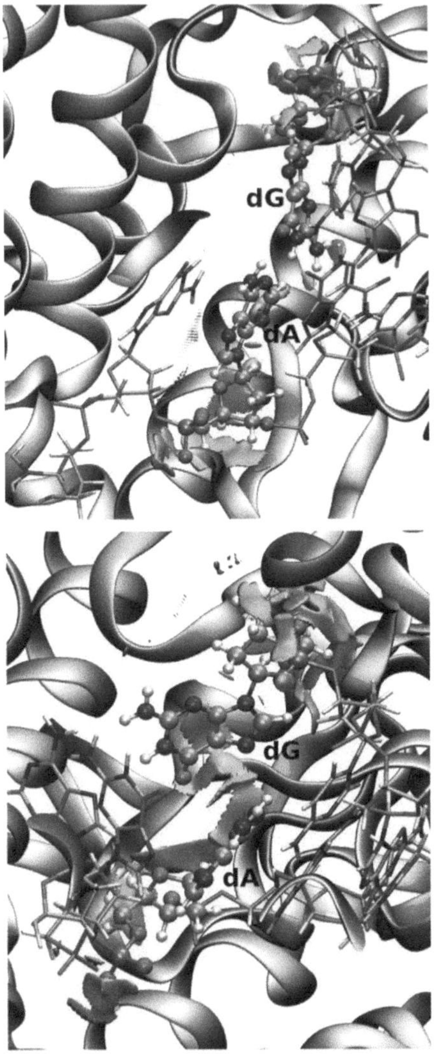

Figure 10 NCI analysis for a selected snapshot for G:A in the preinsertion (top) and the checking (bottom) sites. Colored surfaces denote noncovalent interactions: green (gray in the print version) surfaces denote weak interactions, e.g., Van der Waals, red (dark gray in the print version) surfaces denote repulsive interactions, and blue (black in the print version) surfaces denote attractive interactions, e.g., hydrogen bonds.

in the A:dCTP mispair; however, the corresponding residue in Klentaq (E615) shows a significantly higher $\Delta\Delta E$ for the G:A mispair than the C:A for EDA. The EFER analysis showed significant free energy for E615; however, all three mispairs showed interactions with roughly the same magnitude.

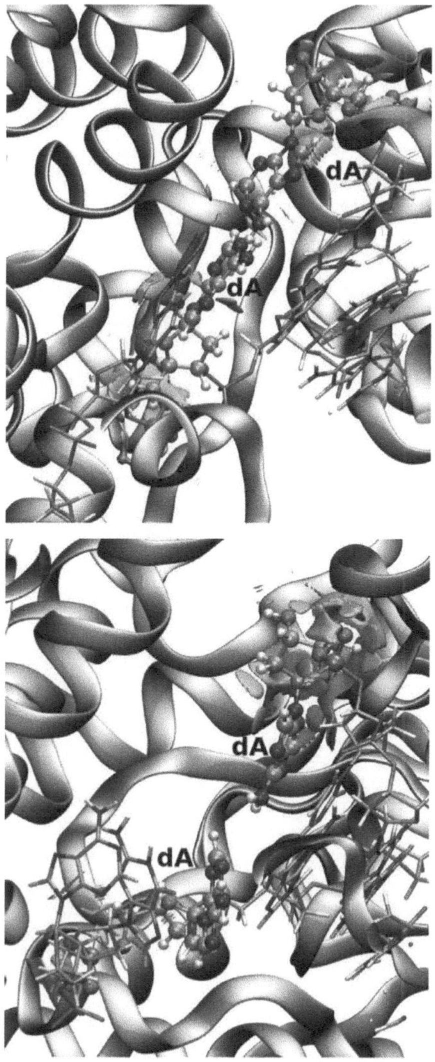

Figure 11 NCI analysis for a selected snapshot for A:A in the preinsertion (top) and the checking (bottom) sites. Colored surfaces denote noncovalent interactions: green (gray in the print version) surfaces denote weak interactions, e.g., van der Waals, red (dark gray in the print version) surfaces denote repulsive interactions, and blue (black in the print version) surfaces denote attractive interactions, e.g., hydrogen bonds.

An increase in the frequency of T:dCTP errors is caused by the R682A and N845A mutants. For the corresponding residue of R682 in Klentaq, the C:A mispair shows a $\Delta\Delta E$ of 10-fold greater than G:A and A:A. For the corresponding residue of N846 in Klentaq, all of the mispairs show

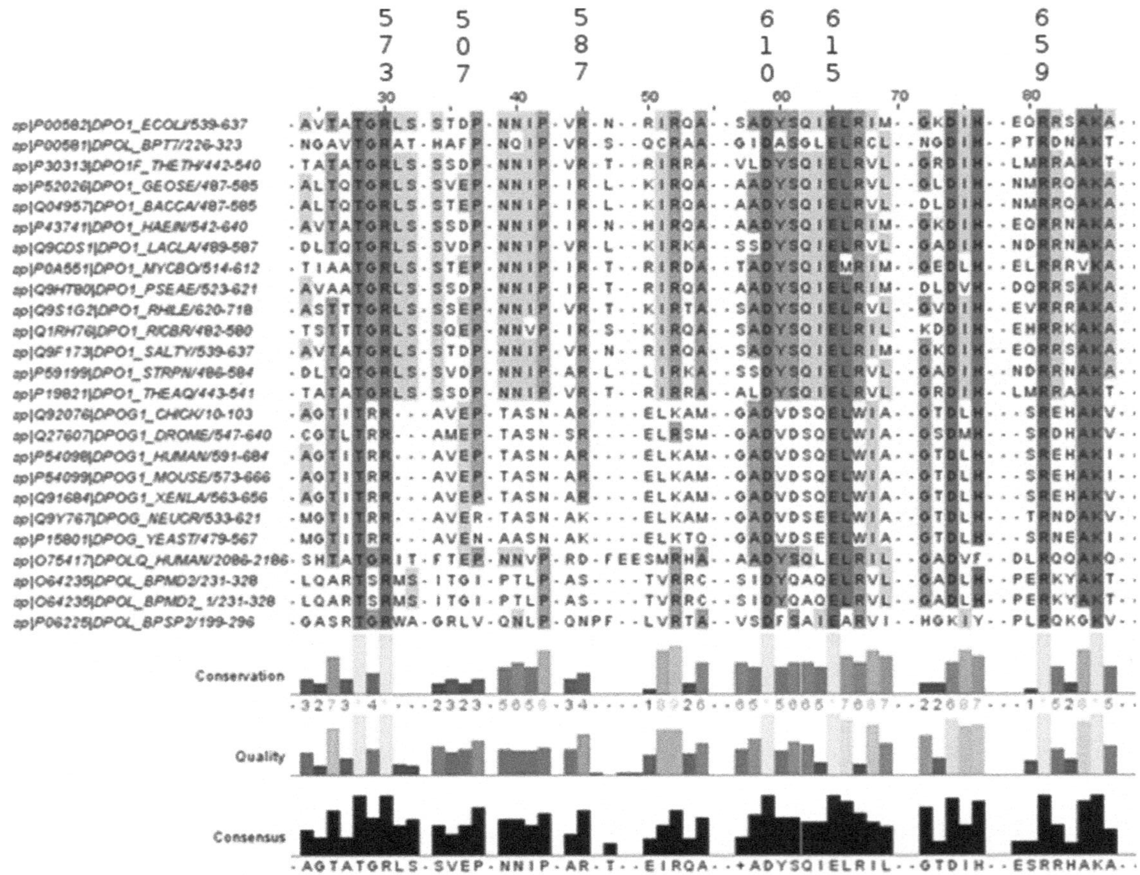

Figure 12 Condensed alignment for family A DNA polymerases. Labeled residues correspond to Klentaq (UniProtKB entry P19821).

interaction energies of greater than 1 kcal/mol and none of them show interaction energy of greater than 10 kcal/mol.

Loh and Loeb have shown that the Q849A mutant in the KF weakened the DNA binding while maintaining the dNTP binding (Loh & Loeb, 2005). Singh and Modak also tested this mutation and reported that additional residues N845, Q849, R668, H881, and Q677 are part of a hydrogen bond track (Singh & Modak, 2003). Residues in Klentaq corresponding to N845, R668, and H881 (750, 573, and 784) showed interactions with all three mispairs.

4. CONCLUSIONS

EDA, EFER, and NCI analyses have been used to determine potential residues that could be involved in a novel postinsertion fidelity-checking mechanism for the Klentaq fragment. A total of eight systems including one properly paired and three mispaired nucleotide bases have been studied for the Klentaq fragment. The results from the eight systems of Klentaq were compared to those of the KF and BF. The results for EDA and EFER show that approximately 30 residues exhibit a change in interaction energy difference for all mispairs. This is consistent with our previous results on the homologous KF and BF DNA polymerases (Graham et al., 2012). Within those residues, five of them are conserved across all three polymerases. NCI analysis gives 28 residues that show noncovalent interactions between the nucleotide pair and residues in the preinsertion and/or putative checking site. A sequence alignment of nonredundant A family polymerases shows that five of the residues that show altered interactions for both EDA and EFER are conserved among all polymerases. Moreover, these five residues are homologous to residues in KF and BF that had previously shown altered interactions (Graham et al., 2012). Comparison of the present results with experimental studies shows that the mutation of several residues identified by these methods result in a decrease in fidelity of DNA synthesis. Our results also agree with the selectivity of misinsertions reported by experimental studies. Our results provide further support for the subsequent experimental mutagenesis studies on the five conserved residues as possible participants in the fidelity-checking mechanism.

REFERENCES

Arora, K., Beard, W. A., Wilson, S. H., & Schlick, T. (2005). Mismatch-induced conformational distortions in polymerase β support an induced-fit mechanism for fidelity. *Biochemistry*, *44*(40), 13328–13341. http://dx.doi.org/10.1021/bi0507682.

Bebenek, K., Garcia-Diaz, M., Zhou, R.-Z., Povirk, L. F., & Kunkel, T. A. (2010). Loop 1 modulates the fidelity of DNA polymerase lambda. *Nucleic Acids Research*, *38*(16), 5419–5431. http://dx.doi.org/10.1093/nar/gkq261.

Bebenek, K., Joyce, C. M., Fitzgerald, M. P., & Kunkel, T. A. (1990). The fidelity of DNA synthesis catalyzed by derivatives of Escherichia coli DNA polymerase I. *Journal of Biological Chemistry*, *265*(23), 13878–13887.

Carroll, S. S., Cowart, M., & Benkovic, S. J. (1991). A mutant of DNA polymerase I (Klenow fragment) with reduced fidelity. *Biochemistry*, *30*(3), 804–813.

Case, D. A., Cheatham, T. E., III, Darden, T., Gohlke, H., Luo, R., Merz, K. M., Jr., et al. (2005). The Amber biomolecular simulation programs. *Journal of Computational Chemistry*, *26*(16), 1668–1688.

Christian, T. D., Romano, L. J., & Rueda, D. (2009). Single-molecule measurements of synthesis by DNA polymerase with base-pair resolution. *Proceedings of the National Academy of Sciences*, *106*(50), 21109–21114.

Cisneros, A., Liu, H., Zhang, Y., & Yang, W. (2003). Ab-initio QM/MM study shows there is no general acid in the reaction catalyzed by 4-Oxalocrotonate Tautomerase. *Journal of the American Chemical Society*, *125*(34), 10384–10393.

Cisneros, G. A., Perera, L., García-Díaz, M., Bebenek, K., Kunkel, T. A., & Pedersen, L. G. (2008). Catalytic mechanism of human DNA polymerase [lambda] with Mg2+ and Mn2+ from ab initio quantum mechanical/molecular mechanical studies. *DNA Repair*, 7(11), 1824–1834.

Cisneros, G. A., Perera, L., Schaaper, R. M., Pedersen, L. C., London, R. E., Pedersen, L. G., et al. (2009). Reaction mechanism of the ε subunit of E. coli DNA polymerase III: Insights into active site metal coordination and catalytically significant residues. *Journal of the American Chemical Society*, *131*(4), 1550–1556. http://dx.doi.org/10.1021/ja8082818.

Cisneros, A., Wang, M., Silinski, P., Fitzgerald, C., & Yang, W. (2004). The protein backbone makes important contributions to 4-Oxalocrotonate Tautomerase enzyme catalysis. *Biochemistry*, *43*(22), 6885–6892.

Contreras-García, J., Johnson, E. R., Keinan, S., Chaudret, R., Piquemal, J.-P., Beratan, D. N., et al. (2011). NCIPLOT: A program for plotting noncovalent interaction regions. *Journal of Chemical Theory and Computation*, 7(3), 625–632. http://dx.doi.org/10.1021/ct100641a.

Cui, Q., Karplus, M., & Valerie, D. (2003). Catalysis and specificity in enzymes: A study of triosephosphate isomerase and comparison with methyl glyoxal synthase. *Advances in Protein Chemistry*, *66*, 315–372.

Darden, T. A., York, D., & Pedersen, L. G. (1993). Particle mesh Ewald: An Nlog(N) method for Ewald sums. *The Journal of Chemical Physics*, *98*, 10089.

Dzantiev, L., & Romano, L. J. (1999). A conformational change in E. coli DNA polymerase I (Klenow fragment) is induced in the presence of a DNTP complementary to the template base in the active site. *Biochemistry*, *39*(2), 356–361. http://dx.doi.org/10.1021/bi991952o.

Essmann, U., Perera, L., Berkowitz, M., Darden, A., Lee, H., & Pedersen, G. (1995). A smooth particle mesh Ewald method. *The Journal of Chemical Physics*, *103*, 8577.

Fijalkowska, I. J., Schaaper, R. M., & Jonczyk, P. (2012). DNA replication fidelity in Escherichia coli: A multi-DNA polymerase affair. *FEMS Microbiology Reviews*, *36*(6), 1105–1121. http://dx.doi.org/10.1111/j.1574-6976.2012.00338.x.

Florian, J., Goodman, M. F., & Warshel, A. (2002). Theoretical Investigation of the binding free energies and key substrate-recognition components of the replication fidelity of human DNA polymerase beta. *The Journal of Physical Chemistry. B*, *106*(22), 5739–5753. http://dx.doi.org/10.1021/jp020790u.

Florian, J., Goodman, M. F., & Warshel, A. (2003). Computer simulation of the chemical catalysis of DNA polymerases: discriminating between alternative nucleotide insertion mechanisms for T7 DNA polymerase. *Journal of the American Chemical Society*, *125*(27), 8163–8177. http://pubs8163.acs.org/acs/journals/doilookup?in_doi=8110.1021/ja028997o.

Florian, J., Goodman, M. F., & Warshel, A. (2005). Computer simulations of protein functions: Searching for the molecular origin of the replication fidelity of DNA polymerases. *Proceedings of the National Academy of Sciences of the United States of America*, *102*(19), 6819–6824. http://dx.doi.org/10.1073/pnas.0408173102.

Graham, S. E., Syeda, F., & Cisneros, G. A. (2012). Computational prediction of residues involved in fidelity checking for DNA synthesis in DNA polymerase I. *Biochemistry*, *51*(12), 2569–2578. http://dx.doi.org/10.1021/bi201856m.

Ishikita, H., & Warshel, A. (2008). Predicting drug-resistant mutations of HIV protease. *Angewandte Chemie, International Edition*, *47*(4), 697–700. http://dx.doi.org/10.1002/anie.200704178.

Johnson, E. R., Keinan, S., Mori-Sánchez, P., Contreras-García, J., Cohen, A. J., & Yang, W. (2010). Revealing noncovalent interactions. *Journal of the American Chemical Society*, *132*(18), 6498–6506. http://dx.doi.org/10.1021/ja100936w.

Jorgensen, W. L., Chandrasekhar, J., Madura, J. D., Impey, R. W., & Klein, M. L. (1983). Comparison of simple potential functions for simulating liquid water. *The Journal of Chemical Physics*, *79*(2), 926–935. http://link.aip.org/link/?JCP/979/926/921.

Joyce, C. M., & Grindley, N. D. (1984). Method for determining whether a gene of Escherichia coli is essential: Application to the polA gene. *Journal of Bacteriology*, *158*(2), 636–643.

Kiefer, J. R., Mao, C., Braman, J. C., & Beese, L. S. (1998). Visualizing DNA replication in a catalytically active Bacillus DNA polymerase crystal. *Nature*, *391*(6664), 304–307. http://dx.doi.org/10.1038/34693.

Kiefer, J. R., Mao, C., Hansen, C. J., Basehore, S. L., Hogrefe, H. H., Braman, J. C., et al. (1997). Crystal structure of a thermostable Bacillus DNA polymerase l large fragment at 2.1 Å resolution. *Structure*, *5*(1), 95–108. http://dx.doi.org/10.1016/S0969-2126(97)00169-X.

Kuchta, R. D., Benkovic, P., & Benkovic, S. J. (1988). Kinetic mechanism whereby DNA polymerase I (Klenow) replicates DNA with high fidelity. *Biochemistry*, *27*(18), 6716–6725.

Kunkel, A. (2003). Considering the cancer consequences of altered DNA polymerase function. *Cancer Cell*, *3*, 105–110.

Li, G., & Cui, Q. (2003). What is so special about Arg 55 in the catalysis of cyclophilin A? Insights from Hybrid QM/MM simulations. *Journal of the American Chemical Society*, *125*(49), 15028–15038. http://dx.doi.org/10.1021/ja0367851.

Loh, E., & Loeb, L. A. (2005). Mutability of DNA polymerase I: Implications for the creation of mutant DNA polymerases. *DNA Repair*, *4*(12), 1390–1398. http://apps.isiknowledge.com/full_record.do?product=WOS&search_mode=GeneralSearch&qid=1391&SID=1391D1393bHpehf1394P1391lE1399cnkm&page=1393&doc=1324.

Lohff, C. J., & Cease, K. B. (1992). PCR using a thermostable polymerase with 3' to 5' exonuclease activity generates blunt products suitable for direct cloning. *Nucleic Acids Research*, *20*(1), 144.

Marti, S., Andres, J., Moliner, V., Silla, E., Tunon, I., & Bertran, J. (2003). Preorganization and reorganization as related factors in enzyme catalysis: The chorismate mutase case. *Chemistry—A European Journal*, *9*(4), 984–991.

Metanis, N., Brik, A., Dawson, E., & Keinan, E. (2004). Electrostatic interactions dominate the catalytic contribution of Arg39 in 4-oxalocrotonate tautomerase. *Journal of the American Chemical Society*, *126*(40), 12726–12727.

Minnick, D. T., Bebenek, K., Osheroff, W. P., Turner, R. M., Jr., Astatke, M., Liu, L., et al. (1999). Side Chains that influence fidelity at the polymerase active site of Escherichia Coli DNA polymerase I (Klenow fragment). *Journal of Biological Chemistry*, *274*(5), 3067–3075. http://dx.doi.org/10.1074/jbc.274.5.3067.

Muegge, I., Tao, H., & Warshel, A. (1997). A fast estimate of electrostatic group contributions to the free energy of protein-inhibitor binding. *Protein Engineering*, *10*(12), 1363–1372. http://dx.doi.org/10.1093/protein/10.12.1363.

Obeid, S., Schnur, A., Gloeckner, C., Blatter, N., Welte, W., Diederichs, K., et al. (2011). Learning from directed evolution: Thermus aquaticus DNA polymerase mutants with translesion synthesis activity. *Chembiochem*, *12*(10), 1574–1580. http://dx.doi.org/10.1002/cbic.201000783.

Radhakrishnan, R., & Schlick, T. (2005). Fidelity discrimination in DNA polymerase β: Differing closing profiles for a mismatched (G:A) versus matched (G:C) base pair. *Journal of the American Chemical Society*, *127*(38), 13245–13252. http://dx.doi.org/10.1021/ja052623o.

Ram Prasad, B., & Warshel, A. (2011). Prechemistry versus preorganization in DNA replication fidelity. *Proteins: Structure, Function, and Bioinformatics*, *79*(10), 2900–2919. http://dx.doi.org/10.1002/prot.23128.

Rucker, R., Oelschlaeger, P., & Warshel, A. (2010). A binding free energy decomposition approach for accurate calculations of the fidelity of DNA polymerases. *Proteins: Structure, Function, and Bioinformatics*, *78*(3), 671–680. http://dx.doi.org/10.1002/prot.22596.

Schaaper, R. M. (1993). Base selection, proofreading, and mismatch repair during DNA replication in Escherichia coli. *Journal of Biological Chemistry*, *268*(32), 23762–23765.

Senn, H. M., O'Hagan, D., & Thiel, W. (2005). Insight into enzymatic C-F bond formation from QM and QM/MM calculations. *Journal of the American Chemical Society*, *127*(39), 13643–13655. http://dx.doi.org/10.1021/ja053875s.

Singh, K., & Modak, M. J. (2003). Presence of 18-A long hydrogen bond track in the active site of Escherichia coli DNA polymerase I (Klenow fragment)—Its requirement in the stabilization of enzyme-template-primer complex. *Journal of Biological Chemistry*, *278*(13), 11289–11302. http://apps.isiknowledge.com/full_record.do?product=WOS&search_mode=GeneralSearch&qid=11281&SID=11281D11283bHpehf11284P11281lE11289cnkm&page=11285&doc=11242.

Szefczyk, B., Claeyssens, F., Mulholland, A. J., & Sokalski, W. A. (2007). Quantum chemical analysis of reaction paths in chorismate mutase: Conformational effects and electrostatic stabilization. *International Journal of Quantum Chemistry*, *107*(12), 2274–2285. http://dx.doi.org/10.1002/qua.21354.

Thompson, E. H. Z., Bailey, M. F., van der Schans, E. J. C., Joyce, C. M., & Millar, D. P. (2002). Determinants of DNA mismatch recognition within the polymerase domain of the Klenow fragment. *Biochemistry*, *41*(3), 713–722. http://dx.doi.org/710.1021/bi0114271.

Tindall, K. R., & Kunkel, T. A. (1988). Fidelity of DNA synthesis by the Thermus aquaticus DNA polymerase. *Biochemistry*, *27*(16), 6008–6013. http://dx.doi.org/10.1021/bi00416a027.

Tsai, Y.-C., & Johnson, K. A. (2006). A new paradigm for DNA polymerase specificity. *Biochemistry*, *45*(32), 9675–9687. http://dx.doi.org/10.1021/bi060993z.

Warshel, A., Sharma, P. K., Kato, M., Xiang, Y., Liu, H., & Olsson, M. H. M. (2006). Electrostatic basis for enzyme catalysis. *Chemical Reviews*, *106*(8), 3210–3235. http://dx.doi.org/10.1021/cr0503106.

Xiang, Y., Goodman, M. F., Beard, W. A., Wilson, S. H., & Warshel, A. (2008). Exploring the role of large conformational changes in the fidelity of DNA polymerase β. *Proteins: Structure, Function, and Bioinformatics*, *70*(1), 231–247. http://dx.doi.org/10.1002/prot.21668.

Yoshida, K., Tosaka, A., Kamiya, H., Murate, T., Kasai, H., Nimura, Y., et al. (2001). Arg660Ser mutation in Thermus aquaticus DNA polymerase I suppresses T−>C transitions: Implication of wobble base pair formation at the nucleotide incorporation step. *Nucleic Acids Research*, *29*(20), 4206–4214.

CHAPTER THREE

New Strategies for Integrative Dynamic Modeling of Macromolecular Assembly

Enrico Spiga*[,1], Matteo Thomas Degiacomi*[,1,2], Matteo Dal Peraro*[,†,3]

*Laboratory for Biomolecular Modeling, Institute of Bioengineering, School of Life Sciences, École Polytechnique Fédérale de Lausanne—EPFL, Lausanne, Switzerland
†Swiss Institute of Bioinformatics—SIB, Lausanne, Switzerland
[1]Authors contributed equally to this work.
[2]Present address: Physical and Theoretical Chemistry Laboratory, South Parks Road, Oxford, United Kingdom.
[3]Corresponding author e-mail address: matteo.dalperaro@epfl.ch

Contents

Abstract

Data reporting on structure and dynamics of cellular constituents are growing with increasing pace enabling, as never before, the understanding of fine mechanistic aspects of biological systems and providing the possibility to affect them in controlled ways. Nonetheless, experimental techniques do not yet allow for an arbitrary level of resolution on cellular processes *in situ*. By consistently integrating a variety of diverse experimental data, molecular modeling is optimally poised to enhance to near-atomistic resolution our understanding of molecular recognition in large assemblies. Within this integrative modeling context, we briefly review in this chapter the recent progresses of molecular simulations at the atomistic and coarse-grained level of resolution to explore protein–protein interactions. In particular, we discuss our recent contributions in this field, which aim at providing a robust bridge between novel optimization algorithms and multiscale molecular simulations for a consistent integration of experimental inputs. We expect that, with the ever-growing sampling ability of molecular simulations and the tireless progress of experimental methods, the impact of such dynamic-based approach could only be more effective with time, contributing to provide detailed description of cellular organization.

Advances in Protein Chemistry and Structural Biology, Volume 96
ISSN 1876-1623
http://dx.doi.org/10.1016/bs.apcsb.2014.06.008

1. INTRODUCTION

We are facing an era in which the accumulation of raw biological data is rapidly exploding, overtaking our capability of analysis in order to extract functional information. As for genomics, proteomic data reporting on structure, dynamics, and interactions of cellular constituents are growing with increasing pace, providing unprecedented functional insights (Babu et al., 2012; Kühner et al., 2009). This is a key ingredient for the thorough understanding of the mechanistic aspects of any biological process and for the possibility to modulate biochemical and biophysical properties in a controlled way for therapeutic and biotechnological purposes. Nonetheless, while progressing rapidly in enabling such quantitative characterization, experimental techniques do not yet allow for an arbitrary level of spatial and temporal resolution on cellular processes *in situ* (i.e., at near-physiological conditions).

The ultimate goal of structural biology is exactly pointing toward this objective, namely the characterization of biomolecular organization at atomic resolution. X-ray crystallography is still the golden standard to capture high-resolution snapshots of biological structures (e.g., Manglik et al., 2012). Unfortunately, it is usually difficult to characterize the structure of large molecular assemblies at atomistic resolution. This is due to both their size and complexity, which make the production of sufficiently pure crystal challenging. The possibility to capture functional conformational changes and fully explore the dynamic properties of biomolecules in native conditions is also usually not accessible by this approach. Spectroscopy techniques, such as nuclear magnetic resonance (NMR), have the great advantage to probe dynamics, but have to face harder limitations when exploring large systems or *in vivo* conditions. A number of techniques have been fortunately flourishing in the past decade, which are able to deliver rich structural and dynamic information at an intermediate-to-low level of resolution, and can be used to complement sparse, incomplete data at higher resolution. Single-particle cryo-electron microscopy (cryo-EM) can currently produce density maps of large assemblies (Bartesaghi & Subramaniam, 2009; Mao et al., 2012). Thanks to recent technical advances, it is revolutionizing the field of structural biology, breaking more and more often the subnanometer resolution limit (Allegretti, Mills, McMullan, Kühlbrandt, & Vonck, 2014; Greber et al., 2014; Li et al., 2013; Liao, Cao, Julius, & Cheng, 2013). Lower resolution information is then offered by other techniques such as small angle X-ray scattering (SAXS) (Lipfert & Doniach, 2007), and high-speed atomic force microscopy (HS-AFM) (Casuso et al., 2012), which have the advantage to probe dynamic properties of large systems at near-physiological conditions. Sparse low-resolution data are also offered by quantitative techniques such as chemical cross-linking (Rinner et al., 2008), ion mobility (Baldwin et al., 2011) or hydrogen–deuterium exchange coupled to mass spectrometry, fluorescence resonance energy transfer (FRET), EPR double electron–electron resonance (DEER) (Barrett et al., 2012), or, for functional genomics, latest generation chromatin conformation capture (Imakaev et al., 2012) and ChIP-seq techniques (Visel et al., 2009).

Molecular modeling is, in this moment, optimally poised to consistently integrate this variety of heterogeneous data, enhancing the potential to deliver structural and dynamic insights at higher resolution and providing a solid bridge to discover the underlying atomistic interactions of biomolecules when they assemble in large, heteromultimeric complexes to perform their specific biological functions (Gavin et al., 2002). For this reason, theoretical and computational methodologies aimed at predicting the structure of molecular assemblies are becoming increasingly effective and promising in the field of structural biology. Currently, a major hurdle for molecular assembly prediction is that molecules are intrinsically dynamic objects and can undergo large and complex conformational changes during assembly (Bahadur & Zacharias, 2008). This greatly increases the complexity of the computational task, making the accurate prediction of macromolecular assembly an open field for research and methods development nowadays. One of the main ways to tackle this challenge is to keep into account the native dynamics of the individual constituents. Molecular Dynamics (MD) simulation has today the potential to serve this scope. In recent years, atomistic MD has unleashed its full potential and can sample with great accuracy the conformational space associated with key processes, such as folding (Lindorff-Larsen, Piana, Dror, & Shaw, 2011; Voelz, Bowman, Beauchamp, & Pande, 2010) and binding (Arkhipov et al., 2013; Kruse et al., 2012). With the guidance provided by experimental inputs, the predictive power of molecular simulations is then enormously enhanced (Trabuco, Villa, Mitra, Frank, & Schulten, 2008) and promises to shed light on mechanistic aspects of molecular organization thus far inaccessible. Meanwhile, various models have been developed to extend even further, at the same computational cost, the boundaries of molecular simulations. Coarse-grained (CG) models for biomolecules, while sacrificing some atomistic degrees of freedom, have been used for studying even larger systems and to address problems such as membrane self-assembly and dynamics and protein folding. These models, although plagued with some drawbacks due to the simplification of the energy potential, have provided numerous insightful results for molecular assembly.

In this domain, we are actively working on two main aspects of the study and prediction of molecular assemblies. On one hand, we have recently developed methods to quickly and robustly predict molecular assembly while keeping monomers native flexibility into account and integrating a variety of low-resolution spatial restraints produced experimentally. On the other, we have been working on the definition of simplified representations of proteins, which can allow simulating large systems for long timescales. The common ground between these topics turned out to be the need of efficient and robust optimization tools. For this reason, we have developed an optimization framework, called *pow*^{er} (i.e., Parallel Optimization Workbench to Enhance the Resolution of biological systems, at http://lbm.epfl.ch) (Degiacomi & Dal Peraro, 2013), currently based on Particle Swarm Optimization (PSO). We have exploited this framework as a workhorse (i) for protein assembly prediction of symmetrical systems characterized by large conformational changes and (ii) for the development of CG models able to faithfully reproduce electrostatic, structural, and dynamical properties. In this chapter, we discuss these newly developed integrative dynamic modeling strategies and their application to specific biological problems. While putting our work in the broader context of the recent progress done by the community actively working in this field, we will give particular emphasis on the capability of MD to describe the intrinsic flexibility of biomolecules, highlighting how simulations can be used to directly study molecular recognition processes.

2. PREDICTING PROTEIN–PROTEIN RECOGNITION AT THE ATOMISTIC LEVEL

Protein–protein assembly prediction usually involves the study of large biomolecular assemblies and of phenomena taking place in a broad range of timescales. Although promising, attempts to study and predict protein–protein interactions solely by means of atomistic MD (Adcock & McCammon, 2006; van Gunsteren et al., 2006) are still at early stages. Deploying an amount of computational power sufficient to study protein binding is not yet routinely possible, nevertheless progress in algorithmic development, construction and exploitation of *ad hoc* hardware, or specific allocations on high performance computing resources can already allow, in some cases, to reach the necessary timescales. For instance, long simulations already allowed to shed light on the properties of the EGF receptor and its coupling to the membrane (Arkhipov et al., 2013), as well as on the dimerization process of the Top7-CFr dimer (Piana, Lindorff-Larsen, & Shaw, 2013).

Enhanced sampling techniques, which already constitute an established methodology for the study of protein–ligand interactions, represent a valid alternative to overcome the limitations of unbiased MD simulations (Buch, Giorgino, & De Fabritiis, 2011; Limongelli, Bonomi, & Parrinello, 2013). These techniques can accurately estimate the free energy of two binding partners and have been already applied for the study of the protein–protein recognition processes for small or medium size assemblies. In this context, replica exchange MD (Sugita & Okamoto, 1999) has been successfully exploited to study protein–peptide binding (Dagliyan, Proctor, D'Auria, Ding, & Dokholyan, 2011). Recently, an accurate procedure to estimate the free energy of binding of protein–protein assemblies has been proposed, returning results in agreement with experimental data (Gumbart, Roux, & Chipot, 2013a, 2013b). Although only applied to the barnase–barstar complex, this theoretical framework is in principle generally applicable to any protein–protein assembly. Metadynamics (Barducci, Bussi, & Parrinello, 2008; Bonomi et al., 2009; Laio & Parrinello, 2002) is another approach suitable for the study of oligomerization processes. For instance, by studying the oligomerization of the fibrin foldon domain, it was observed that fast assembly kinetics could be due to an interplay between thermodynamic bias toward native structure formation and residual flexibility (Barducci, Bonomi, Prakash, & Parrinello, 2013). In another work, this enhanced sampling technique allowed to elucidate the effect of pH on a large conformational change in the dengue virus envelope protein (Prakash, Barducci, & Parrinello, 2010).

3. TACKLING PROTEIN–PROTEIN INTERACTIONS AT COARSE-GRAINED RESOLUTION

Despite successful in recent applications, some of which highlighted above, atomistic MD simulations still seem to fall short on providing an effective solution to protein assembly prediction. One of the main reasons lies in the extremely large computational cost involved in the simulation of such systems. A valid alternative to atomistic simulations is the adoption of

CG models, where a nearly atomistic representation of biomolecules is adopted. CG models have a long history (Levitt & Warshel, 1975), but only in recent years they have been gaining an increasing interest. In a CG representation, atoms are grouped in order to create so-called beads, the interactions of which are described by effective potentials. This simplified representation has, first, the merit to reduce the number of particles to account for during force calculations. Second, the softer nature of the potentials allows for larger time steps, increasing further the time and spatial scales thus accessible. Several excellent reviews have been recently published on the variety of multiscale models available for biomolecular simulations (Baaden & Marrink, 2013; Ingolfsson et al., 2014; Tozzini, 2010). Here, we will mainly focus on CG models especially developed to study protein–protein assemblies, which could thus be also applicable in the context of integrative modeling of macromolecular assemblies (see next section).

Some of the existing CG force fields retain an almost atomistic representation of the backbone and use from one to four beads to represent side chains. In this context, a hybrid CG/atomistic model using implicit solvent to speed up calculations has been recently proposed by Zacharias (2013). This model was successfully used to study the self-assembly process of the axin-peptide– ADC receptor system and the α-helical p53 peptide binding to the MDM2 receptor protein. Another CG force field, OPEP, was used in conjunction with replica exchange to simulate the self-assembly process of amyloid peptides (Derreumaux, 1999; Maupetit, Tuffery, & Derreumaux, 2007; Santini, Mousseau, & Derreumaux, 2004). The PaLaCe force field, recently introduced by Lavery and coworkers, has shown promising results for the study of protein–protein complexes (Ceres, Pasi, & Lavery, 2012; Pasi, Lavery, & Ceres, 2013) where, although implicitly modeled, the solvent is carefully calibrated to account for surface effects. The Primo force field, that also exploits an implicit solvent model, is capable to reproduce protein–protein complexes dynamics, performing in fact well also in comparison to atomistic force fields (Kar, Gopal, Cheng, Predeus, & Feig, 2013). Finally, the Plum force field was able to address problems such as small peptides aggregation (Bereau & Deserno, 2009) and protein–membrane interactions (Tristan Bereau, Wang, & Deserno, 2014).

Other CG models feature one single bead representing the backbone and from one to four beads for the side chains. Within this class of mappings, one of the most relevant examples is represented by the Martini force field. Although mainly developed to simulate protein–protein complexes in phospholipid bilayers (Marrink, Risselada, Yefimov, Tieleman, & de Vries, 2007;

Monticelli et al., 2008), its recent application to soluble proteins showed promising results and room for improvement (May et al., 2014; Stark, Andrews, & Elcock, 2013). Among the most successful applications, we find the self-assembly of GPCR proteins (Periole, Huber, Marrink, & Sakmar, 2007, 2012). Sansom and coworkers have used the same approach to study the self-assembly of cohesin and dockerin, demonstrating that CG models are even able to capture the influence of point mutations affecting binding (Hall & Sansom, 2009), maintaing a good agreement with experimental results (Parton, Kilngelhoefer, & Sansom, 2011; Scott et al., 2008). Scorpion, another CG force field of this mapping class and featuring an explicit solvent model, allows to simulate solvated protein complexes (Basdevant, Borgis, & Ha-Duong, 2004; Ha-Duong, 2009; Ha-Duong, Basdevant, & Borgis, 2009). In the case of the barnase–barstar, self-assembled structures were closely resembling the known target complex (Basdevant, Borgis, & Ha-Duong, 2013). Finally, the UNRES force field (Liwo, Czaplewski, Pillardy, & Scheraga, 2001), originally developed to simulate folding and aggregation of small proteins (Czaplewski, Kalinowski, Liwo, & Scheraga, 2009), has been recently used to study the recognition process of Hsp70 chaperones (Gołaś et al., 2012), and the aggregation of the β-amyloid fragments ($A\beta_{1-28}$), demonstrating that helical intermediate secondary structure elements may play a role during aggregation (Rojas, Liwo, & Scheraga, 2011).

Within this mapping class, we have recently developed an alternative CG model based on an improved description of protein electrostatics. Besides the net charges for basic and acidic amino acids, the permanent dipolar contributions for both the backbone and polar side chains are considered as based on a topologically reconstructed multipolar expansion of all-atom centers of charge (Alemani, Collu, Cascella, & Dal Peraro, 2010; Cascella, Neri, Carloni, & Dal Peraro, 2008; Spiga, Alemani, Degiacomi, Cascella, & Dal Peraro, 2013). We initially showed that CG models holding such a simplified description of the electrostatics are sufficiently accurate to describe the electrostatic field at the active site of an enzyme (Cascella et al., 2008). At the same time, the potential terms mimicking the dipole–dipole interactions (Fig. 1A) are able to naturally stabilize elementary secondary structure motifs, such as α-helices and β-sheets, and to modulate transitions to super-secondary structure assemblies (Fig. 1C) (Alemani et al. 2010). In particular, the directionality of the backbone structure is directly connected to the CG bending angle (Fig. 1AB). Thus, the secondary structure propensity of amino acids, which is chemically encoded in the side chain, can be elegantly controlled by using the backbone bending potential as order

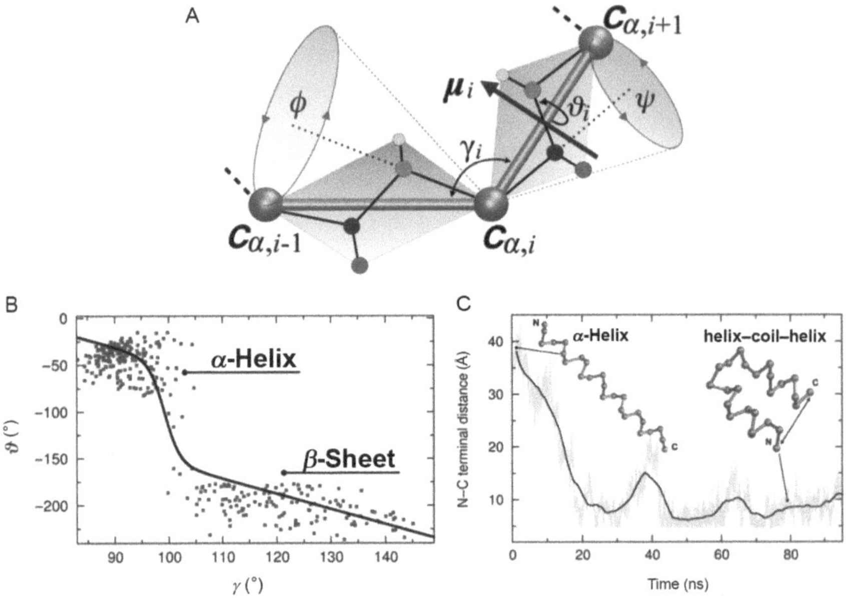

Figure 1 Dipole description in a CG representation. (A) Relationship between the CG angle γ_i and the all-atom Ramachandran angles ϕ and ψ. Rotation of any Ramachandran angle corresponds to a precession (ϑ_i) of the respective CG backbone dipole moment μ_i, resulting in a variation of the angle γ_i. (B) Statistical distribution of γ_i as a function of ϑ_i extracted from the PDB (in blue (dark gray in print version)) and its analytical fitting function (solid line). (C) Nonspecific CG force fields obtained using this scheme can stabilize secondary structure motifs and sample conformational transitions in MD simulations (e.g., the initial helical structure rearranges into a super-secondary helix–coil–helix motif). *Adapted with permission from Alemani et al.; J. Chem. Theory Comput., 2010, 6 (1), pp 315–324. Copyright 2010 American Chemical Society.*

parameter. Based on this CG mapping, we were able to define a consistent functional form of the CG potential and a first generation of CG parameters produced using a force-matching procedure based on an extended ensemble of all-atom MD simulations.

In particular, our CG model for proteins is based on an approximately four-to-one atoms-to-bead mapping as adopted by others (DeVane, Shinoda, Moore, & Klein, 2009; Marrink et al., 2007; Monticelli et al., 2008), with the only exception of aromatic side chains, where a higher resolution was used to map their geometric specificity (Fig. 2). All the amino acids are described by a backbone bead placed on top of the C_α atom and one or more beads for the side chains, which are placed at the center of mass of the constituent heavy atoms. Alanine and glycine amino acids are composed of a single backbone bead. The mass of each bead is the total mass of the

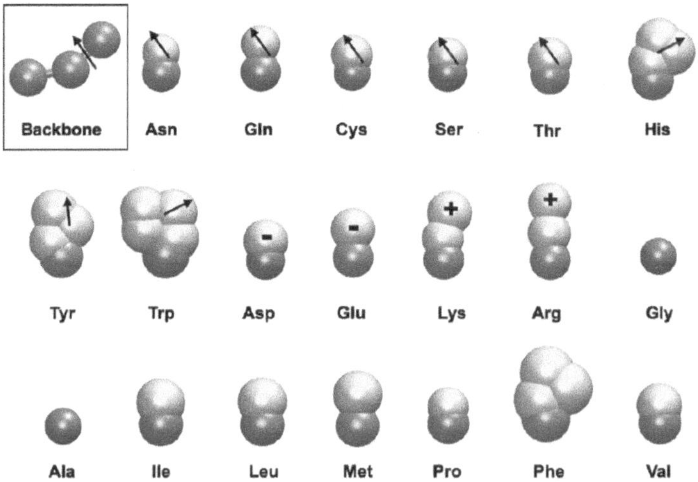

Figure 2 Molecular mapping of the dipolar CG force field. Schematic view of the atomistic to CG mapping in our force field. Backbone (iceblue (gray in print version)) and side chains (yellow (light gray in print version)) beads are represented by different CG potentials. Backbone and polar amino-acids also retain an electrostatic contribution described by their permanent dipoles (arrows). *Adapted with permission from Spiga et al.; J. Chem. Theory Comput., 2013, 9 (8), pp 3515–3526. Copyright 2013 American Chemical Society.*

constituent atoms (Fig. 2; Spiga et al., 2013). Amino- and carboxy-terminal backbone beads are represented with their respective zwitterionic charge and mass. The Hamiltonian associated with our CG model is an additive potential function similar to that used in atomistic simulations, adopted as a good compromise between accuracy and computational efficiency (Tozzini, 2010). The total potential function is thus given by:

$$\begin{aligned} V_{total} = & \sum_{bonds} k_b(|\vec{r}_{ij}| - r_0)^2 + \sum_{bendings} \sum_{n=2}^{4} k_{a,n}(\theta_{ijk} - \theta_0)^n \\ & + \sum_{dihe} \sum_{n=1}^{3} k_{d,n}[1 + \cos(n\phi_{ijkl} - \phi_0)] + \sum_{improper} \sum_{n=2}^{4} k_{i,n}(\psi_{ijkl} - \psi_0)^n \\ & + \sum_{pairs} 4\epsilon_{ij}\left[\left(\frac{\sigma_{ij}}{|\vec{r}_{ij}|}\right)^{12} - \left(\frac{\sigma_{ij}}{|\vec{r}_{ij}|}\right)^{6}\right] + V_{el}(\vec{r}_{ij}) \end{aligned} \quad (1)$$

where the first four terms describe bonded interactions and the remaining are used to describe nonbonded ones. In particular, the first term describes pseudobonds between backbone beads and beads belonging to the same residue, having r_0 as the equilibrium value and k_b as the constant force. The second term accounts for pseudobending for backbone and side chain beads (Tozzini, Rocchia, & McCammon, 2006), with θ_0 as equilibrium value and $k_{a,n}$ as constant forces. The third term, for torsion potential of pseudodihedrals (Tozzini et al., 2006) for backbone and multibead side chains, has ϕ_0 as the equilibrium values and $k_{d,n}$ as constant forces. The final term of the bonded potential describes improper torsion potentials used to force the L-chirality to the side chains or to force the planarity of aromatic side chains having ψ_0 as equilibrium value and $k_{i,n}$ as force constant. The last two terms represent the nonbonded part of the total potential function: a common 6–12 Lennard-Jones potential is used to account for effective nonbonded interactions not explicitly included in the electrostatics potential term. The electrostatic potential, instead, reads:

$$V_{\mathrm{el}}(\vec{r}_{ij}) = C(|\vec{r}_{ij}|)[V_{q_iq_j} + V_{q_i\mu_j} + V_{\mu_iq_j} + V_{\mu_i\mu_j}] \tag{2}$$

where all charge–charge, charge–dipole, and dipole–dipole interactions are considered. In our model, the solvation effects can be treated both implicitly and explicitly (Spiga et al., 2013). The implicit solvent model consists of a distance-dependent dielectric constant (Rubinstein & Sherman, 2004; Spiga et al., 2013), whereas the explicit solvent model is based on the work of Warshel (Florian & Warshel, 1997) and Borgis (Basdevant et al., 2004, 2013; Ha-Duong et al., 2009; Ha-Duong, Phan, Marchi, & Borgis, 2002).

We showed earlier that the explicit introduction of electrostatic dipolar terms (Cascella et al., 2008) contributes to enhance the stability of secondary structure elements without the use of *ad hoc* bias potentials (Alemani et al., 2010). Backbone dipoles μ_i are in fact univocally defined by the C_α trace of the protein (Alemani et al., 2010), associated to a triplet of consecutive C_α beads with their orientation determined by the angle of the bead triplet (Alemani et al., 2010; Cascella et al., 2008) (Fig. 1C). Remarkably, simple conformational changes, such as α-helix to helix–coil–helix transitions and α-helix to β-hairpin transitions could be also reproduced (Alemani et al., 2010). Moreover, we showed that this multipolar CG representation is able to reproduce peculiar electrostatic properties of proteins. For instance, protein-induced electrostatic field on *Pseudomonas aeruginosa* actin and azurin binding sites (Cascella et al., 2008; Neri, Anselmi, Cascella, Maritan, & Carloni, 2005) was well recapitulated using this approach. The calculated electrostatic potential on the active site agreed with its atomistic representation, demonstrating a striking match of the mutual orientation and module of the electric field at the two levels of resolution. Furthermore, we verified the ability of our representation to describe electrostatic tuning upon conformational changes in α-helices (Cascella et al., 2008).

Prompted by these promising results, we have more recently developed a protocol to derive electrostatic consistent CG parameters within this Hamiltonian form. Based on the adopted CG mapping scheme (Fig. 2), the individual CG distributions on a nonredundant subset of structures extracted from the PDB databank for bond lengths $\{r\}$, bending angles $\{\theta\}$ and torsions $\{\phi\}$, $P^{CG}(\chi_i, T)$ were Boltzmann inverted (Rühle, Junghans, Lukyanov, Kremer, & Andrienko, 2009; Tschöp, Kremer, Batoulis, Bürger, & Hahn, 1998) to obtain initial seeding potentials for the generic degree of freedom $\chi_i = r_i, \theta_i, \phi_i$ (Spiga et al., 2013), which were subsequently optimized using a force matching procedure for bonded and nonbonded potentials.

Force matching schemes have been widely discussed in several publications (Ercolessi & Adams, 1994; Izvekov & Voth, 2005; Noid, Liu, et al., 2008; Thorpe, Zhou, & Voth, 2008). Here, let $\{\omega\}$ the entire set of L parameters $\{\omega_1, \ldots \omega_L\}$ used to define the potential function adopted for the CG representation. The optimal $\{\omega\}$ set defining the CG potential function is the one minimizing the fitness function $Z_F(\omega)$:

$$Z_F(\omega) = \sqrt{\left(3\sum_{k=1}^{M} N_k\right)^{-1} \sum_{k=1}^{M}\sum_{i=1}^{N_k} \left| F_{ki}(\omega) - F_{ki}^{0} \right|^2} \tag{3}$$

where M is the number of sets of atomic configurations available, N_k is the number of beads in configuration k, $F_{ki}(\omega)$ is the force on the i-th bead in set k obtained with parametrization ω, and F_{ki}^0 is the reference force acting on the bead as given by the following formula:

$$F_{ki}^{0} = \sum_{j=1}^{L_i} F_{jki}^{0} \tag{4}$$

that is the sum of forces acting on the atoms belonging to the i-th bead. All quantities are averaged for a large set of different configurations, sampled from a preceding all-atom MD run.

The set of parameters $\{\omega\}$ minimizing the fitness function $Z_F(\omega)$ were obtained by means of a PSO search (Kennedy & Eberhart, 1995). PSO is a distributed heuristic optimization technique, that we implemented in a modular optimization environment called *pow*^{er} (Degiacomi & Dal

Peraro, 2013; Spiga et al., 2013), which we also used to predict molecular assemblies (further information is provided in the next section). In PSO, an ensemble of solutions (also called particles p) have their position $\omega(\mathrm{p})$ and velocity $\nu(\mathrm{p})$ randomly initialized in the multidimensional search space identified by boundaries. Along the whole optimization process, every particle will keep track of the position $\omega(\mathrm{p})$ associated with the best objective (fitness) function value $Z_F\,(\omega(\mathrm{p}))$. The boundaries associated with each parameter in the search space are guided by previous values obtained by Boltzmann inversion (as in the case of side chain bonded terms), by physically plausible values, or from literature (DeVane et al., 2009; Marrink et al., 2007; Monticelli et al., 2008). In this context, *pow^er^* was exploited to solve the force-matching problem on several atomistic references extracted from all-atom MD simulations carried out for five proteins representative of distinct SCOP families and covering single molecules in solution and protein–protein complexes (i.e., α_3W (1lq7; Dai et al., 2002), the β-protein Cox11 (1sp0; Banci et al., 2004), the α/β-protein LysM (1e0g; Bateman & Bycroft, 2000), the coiled-coil engineered water soluble phospholamban (Slovic, Stayrook, North, & DeGrado, 2005; Slovic, Summa, Lear, & DeGrado, 2003), and the barnase-barstar complex (1brs; Buckle, Schreiber, & Fersht, 1994)). At completion of every force matching, the difference between reference and calculated forces was in the order of 1 kcal mol^{-1} $\mathrm{\AA}^{-1}$ per degree of freedom (Spiga et al., 2013). During the optimization process, some parameters converged sooner than others. For instance, backbone bonded and nonbonded parameters converged to similar values (Spiga et al., 2013), hinting that such values may be transferable and used for a general CG force field. PSO demonstrated to be able to optimize simultaneously the large amount of parameters controlling the force matching error function. By parallelizing this algorithm on multiple processors, we therefore obtained an accurate, simple, and fast (i.e., few hours using 8 core processors) pipeline for the parameterization of *ad hoc* CG force fields for proteins.

All the proteins simulated at the CG level with our optimized force fields conserved their fold as during atomistic simulations, showing very good agreement with their all-atom description (Table 1) (Spiga et al., 2013). Secondary structures were conserved without the use of any additional *ad hoc* biases on the bending and torsional potential terms. Only minor discrepancies are observed on the loop regions connecting secondary structure elements in α^3W and Cox11 (Spiga et al., 2013). RMSD values reach convergence in around 5–10 ns (similarly to atomistic MD),

Table 1 Summary of the structural and dynamic properties of our CG model for different proteins. In square brackets are reported the all-atom values for RMSD and gyration radius in Å.

Protein	RMSD	R_g
α_3W	2.6±0.2 [2.5±0.2]	11.0±0.2 [12.2±0.2]
Cox11	3.2±0.2 [3.0±0.3]	15.1±0.1 [16.8±0.2]
LysM Domain	2.6±0.2 [2.7±0.4]	8.7±0.3 [9.9±0.2]
Water soluble phospholamban	5.4±0.6 [2.7±0.5]	14.8±0.4 [16.4±0.2]
Barnase–Barstar	3.5±0.2 [1.1±0.2]	15.7±0.1 [17.2±0.1]

fluctuating to values as low as 3 Å for ~1 μs for all three proteins (Table 1) (Spiga et al., 2013). The absolute values observed for RMSD are in line with results reported using other CG models (Chebaro, Pasquali, & Derreumaux, 2012). The gyration radius is systematically slightly higher at all-atom level with respect to the CG representation, the difference being, however, in the order of 1 Å(Table 1). This slight collapse is likely to be intrinsically dependent on the CG representation. The adopted mapping may not be able to completely reproduce the steric effects of all the side chains, and buried cavities accommodating few water molecules cannot be filled by water beads having larger hindrance at CG granularity (Takada, 2012).

In general, dynamical features were also in good agreement with the atomistic MD simulations. The RMSF calculated at CG level is systematically lower than for the all-atom one, as already observed using other models (Ha-Duong, 2009). Major differences are on the loop regions. For instance, in the case of the α_3W protein, the loops are composed by glycines and are therefore very flexible. The decrease in flexibility observed for the RMSF is confirmed also when calculating the S^2 order parameter of the backbone. The difference in flexibility observed for RMSF and S^2 is attributed (i) to the simple fact that at CG level the lower number of degrees of freedom does not intrinsically allow for a complete description of structural fluctuations, and (ii) to the potential form of the bending terms that, at this stage, is not parameterized to be sequence dependent, but has a general form meant to describe at the same time α, β, and coil structures. Backbone and side chains charges and dipolar terms are able to reproduce well the electrostatic potential of the proteins. Comparison with values obtained via atomistic simulations revealed high similarity indices (Table 2).

Table 2 Summary of the electrostatics properties

Protein	$\|P_{AA}\|$	$\|P_{CG}\|$	SI_{ele}
α_3W	88.6	119.3	0.97
Cox11	50.6	55.4	0.99
LysM	39.0	44.2	0.99
Water soluble phospholamban	630.7	630.2	0.95
Barnase–Barstar	200.2	218.8	0.93
L25	47.3	48.3	0.97
B1 Immunoglobulin binding	63.8	94.6	0.93

We could also show that these models already have a reasonable degree of transferability. In fact, simply averaging force field parameters obtained for different systems was sufficient to have a preliminary version of a quasi-transferable set of electrostatic-consistent CG potentials that produced structural and dynamic results in reasonable agreement with all-atom MD simulations. For improving transferability, a broader ensemble of MD simulations including all the representative folding classes might be used, exploiting the capabilities of PSO to work on high dimensional search spaces.

Importantly, once we extended our investigation to protein–protein complexes, we could show that this approach also produces dynamically stable quaternary complexes. Like in the case of the engineered soluble phospholamban (Slovic et al., 2005) and barnase-barstar (Buckle et al., 1994), the main interface interactions were preserved (Fig. 3A). These results are extremely promising and suggest that electrostatic-consistent CG potentials of this kind can be efficiently used to explore protein–protein molecular recognition using MD sampling. Our results are, in fact, in good agreement with all-atom MD simulations (Fig. 3B and C) and, when directly compared with previously reported CG force fields, showed similar or better performances in describing structural and dynamic determinants of soluble proteins. The extension of this optimization procedure to a larger data set may prelude to the generation of a fully transferable CG force field that will be applied in principle to any protein or, more interestingly, to explore the assembly mechanism of large macromolecular systems for which direct, long all-atom MD simulations may not be easily affordable. In addition, this CG force field may provide an energy scoring function at intermediate molecular resolution for the prediction and screening of protein–protein complexes during integrative modeling (see next section).

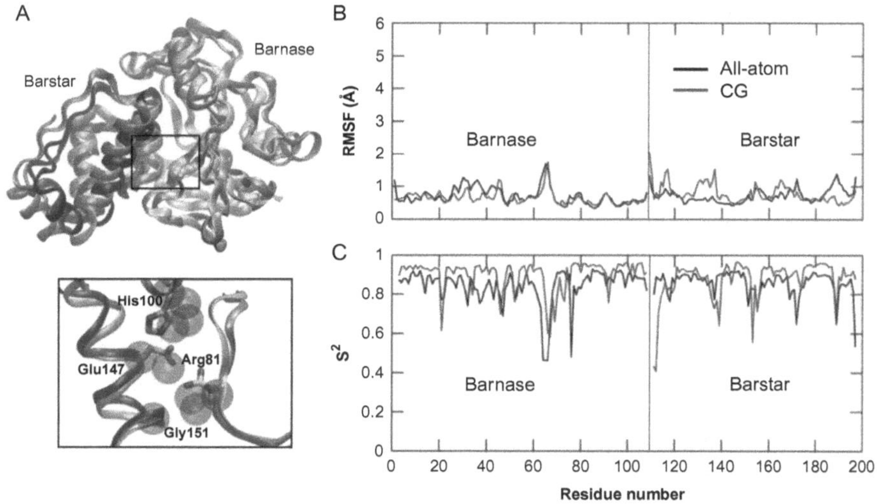

Figure 3 CG description of protein–protein complexes. (A) Backbone superimpositions of the structure obtained from all-atom and CG MD simulations for the barnase–barstar complex. In the inset, the interface of barnase–barstar is indicated with all-atom residues in licorice representation and CG ones in transparent van der Waals representations. (B and C) Comparison between dynamic properties of all-atom and CG MD simulations of molecular complexes. RMSF (B) and S^2 (C) are, respectively, reported for the barnase–barstar complex. *Adapted with permission from Spiga et al.; J. Chem. Theory Comput., 2013, 9 (8), pp 3515–3526. Copyright 2013 American Chemical Society.*

4. PREDICTING PROTEIN ASSEMBLY USING INTEGRATIVE MODELING

We have previously discussed as the prediction from first principles of protein assembly at nearly atomistic resolution is an extremely complex task. Since 2001 the CAPRI benchmark for blind protein assembly prediction (Janin et al., 2003; Lensink & Wodak, 2013) assesses the performance of methodologies targeting this challenge. Despite a steady progress in predictions accuracy, it is apparent that this problem is far from being robustly solved. Several outstanding reviews have been written about this topic (Alber, Förster, Korkin, Topf, & Sali, 2008; Russel et al., 2009; Ward, Sali, & Wilson, 2013). Here, we will briefly overview different available methodologies, and describe our own approach to tackle this challenge.

The prediction of protein assemblies can be seen as an optimization problem within a large and multidimensional search space, with the final goal to produce an exhaustive model of the multiprotein complex using known structures of its subunits as building blocks. The search for such a structure is increasingly difficult when more than two binding partners are present. To make this problem computationally tractable, a majority of existing methods feature some strategy to either reduce its size, or bias the search process toward specific regions of interest. A means to reduce the search space dimensions is to impose a specific symmetry (Andre, Bradley, Wang, & Baker, 2007; Berchanski, Segal, & Eisenstein, 2005; Goodsell & Olson, 2000; Pierce, Tong, & Weng, 2005; Plaxco & Gross, 2009; Schneidman-Duhovny, Inbar, Nussinov, & Wolfson, 2005). Other strategies consist in exploiting statistical information collected from a database of known assemblies to generate a scoring function biased toward favorable protein– protein arrangements or providing an initial seed to limit the search around a specific region of interest, or hotspot.

Another viable strategy to bias the search toward relevant regions has been more recently emerging. Methods, which are commonly defined integrative, directly exploit low-resolution experimental data to guide their structural predictions (Lensink & Wodak, 2010). In this context, density maps produced by electron microscopy constitute a good example of low-resolution data. 2D ensemble averages and 3D reconstructions can in fact provide information about the coarse arrangement of molecules within the complex. This growing wealth of data has already attracted the attention of several groups, that have successfully exploited density maps coupled to MD simulations as a source of information for protein assembly prediction and refinement. In this context, Grubmüller and coworkers characterized structure and kinetic pathways of 13 intermediate-translocation state models based on cryo-EM data of the *E. coli* ribosome (Bock et al., 2013). Another example is represented by the Attract method, that has been recently extended to allow the prediction of protein–protein assemblies for which cryo-EM data is available (de Vries & Zacharias, 2012; Zacharias, 2003). A final remarkable example is the MD flexible fitting (MDFF) method developed by Schulten and coworkers, which relies on steered MD to drive atoms of a structure toward areas having a high electron density in the map. MDFF has been successfully applied to a growing list of systems, such as the ribosome in order to interpret different stages of translation, the structure of a membrane-curvature-inducing photosynthetic complex, and the mature HIV-1 capsid (Trabuco et al., 2011; Zhao et al., 2013).

Besides density maps, structural data to be integrated can come in several other forms. These can range from interatomic distances (e.g., from cross-linking, Förster resonance energy transfer) to the location of specific amino acids in the assembly structure (e.g., gold labelling or alanine scanning), and macroscopic information (e.g., AFM or SAXS). Integrating all these different sources of information is desirable, a challenge lies, however, in the heterogeneity and possible ambiguity of these data. One of the most relevant efforts toward the consistent integration of these different spatial restraints is currently represented by the Integrative Modelling Platform, IMP (Lasker, Topf, Sali, & Wolfson, 2009; Russel et al., 2012). This approach mainly relies on a Monte Carlo protocol to model any kind of heterogeneous assembly at various granularities by exploiting a wide range of experimental information, as recently shown for the cases of the nuclear pore complex (Alber et al., 2007) and of the 26S proteasome (Lasker et al., 2012).

Another major challenge lies in accounting for protein flexibility, while integrating available experimental inputs. A way to deal with it is to first perform a rigid docking of constituents, and subsequently relax the resulting complex. This is the strategy followed, among others, by HADDOCK (Dominguez, Boelens, & Bonvin, 2003; Karaca & Bonvin, 2013). This integrative method can deal with a wide range of experimental restraints, and exploits MD simulations to refine the interfaces of produced models. An alternative strategy is to keep into account protein conformational space by docking ensembles of structures. Improved performances have been reported by methods docking ensembles generated via linear combinations of eigenvectors extracted by essential dynamics analysis (Moal & Bates, 2010; Mustard & Ritchie, 2005), NMR experiments (Chaudhury & Gray, 2008) or MD simulations (Boehr, Nussinov, & Wright, 2009; Di Nola, Roccatano, & Berendsen, 1994; Grünberg, Leckner, & Nilges, 2004; Mangoni, Roccatano, & Berendsen, 1999; Pak & Wang, 2000; Smith, Sternberg, & Bates, 2005).

In fact, MD simulations offer the important advantage of allowing the assessment of both the structural and mechanistic features of a protein, treating accurately its intrinsic dynamics. Furthermore, MD simulation can be easily interfaced with experimental data, as we recently showed predicting the key transmembrane domain of the two-component system PhoQ histidine kinase from *E. coli* by exploiting MD simulations guided by cross-linking data. In agreement with experimental findings, this work supported a possible signal transduction mechanism mediated by PhoQ, namely a combination of scissoring and rotational movements of the transmembrane helices (Lemmin, Soto, Clinthorne, DeGrado, & Dal Peraro, 2013). In another work following the same conceptual strategy, a synergistic approach combining MD simulations with double electron–electron resonance (DEER) EPR data permitted to clarify how the amyloid precursor protein prefers a straight α-helical conformation once embedded in a membrane bilayer (Lemmin, Dimitrov, Fraering, & Dal Peraro, 2013), allowing to better rationalize the mechanism of proteolytic process by the γ-secretase enzyme, and the interaction with drug modulators of this enzymatic activity (Dimitrov et al., 2013).

In our group, we have recently developed a novel and alternative integrative method capable to predict at near-atomistic level protein–protein assemblies of heterodimers and homomultimers. Our approach is based on three distinctive points. First, a correct coarse assembly arrangement can be quickly produced by exploiting a limited amount of structural information. Refinement of protein–protein interfaces, that is usually computationally expensive, can be postponed to a second phase and applied only on the most promising candidates. Second, a conformational ensemble can be dynamically explored during the protein assembly prediction process. Third, an optimization technique geared to tackle hard optimization problems is suitable for solving protein assembly prediction problems.

When dealing with heterodimers, one of the two proteins (the receptor) is kept fixed, whereas the other (the ligand) is freely displaced and rotated, leading to a 6D search space. If ensembles of structures are available, typically obtained from a MD trajectory, flexibility can be introduced as set of further dimensions in the search space. To do so, a principal component analysis (PCA) is initially performed in order to highlight the most relevant motions within the ensembles, i.e., the eigenvectors having the largest associated eigenvalue. In order to define the minimal number of eigenvectors required to represent protein motion, we first compute their associated eigenvalues cumulative energy g.

$$g(m) = \sum_{q=0}^{m} V(q,q), \forall m \in [0, 3n] \subset \mathbb{N} \tag{5}$$

Where the diagonal matrix V contains the eigenvalues ranked from the largest to the smallest, and n represents the number of atoms in the protein (corresponding therefore to $3n$ degrees of freedom). The desired amount of eigenvectors m_{min} is subsequently selected by means of the following condition:

$$\frac{g(m)}{g(3n)} \geq p \tag{6}$$

where p represents a user defined percentage of protein's motion. Finally, the projection value P of every structure in the ensemble along the first m_{min} eigenvectors is computed:

$$P = E(0 \leq m \leq m_{min}, 3n)^T \dot{T} \tag{7}$$

The obtained projection values P, also called fluctuations, are used as a way to index the structures within the ensemble, which we consider as a conformation database. The search space is therefore characterized by three rotations, three translations, and n fluctuations. In order to produce a complete assembly, the ligand (and the receptor) in the database having its eigenvector projection being the closest to the desired fluctuation values is first extracted. Subsequently, the rigid geometric operations are performed. The main advantage of using a MD trajectory with this approach is that the protein conformations used to assemble the multimer will respect protein natural (and physically plausible) flexibility. This methodology can be naturally extended to larger assemblies and has been applied to homomultimers arranged according to a circular symmetry, where the search space is defined by the three rotation angles of a single monomer with respect to a center of symmetry aligned along the z axis, and a tangential displacement with respect to it, which represents the radius of the assembly in its narrowest point (i.e., a 4D search space). As an additional extension, we geared our method to allow the docking of a symmetrical assembly on an given substrate. In this case, the substrate is kept fixed, while the obtained assembly is rototranslated around it.

To score the quality of an assembly, a lightweight fitness function is exploited, depending on two factors: geometry and energy. For geometric contribution, specific measures of the current multimer are compared to values being experimentally known. The aim is to minimize the difference within obtained multimer and desired measures. Target measures can be as diverse as assembly width or height obtained from cryo-EM maps to atomic distances obtained with FRET or cross-linking experiments. The energetic term of the fitness function aims at avoiding clashes. Two flavors of this energy term are available. The first is "minimalistic", it is indeed constituted by a 6–9 Lennard–Jones potential computed on all C_α and C_β atoms:

$$E(m) = 4\epsilon\left[\left(\frac{\sigma}{r}\right)^9 - \left(\frac{\sigma}{r}\right)^6\right] \tag{8}$$

where r are all the distances within couples of atoms being at a distance smaller than 12 Å, $\epsilon = 1$ kcal mol^{-1} (i.e., kcal/mol) and $\sigma = 4.7$ Å. A second energy function, more precise but also computationally more expensive, is available when just two binding partners (ligand and receptor) are docked. This is based on the estimation of the molecular mechanics contribution to the binding free energy of two monomers via the electrostatic consistent CG representation of the complex discussed in the previous section (Spiga et al., 2013).

The structural ensembles and defined scoring function are submitted to PSO engine, a particularly robust distributed heuristic optimization technique, which he have briefly introduced in the previous section to approach the problem of CG parameterization via a force-matching scheme. In the past 20 years, PSO has been enjoying popularity in fields requiring the resolution of hard optimization problems. A wealth of applications, as well as a wealth of variations of the basic algorithm, have been proposed (Poli, Kennedy, & Blackwell, 2007). In this specific context, we developed a new flavor of PSO, called "kick and reseed" (PSO-KaR) specifically conceived to avoid early convergence (Degiacomi & Dal Peraro, 2013) in hard optimization problems. All fitness evaluations produced by PSO-KaR are collected, and solutions having a fitness lower than a predefined threshold are retained. Since several solutions usually represent similar conformations, clustering on the basis of structures' reciprocal RMSD is subsequently performed. The complete pipeline described here is represented in Fig. 4.

We showed in a first benchmarking phase that our method can predict the rigid assembly of a large set of bound complexes spanning several symmetry classes, namely Chorismate Mutase (C3 symmetry), Acyl Carrier Protein Synthase (C3), Lumazine Synthase (C5), SM Archeal Protein (C7) and EscJ (C24). In all the cases, the reconstruction of the original multimer is attempted on the basis of a known stoichiometry by providing various combinations of experimentally plausible geometric restraints, and obtained a small ensemble of models (typically less than 10) having RMSD as low as ≤ 2 Å) with respect to the known bound conformation.

This approach gives, however, its best when conformational changes are associated with the assembly of the final complex. To assess the capacity of our methodology to account for protein intrinsic flexibility, three cases were studied where both the unbound and bound structures were known, namely phospholipase A2 (having an RMSD within bound and unbound conformation equal to 0.8 Å), flavivirus trimeric envelope glycoprotein (4.4 Å) and HIV-1 hexameric capsomer (10.5 Å). Within our protocol, first, MD simulations in the multi-ns timescale were able to sufficiently explore the conformational space of the unbound state; then PSO was successful in capturing the sampled configurations most closed to the bound state to predict the final assembly. In particular, we observed that when the difference between bound and unbound states is mild, such in the case of phospholipase

A2, no major difference between the results obtained with a rigid and flexible approach can be observed. However, when the conformational difference starts to be relevant, as for the flavivirus envelope glycoprotein, *pow*^{er}, our dynamic modeling approach outperformed a rigid docking strategy, producing models as close as 3.8 Å in RMSD with respect to the known assembled crystal. By comparison with the unbound structure, the selected frames had the important advantage of better capturing the protein-wide motion involved in the binding process. A more striking case is, however,

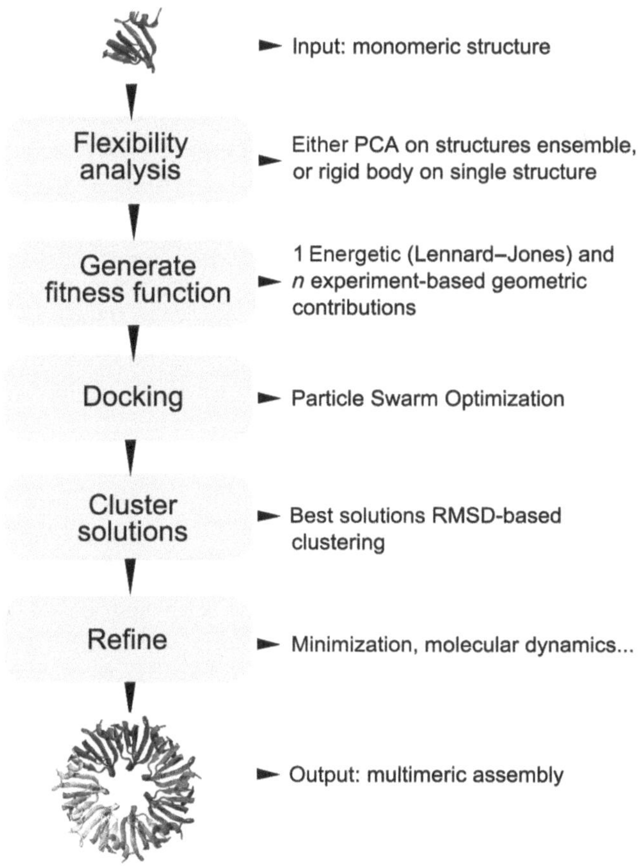

Figure 4 Macromolecular assembly prediction workflow as implemented in *pow*^{er} (at http://lbm.epfl.ch). When a structural ensemble is provided from a MD trajectory, fluctuations of main eigenvectors are added to the search space. Every protein assembly is evaluated by a fitness function depending on energy and geometry typically based on experimental evidence. Fitness function minima in the search space are sampled via a PSO algorithm. *Figure from Degiacomi and Dal Peraro (2013).*

constituted by the HIV-1 hexameric capsomer, which is composed of two domains connected by a flexible linker, and an extremely large RMSD between unbound and bound state. By exploiting a $\simeq$500 ns long MD simulation of the unbound conformation, *pow*^{er} produced a dozen of models, the best of which had a RMSD equal to 3.7 Å with respect to the known bound structure (Fig. 5). These results clearly highlight how our method is able to pinpoint the most suitable structure for assembly within a conformational ensemble, outperforming a purely rigid docking approach and producing states that can be further refined by other techniques.

The capabilities of this approach assume a broader reach within a conformational selection scenario of molecular assembly (Boehr et al., 2009; Lange et al., 2008; Peters & de Groot, 2012). If the bound conformational states of a molecular assembly are expected to already populate the accessible conformational space explored by the individual components alone, then the ensemble of states that are relevant for assembly prediction can be accessed by a thorough exploration of the conformational space of single subunits. If the complete conformational space could be sampled by MD, one might expect that the *pow*^{er} -based search would eventually find the optimal states to assemble the multimeric complex. Therefore, with the ever-growing sampling ability of MD simulations and the continuous progress of experimental methods the impact of such integrative and dynamic approach can only keep increasing with time.

We have recently exploited this approach to model the heptameric conformations of aerolysin, a pore-forming toxin from *Aeromonas hydrophila* (Degiacomi et al., 2013). In this case, protein flexibility plays an extremely important role, aerolysin assembles in fact first into a multimeric prepore complex upon proteolytic cleavage of its C-terminal region. Via a concerted movement, the prepore subsequently inserts into the lipid bilayer by forming a transmembrane β-barrel. Cryo-EM images of mutant species at 16 Å provided low-resolution information about both the prepore and pore states, which appeared to be markedly different. To model the prepore state, we first characterized by MD the dynamic features of activated aerolysin (i.e., with its C-terminal region removed; Fig. 6A), and used this conformational space to produce an assembly matching the prepore dimensions as observed in the available density map (Fig. 6B). We obtained six models, which we ranked by their cross-correlation coefficient (CCC) with respect to the density map. The highest ranked structure had a CCC of 0.79, a single subunit of this model having an RMSD equal to 4.3 Å with respect to the original

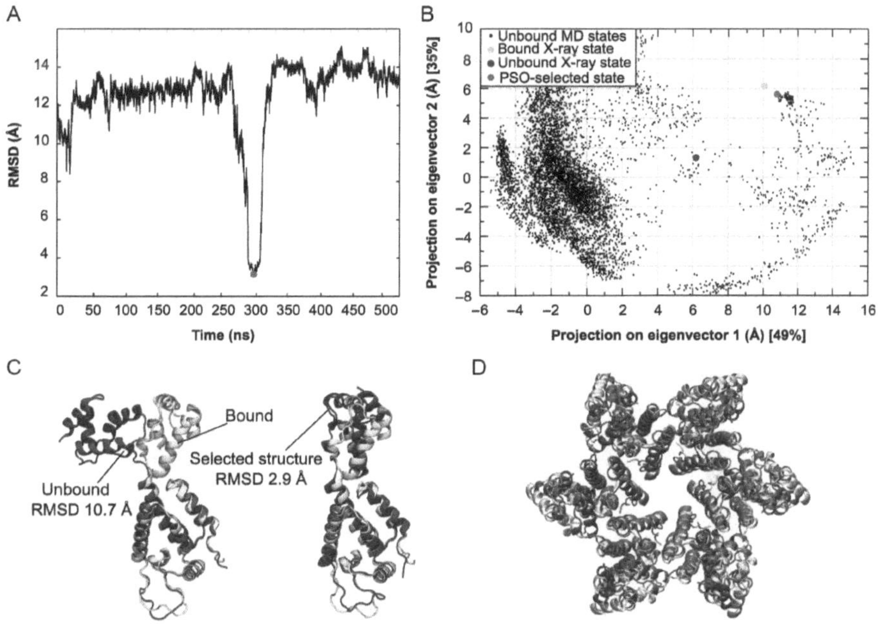

Figure 5 Dynamic conformational capturing. (A) Time evolution of monomeric HIV-1 hexameric capsomer RMSD with respect to initial unbound reference X-ray state. At around 300 ns, the protein visits conformations with low RMSD (less than 3 Å, indicated by a red dot (gray in print version)) with respect to the bound state. (B) Projection of the first two eigenvectors calculated by PCA reveals that, in the MD eigenspace of the unbound state, some structures appear close to the bound state. (C) Superposition of the unbound structure and the best frame extracted from MD simulation that is close to the bound state. (D) Best model produced by our method (3.7 Å RMSD, in yellow (light gray in print version)) is superimposed to known X-ray crystal structure of the complex (in blue (dark gray in print version)). A frame having a low RMSD in the conformational database (2.9 Å, indicated by the red dot in panels A and B (gray in print version)) was automatically selected as building block. *Figure from Degiacomi and Dal Peraro (2013).*

aerolysin crystal. It is also noteworthy to mention that by attempting to assemble a heptamer by solely exploiting the crystal structure no suitable candidate could be produced. In a last stage, the model was refined by means of an MDFF protocol (Trabuco et al., 2008), which led to a structure optimally fitting it (CCC equal to 0.87). The same pipeline was adopted to model aerolysin in its membrane inserted conformation (Fig. 6C). In this case, in order to produce a conformational space representative of the membrane-inserted state, we mimicked the second activation step by inducing the conformational rearrangement required for pore formation by means of steered MD simulations. The region forming

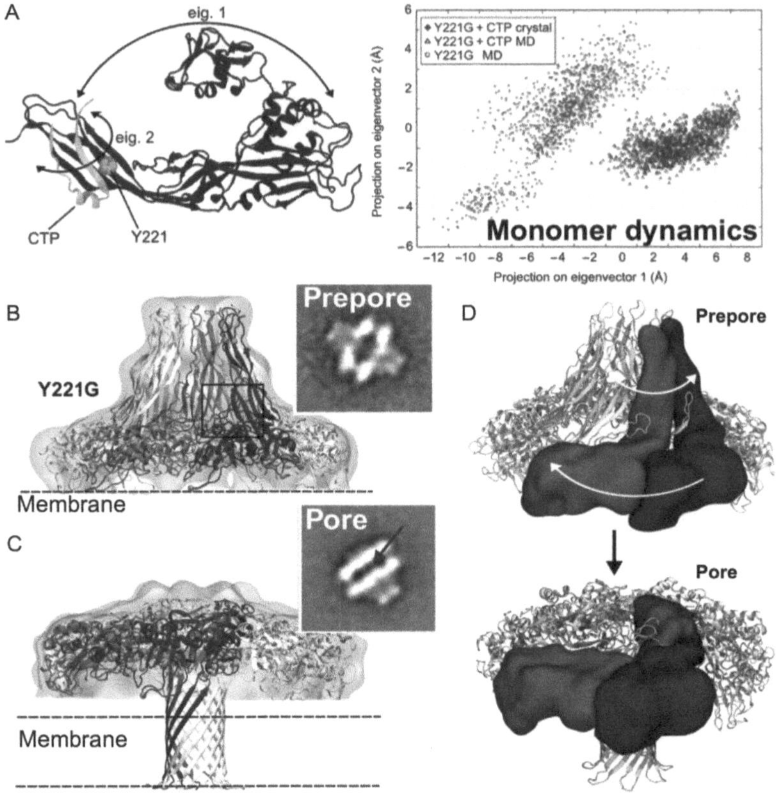

Figure 6 Dynamic modeling of the assembly of pore-forming toxin aerolysin. (A) Aerolysin monomers undergo some large conformational changes upon activation (i.e., removal of the C term peptide, CTP) as observed by a principal component analysis (PCA). (B and C) Cryo-EM maps of aerolysin trapped at the Y221G prepore and K246C–E258C quasi-pore states were used as spatial restraints for the construction of near-atomistic pore models. The dynamic ensemble in (A) was taken into account during the modeling step. (D) The comparison between the two states revealed a concerted swirling mechanism leading to pore insertion. *Adapted from Degiacomi et al. (2013).*

the β-barrel, called prestem loop, was extracted from its resting place using a rotational force, and ensemble of conformations thus obtained was used to assemble the final structure of the pore. The best ranked model had a CCC equal to 0.72, that was further improved to 0.90 by MDFF simulations. Importantly, the obtained prepore and pore conformation recapitulated all known biophysical and biochemical data, allowed to predict a completely new orientation of the pore at the target membrane, and to pinpoint relevant point mutations impairing the assembly process. The comparison between the prepore and pore state allowed proposing a concerted swirling motion driving a vertical collapse on the membrane and the formation of the transmembrane β-barrel (Fig. 6D). This mechanism could not have been observed without accounting with simulations for the native flexibility of aerolysin subunits.

Using the same strategy, the basal body of type III secretion system (a.k.a. injectisome) from *Yersinia enterocolitica* has been also determined (Kudryashev et al., 2013; Wagner et al., 2009; Wagner, Stenta, Metzger, Dal Peraro, & Cornelis, 2010). To this aim, we combined the first cryo-electron tomography (ET) map of an injectisome *in situ* (Fig. 7A) with new X-ray structures of basal body subunits (YscD) to build the periplasmic portion of the injectisome (Fig. 7B). Interestingly, YscD is composed by three independent, modular domains connected by flexible loops, and showed the capability to explore states ranging from a very compact to completely extended conformation (Fig. 7C). This conformational ensemble was subsequently used to assemble a 24-mer near-atomistic structure of the basal body, for which only a fully extended YscD could be fitted in the *in situ* cryo-ET map. When compared to isolated forms of injectisomes from other organisms not embedded in the bacterial membranes, like *Salmonella* or *Shighella*, *Yersinia* basal body showed a much larger flexibility and elongation (40%), which is likely promoted by the set of multidomain proteins constituting the basal body (Fig. 7B and C). This global elasticity is expected to be critical for injectisomes to resist the stress induced by thickness variations of the bacterial periplasmic space.

Altogether, these initial applications (Degiacomi et al., 2013; Hofmeyer et al., 2013; Kudryashev et al., 2013) show the effectiveness of our newly developed *pow^{er}* framework to model protein assembly. In particular, exploiting an ensemble of structures representative of the native flexibility of individual complex constituents leads eventually to improved performances in assembly prediction, and to the generation of more biologically sound structural models. In this context, the exploration of the conformational space of individual subunits provided by extended MD simulations has shown to be critical, especially to take into account activated states, which are not easy to be observed and characterized by other experimental approaches.

5. CONCLUSIONS AND PERSPECTIVES

In this chapter, we presented our recent efforts to predict and study large molecular assemblies using a computational strategy that proposes a

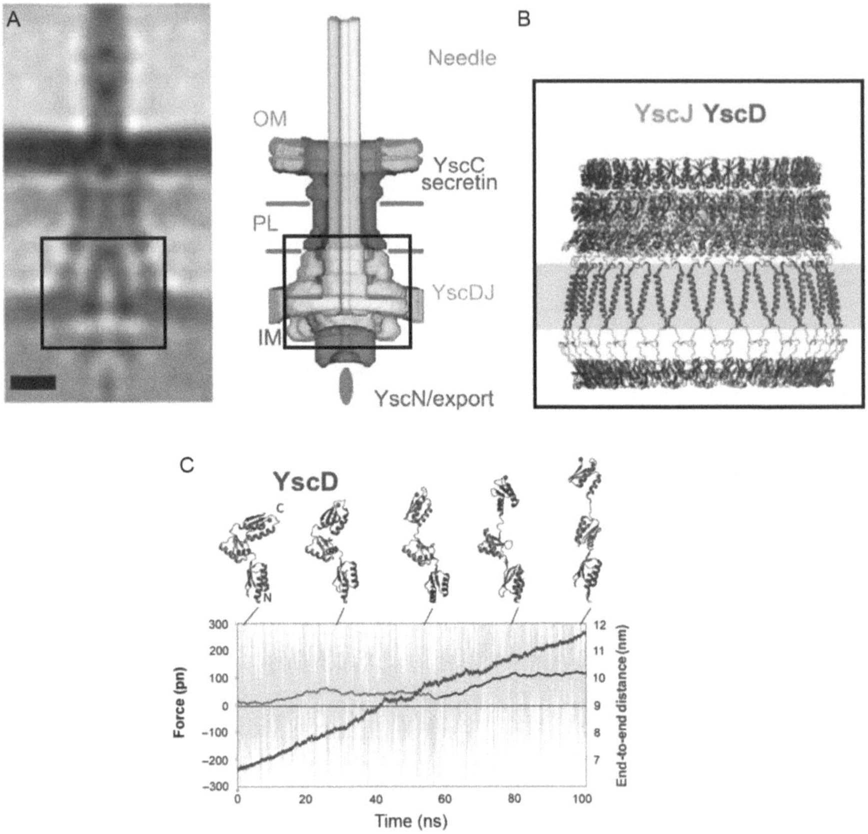

Figure 7 Assembly of the basal body of type III secretion system from Yersinia. (A) Slice through the cryo-electron tomography (cryo-ET) average structure of the injectisome and a correspondent model representation with indicated components. OM, outer membrane; PL, peptidoglycan layers; IM, inner membrane. (B) 24-mer ring model of YscD (blue (gray in print version)) and YscJ (yellow (light gray in print version)) proteins arranged at the basal body, as fitted on the injectisome cryo-ET map reported in (A). (C) Force-extension profile from steered MD simulations stretching an YscD monomer, respectively; end-to-end distance is reported in blue (dark gray in print version). Only fully extended conformations consistent with YscD X-ray crystals could fit the ET map in (A) to produce the basal body model reported in (B). *Adapted from Kudryashev et al. (2013).*

robust bridge among novel optimization algorithms, integration of experimental inputs and multiscale MD simulations.

While integrative modeling in biology can be dated back to the prediction of protein secondary structures by Pauling and Corey (1950, 1951) and DNA by Watson and Crick (1953), during the past decade this field has been displaying a constant progress, largely widening the scope and accuracy of its predictions. Theoretical and technological innovation on both the computational and experimental side has allowed to shed light on the mechanistic functioning of increasingly large and complex assemblies, by effectively enhancing to near-atomistic resolution our view on systems that are relevant for cellular biology. Still today, experimental techniques alone can rarely reach such resolution limit, and often, aiming at doing that, they perturb the native structural and dynamic determinants relevant at physiological conditions. Therefore, the development of computational strategies able to consistently integrate the growing wealth of functional data is a timely solution to close the spatial and temporal resolution gap existent till just few years ago between these approaches.

A key aspect, while modeling large molecular assemblies, is the emerging role of the native dynamics of the individual components in determining specific function. We think that one of the most crucial challenges for the years to come lies in a correct description of these dynamic determinants. Here, we discussed some effective strategies to directly include these ingredients during model building. First, the development of more accurate CG models able to sample conformational changes and describe more precisely electrostatic properties can help exploring from first principles the assembly mechanism of large molecular complexes via MD simulations. Second, the same CG models along with their atomistic counterparts can be employed to include the native dynamic contribution of individual subunits when models are built based on the integration of a variety of low resolution, sparse experimentally derived spatial restraints.

The future developments in this field will contribute to advance our understanding of the physicochemical principles underlying molecular recognition, and will hopefully foster the routinely adoption of dynamic and integrative strategies for the study of biological systems.

ACKNOWLEDGMENTS

The authors thank the entire past and present members of the Laboratory for Biomolecular Modeling (LBM) for useful discussions and suggestions. Their propositions, through kind and honest contributions in internal debates, have been of paramount importance. This work was supported by the Swiss National Science Foundation (SNSF) (grant numbers 200021 122120, 200020 138013 to M. D. P.).

REFERENCES

Adcock, S. A., & McCammon, J. A. (2006). Molecular dynamics: Survey of methods for simulating the activity of proteins. *Chemical Reviews*, *106*(5), 1589–1615.

Alber, F., Dokudovskaya, S., Veenhoff, L., Zhang, W., Kipper, J., Devos, D., et al. (2007). Determining the architectures of macromolecular assemblies. *Nature*, *450*(7170), 683–694.

Alber, F., Förster, F., Korkin, D., Topf, M., & Sali, A. (2008). Integrating diverse data for structure determination of macromolecular assemblies. *Annual Review of Biochemistry*, *77*, 443–477.

Alemani, D., Collu, F., Cascella, M., & Dal Peraro, M. (2010). A nonradial coarse-grained potential for proteins produces naturally stable secondary structure elements. *Journal of Chemical Theory and Computation*, *6*(1), 315–324. http://dx.doi.org/10.1021/ct900457z.

Allegretti, M., Mills, D. J., McMullan, G., Kühlbrandt, W., & Vonck, J. (2014). Atomic model of the f420-reducing [nife] hydrogenase by electron cryo-microscopy using a direct electron detector. *eLife*, *3*.

Andre, I., Bradley, P., Wang, C., & Baker, D. (2007, October). Prediction of the structure of symmetrical protein assemblies. *Proceedings of the National Academy of Sciences*. http://dx.doi.org/10.1073/pnas.0702626104. Retrieved from, http://dx.doi.org/10.1073/pnas.0702626104.

Arkhipov, A., Shan, Y., Das, R., Endres, N. F., Eastwood, M. P., Wemmer, D. E., et al. (2013). Architecture and membrane interactions of the egf receptor. *Cell*, *152*(3), 557–569.

Baaden, M., & Marrink, S. J. (2013). Coarse-grain modelling of proteinprotein interactions. *Current Opinion in Structural Biology*, *23*(6), 878–886.

Babu, M., Vlasblom, J., Pu, S., Guo, X., Graham, C., Bean, B. D., et al. (2012). Interaction landscape of membrane-protein complexes in saccharomyces cerevisiae. *Nature*.

Bahadur, R., & Zacharias, M. (2008). The interface of protein-protein complexes: Analysis of contacts and prediction of interactions. *Cellular and Molecular Life Sciences*, *65*, 1059–1072.

Baldwin, A. J., Lioe, H., Hilton, G. R., Baker, L. A., Rubinstein, J. L., Kay, L. E., et al. (2011). The polydispersity of αb-crystallin is rationalized by an interconverting polyhedral architecture. *Structure*, *19*(12), 1855–1863.

Banci, L., Bertini, I., Cantini, F., Ciofi-Baffoni, S., Gonnelli, L., & Mangani, S. (2004). Solution structure of cox11, a novel type of β-immunoglobulin-like fold involved in cu_B site formation of cytochrome c oxidase. *The Journal of Biological Chemistry*, *279*(33), 34833–34839.

Barducci, A., Bonomi, M., Prakash, M. K., & Parrinello, M. (2013). Free-energy landscape of protein oligomerization from atomistic simulations. *PNAS*, *19*, E4708–E4713.

Barducci, A., Bussi, A., & Parrinello, M. (2008). Well-tempered metadynamics: A smoothly converging and tunable free-energy method. *Physical Review Letters*, *100*(2), 020603.

Barrett, P. J., Song, Y., Van Horn, W. D., Hustedt, E. J., Schafer, J. M., Hadziselimovic, A., et al. (2012). The amyloid precursor protein has a flexible transmembrane domain and binds cholesterol. *Science*, *336*(6085), 1168–1171.

Bartesaghi, A., & Subramaniam, S. (2009). Membrane protein structure determination using cryo-electron tomography and 3d image averaging. *Current Opinion in Structural Biology*, *19*(4), 402–407.

Basdevant, N., Borgis, D., & Ha-Duong, T. (2004). A semi-implicit solvent model for the simulation of peptides and proteins. *Journal of Computational Chemistry*, *25*(8), 1015–1029.

Basdevant, N., Borgis, D., & Ha-Duong, T. (2013). Modeling protein-protein recognition in solution using the coarse-grained force field scorpion. *Journal of Chemical Theory and Computation*, *9*(1), 803–813.

Bateman, A., & Bycroft, M. (2000). The structure of a lysm domain from e. coli membranebound lytic murein transglycosylase. *Journal of Molecular Biology*, *209*, 1113–1119. http://dx.doi.org/10.1006/jmbi.2000.3778.

Berchanski, A., Segal, D., & Eisenstein, M. (2005). Modeling oligomers with cn or dn symmetry: Application to capri target 10. *Proteins: Structure, Function, and Bioinformatics*, *60*(2), 202–206.

Bereau, T., & Deserno, M. (2009). Generic coarse-grained model for protein folding and aggregation. *The Journal of Chemical Physics*, *130*(235106), 1–15.
Bock, L. V., Blau, C., Schröder, G. F., Davydov, I. I., Fischer, N., Stark, H., et al. (2013). Energy barriers and driving forces in trna translocation through the ribosome. *Nature Structural and Molecular Biology*, *20*(12), 1390–1397.
Boehr, D. D., Nussinov, R., & Wright, P. E. (2009). The role of dynamic conformational ensembles in biomolecular recognition. *Nature Chemical Biology*, *5*, 789–796.
Bonomi, M., Branduardi, D., Bussi, G., Camilloni, C., Provasi, D., Raiteri, P., et al. (2009). Plumed: A portable plugin for free-energy calculations with molecular dynamics. *Computer Physics Communications*, *180*(10), 1961.
Buch, I., Giorgino, T., & De Fabritiis, G. (2011). Complete reconstruction of an enzyme inhibitor binding process by molecular dynamics simulations. *Proceedings of the National Academy of Sciences*, *108*(25), 10184–10189.
Buckle, A. M., Schreiber, G., & Fersht, A. R. (1994). Protein-protein recognition: Crystal structural analysis of a barnase-barstar complex at 2.0-å resolution. *Biochemistry*, *33*(30), 8878–8889.
Cascella, M., Neri, M. A., Carloni, P., & Dal Peraro, M. (2008). Topologically based multipolar reconstruction of electrostatic interactions in multiscale simulations of proteins. *Journal of Chemical Theory and Computation*, *4*(8), 1378–1385.
Casuso, I., Khao, J., Chami, M., Paul-Gilloteaux, P., Husain, M., Duneau, J.-P., et al. (2012). Characterization of the motion of membrane proteins using high-speed atomic force microscopy. *Nature Nanotechnology*, 7(8), 525–529.
Ceres, N., Pasi, M., & Lavery, R. (2012). A protein solvation model based on residue burial. *Journal of Chemical Theory and Computation*, *8*(6), 2141–2144.
Chaudhury, S., & Gray, J. J. (2008). Conformer selection and induced fit in flexible backbone protein-protein docking using computational and nmr ensembles. *Journal of Molecular Biology*, *381*, 1068–1087.
Chebaro, Y., Pasquali, S., & Derreumaux, P. (2012). The coarse-grained opep force field for non-amyloid and amyloid proteins. *The Journal of Physical Chemistry B*, *116*(30), 8741–8752.
Czaplewski, C., Kalinowski, S., Liwo, A., & Scheraga, H. A. (2009). Application of multiplexed replica exchange molecular dynamics to the unres force field: Tests with α and $\alpha+\beta$ proteins. *Journal of Chemical Theory and Computation*, *5*(3), 627–640.
Dagliyan, O., Proctor, E. A., D'Auria, K. M., Ding, F., & Dokholyan, N. V. (2011). Structural and dynamic determinants of protein-peptide recognition. *Structure*, *19*, 1837–1845.
Dai, Q. H., Tommos, C., Fuentes, J. E., Blomberg, M. R. A., Dutton, P. L., & Wand, A. J. (2002). Structure of a de novo designed protein model of radical enzymes. *Journal of the American Chemical Society*, *124*(37), 10952–10953. http://dx.doi.org/10.1021/ja0264201.
de Vries, S. J., & Zacharias, M. (2012). Attract-em: A new method for the computational assembly of large molecular machines using cryo-em maps. *PLoS One*, 7(12), e49733.
Degiacomi, M. T., & Dal Peraro, M. (2013). Macromolecular symmetric assembly prediction using swarm intelligence dynamic modelling. *Structure*, *21*, 1097–1106.
Degiacomi, M. T., Iacovache, I., Pernot, L., Chami, M., Kudryashev, M., Stahlberg, H., et al. (2013). Molecular assembly of the aerolysin pore reveals a swirling membrane-insertion mechanism. *Nature Chemical Biology*, *9*(10), 623–629.
Derreumaux, P. (1999). From polypeptide sequences to structures using monte carlo simulations and an optimized potential. *The Journal of Chemical Physics*, *111*(5), 2301–2310.
DeVane, R., Shinoda, W., Moore, P. B., & Klein, M. L. (2009). Transferable coarse grain non-bonded interaction model for amino acids. *Journal of Chemical Theory and Computation*, *5*(9), 2115–2124.

Di Nola, A., Roccatano, D., & Berendsen, H. J. C. (1994). Molecular dynamics simulation of the docking of substrates to proteins. *Proteins: Structure, Function, and Bioinformatics, 19*, 174–182.

Dimitrov, M., Alattia, J.-R., Lemmin, T., Lehal, R., Fligier, A., Houacine, J., et al. (2013). Alzheimers disease mutations in app but not γ-secretase modulators affect epsilon cleavage-dependent aicd production. *Nature Communications, 4*.

Dominguez, C., Boelens, R., & Bonvin, A. M. J. J. (2003). Haddock: A protein-protein docking approach based on biochemical or biophysical information. *Journal of the American Chemical Society, 125*(9), 1731–1737.

Ercolessi, F., & Adams, J. B. (1994). Interatomic potentials from first-principles calculations: The force-matching method. *Europhysics Letters, 26*(8), 583–588.

Florian, J., & Warshel, A. (1997). Langevin dipoles model for ab initio calculations of chemical processes in solution: Parametrization and application to hydration free energies of neutral and ionic solutes and conformational analysis in aqueous solution. *The Journal of Physical Chemistry B, 101*(28), 5583–5595.

Gavin, A. C., Bösche, M., Krause, R., Grandi, P., Marzioch, M., Bauer, A., et al. (2002). Functional organization of the yeast proteome by systematic analysis of protein complexes. *Nature, 415*, 141–147.

Gołaś, E., Maisuradze, G. G., Senet, P., Ołdziej, S., Czaplewski, C., Scheraga, H. A., et al. (2012). Simulation of the opening and closing of hsp70 chaperones by coarse-grained molecular dynamics. *Journal of Chemical Theory and Computation, 8*(5), 1750–1764.

Goodsell, D. S., & Olson, A. J. (2000). Structural symmetry and protein function. *Annual Review of Biophysics and Biomolecular Structure, 29*(1), 105–153.

Greber, B. J., Boehringer, D., Leitner, A., Bieri, P., Voigts-Hoffmann, F., Erzberger, J. P., et al. (2014). Architecture of the large subunit of the mammalian mitochondrial ribosome. *Nature, 505*(7484), 515–519.

Grünberg, R., Leckner, J., & Nilges, M. (2004). Complementarity of structure ensembles in protein-protein binding. *Structure, 12*, 2125–2136.

Gumbart, J. C., Roux, B., & Chipot, C. (2013a). Efficient determination of protein–protein standard binding free energies from first principles. *Journal of Chemical Theory and Computation, 9*(8), 3789–3798.

Gumbart, J. C., Roux, B., & Chipot, C. (2013b). Standard binding free energies from computer simulations: What is the best strategy? *Journal of Chemical Theory and Computation, 9*(1), 794–802.

Ha-Duong, T. (2009). Protein backbone dynamics simulations using coarse-grained bonded potentials and simplified hydrogen bonds. *Journal of Chemical Theory and Computation, 5*(12), 3211–3223.

Ha-Duong, T., Basdevant, N., & Borgis, D. (2009). A polarizable coarse-grained water model for coarse-grained proteins simulations. *Chemical Physics Letters, 468*, 79–82.

Ha-Duong, T., Phan, S., Marchi, M., & Borgis, D. (2002). Electrostatics on particles: Phenomenological and orientational density functional approach. *The Journal of Chemical Physics, 117*(2), 541–556.

Hall, B. A., & Sansom, M. S. P. (2009). Coarse-grained md simulations and protein-protein interactions: The cohesin-dockerin system. *Journal of Chemical Theory and Computation, 5*(9), 2465–2471.

Hofmeyer, T., Schmelz, S., Degiacomi, M. T., Dal Peraro, M., Daneschdar, M., Scrima, A., et al. (2013). Arranged sevenfold: Structural insights into the c-terminal oligomerization domain of human c4b-binding protein. *Journal of Molecular Biology, 425*(8), 1302–1317.

Imakaev, M., Fudenberg, G., McCord, R. P., Naumova, N., Goloborodko, A., Lajoie, B. R., et al. (2012). Iterative correction of hi-c data reveals hallmarks of chromosome organization. *Nature Methods, 9*(10), 999–1003.

Ingolfsson, H. I., Lopez, C. A., Uusitalo, J. J., de Jong, D. H., Gopal, S. M., Periole, X., et al. (2014). The power of coarse graining in biomolecular simulations. *WIREs Computational Molecular Science*, *4*, 225–248.

Izvekov, S., & Voth, G. A. (2005). A multiscale coarse-graining method for biomolecular systems. *The Journal of Physical Chemistry B*, *109*(7), 2469–2473.

Janin, J., Henrick, K., Moult, J., Eyck, L. T., Sternberg, M. J., Vajda, S., et al. (2003). Capri: A critical assessment of predicted interactions. *Proteins: Structure, Function, and Bioinformatics*, *52*(1), 2–9.

Kar, P., Gopal, S. M., Cheng, Y. M., Predeus, A., & Feig, M. (2013). Primo: A transferable coarse-grained force field for proteins. *Journal of Chemical Theory and Computation*, *9*(8), 3769–3788.

Karaca, E., & Bonvin, A. M. J. J. (2013). Advances in integrative modeling of biomolecular complexes. *Methods*, *59*, 372–381.

Kennedy, J., & Eberhart, R. (1995). Particle swarm optimization. In *Proceedings of the ieee international conference on neural networks, 1995: Vol. 4.* (pp. 1942–1948).

Kruse, A. C., Hu, J., Pan, A. C., Arlow, D. H., Rosenbaum, D. M., Rosemond, E., et al. (2012). Structure and dynamics of the m3 muscarinic acetylcholine receptor. *Nature*, *482*(7386), 552–556.

Kudryashev, M., Stenta, M., Schmelz, S., Amstutz, M., Wiesand, U., Castao-Díez, D., et al. (2013). In situ structural analysis of the yersinia enterocolitica injectisome. *eLife*, *2*, e00792.

Kühner, S., van Noort, V., Betts, M. J., Leo-Macias, A., Batisse, C., Rode, M., et al. (2009). Proteome organization in a genome-reduced bacterium. *Science*, *326*(5957), 1235–1240.

Laio, A., & Parrinello, M. (2002). Escaping free-energy minima. *PNAS*, *99*(22), 12562–12566.

Lange, O. F., Lakomek, N. A., Fares, C., Schroder, G. F., Walter, K. F., Becker, S., et al. (2008). Recognition dynamics up to microseconds revealed from an rdcderived ubiquitin ensemble in solution. *Science*, *320*(5882), 1471–1475.

Lasker, K., Förster, F., Bohn, S., Walzthoeni, T., Villa, E., Unverdorben, P., et al. (2012). Molecular architecture of the 26s proteasome holocomplex determined by an integrative approach. *Proceedings of the National Academy of Sciences*, *109*(5), 1380–1387.

Lasker, K., Topf, M., Sali, A., & Wolfson, H. (2009). Inferential optimization for simultaneous fitting of multiple components into a cryoem map of their assembly. *Journal of Molecular Biology*, *388*(1), 180–194.

Lemmin, T., Dimitrov, M., Fraering, P. C., & Dal Peraro, M. (2013). Perturbation of the straight transmembrane α-helical structure of the amyloid precursor protein affect its processing by γ-secretase. *The Journal of Biological Chemistry*. http://dx.doi.org/10.1074/jbc.M113.479781.

Lemmin, T., Soto, C. S., Clinthorne, G., DeGrado, W. F., & Dal Peraro, M. (2013). Assembly of the transmembrane domain of e. coli phoq histidine kinase: Implications for signal transduction from molecular simulations. *PLoS Computational Biology*, *9*, e1002878.

Lensink, M. F., & Wodak, S. J. (2010). Docking and scoring protein interactions: Capri 2009. *Proteins: Structure, Function, and Bioinformatics*, *78*(15), 3073–3084.

Lensink, M. F., & Wodak, S. J. (2013). Docking, scoring, and affinity prediction in capri. *Proteins: Structure, Function, and Bioinformatics*, *81*(12), 2082–2095.

Levitt, M., & Warshel, A. (1975). Computer simulation of protein folding. *Nature*, *253*(5494), 94–98.

Li, X., Mooney, P., Zheng, S., Booth, C. R., Braunfeld, M. B., Gubbens, S., et al. (2013). Electron counting and beam-induced motion correction enable near-atomic-resolution single-particle cryo-em. *Nature Methods*, *10*(6), 584–590.

Liao, M., Cao, E., Julius, D., & Cheng, Y. (2013). Structure of the trpv1 ion channel determined by electron cryo-microscopy. *Nature*, *504*(7478), 107–112.

Limongelli, V., Bonomi, M., & Parrinello, M. (2013). Funnel metadynamics as accurate binding free-energy method. *Proceedings of the National Academy of Sciences*, *110*(16), 6358–6363.

Lindorff-Larsen, K., Piana, S., Dror, R. O., & Shaw, D. E. (2011). How fast-folding proteins fold. *Science*, *334*(6055), 517–520.

Lipfert, J., & Doniach, S. (2007). Small-angle x-ray scattering from rna, proteins, and protein complexes. *Annual Review of Biophysics and Biomolecular Structure*, *36*, 307–327.

Liwo, A., Czaplewski, C., Pillardy, J., & Scheraga, H. A. (2001). Cumulant-based expressions for the multibody terms for the correlation between local and electrostatic interactions in the united-residue force field. *The Journal of Chemical Physics*, *115*(5), 2323–2357.

Manglik, A., Kruse, A. C., Kobilka, T. S., Thian, F. S., Mathiesen, J. M., Sunahara, R. K., et al. (2012). Crystal structure of the [micro]-opioid receptor bound to a morphinan antagonist. *Nature*, *485*(7398), 321–326.

Mangoni, M., Roccatano, D., & Berendsen, H. J. C. (1999). Docking of flexible ligands to flexible receptors in solution by molecular dynamics simulation. *Proteins: Structure, Function, and Bioinformatics*, *35*, 153–162.

Mao, Y., Wang, L., Gu, C., Herschhorn, A., Xiang, S.-H., Haim, H., et al. (2012). Subunit organization of the membrane-bound hiv-1 envelope glycoprotein trimer. *Nature Structural & Molecular Biology*, *19*(9), 893–899.

Marrink, S. J., Risselada, H. J., Yefimov, S., Tieleman, D. P., & de Vries, A. H. (2007). The martini force field: Coarse grained model for biomolecular simulations. *The Journal of Physical Chemistry B*, *111*(27), 7812–7824.

Maupetit, J., Tuffery, P., & Derreumaux, P. (2007). A coarse-grained protein force field for folding and structure prediction. *Proteins*, *69*, 394–408.

May, A., Pool, R., van Dijk, E., Bijlard, J., Abeln, S., Heringa, J., et al. (2014). Coarse-grained versus atomistic simulations: Realistic interaction free energies for real proteins. *Bioinformatics*, *30*(3), 326–334.

Moal, I. H., & Bates, P. A. (2010). Swarmdock and the use of normal modes in protein-protein docking. *International Journal of Molecular Sciences*, *11*(10), 3623–3648.

Monticelli, L., Kandasamy, S. K., Periole, X., Larson, R. G., Tieleman, D. P., & Marrink, S. J. (2008). The martini coarse-grained force field: Extension to proteins. *Journal of Chemical Theory and Computation*, *4*(5), 819–834.

Mustard, D., & Ritchie, D. W. (2005). Docking essential dynamics eigenstructures. *Proteins*, *60*, 269–274.

Neri, M. A., Anselmi, C., Cascella, M., Maritan, A., & Carloni, P. (2005). Coarse-grained model of proteins incorporating atomistic detail of the active site. *Physical Review Letters*, *95*, 218102.

Noid, W. G., Liu, P., Wang, Y., Chu, J. W., Ayton, G. S., Izvekov, S., et al. (2008). The multiscale coarse-graining method. 2. numerical implementation for coarse-grained molecular models. *The Journal of Chemical Physics*, *128*(244115), 1–20.

Pak, Y., & Wang, S. (2000). Application of a molecular dynamics simulation method with a generalized effective potential to the flexible molecular docking problems. *The Journal of Physical Chemistry*, *104*, 354–359.

Parton, D. L., Kilngelhoefer, J. W., & Sansom, M. S. P. (2011). Aggregation of model membrane proteins, modulated by hydrophobic mismatch, membrane curvature, and protein class. *Biophysical Journal*, *101*(3), 691–699.

Pasi, M., Lavery, R., & Ceres, N. (2013). Palace: A coarse-grain protein model for studying mechanical properties. *Journal of Chemical Theory and Computation*, *9*(1), 785–793.

Pauling, L., & Corey, R. B. (1950). Two hydrogen-bonded spiral configurations of the polypeptide chain. *Journal of the American Chemical Society*, *72*(11), 5349.

Pauling, L., & Corey, R. B. (1951). The pleated sheet, a new layer configuration of polypeptide chains. *Proceedings of the National Academy of Sciences of the United States of America*, *37*(5), 251.

Periole, X., Huber, T., Marrink, S. J., & Sakmar, T. P. (2007). G protein-coupled receptors self-assemble in dynamics simulations of model bilayers. *Journal of the American Chemical Society*, *129*(33), 10126–10132.

Periole, X., Huber, T., Marrink, S. J., & Sakmar, T. P. (2012). Structural determinants of the supramolecular organization of g protein-coupled receptors in bilayers. *Journal of the American Chemical Society*, *134*(26), 10959–10965.

Peters, J. H., & de Groot, B. L. (2012). Ubiquitin dynamics in complexes reveal molecular recognition mechanisms beyond induced fit and conformational selection. *PLoS Computational Biology*, *8*(10), e1002704.

Piana, S., Lindorff-Larsen, K., & Shaw, D. E. (2013). Atomistic description of the folding of a dimeric protein. *The Journal of Physical Chemistry B*, *117*(42), 12935–12942.

Pierce, B., Tong, W., & Weng, Z. (2005). M-zdock: A grid-based approach for cn symmetric multimer docking. *Bioinformatics*, *21*(8), 1472–1478.

Plaxco, K. W., & Gross, M. (2009). Protein complexes: The evolution of symmetry. *Current Biology*, *19*(1), R25–R26.

Poli, R., Kennedy, J., & Blackwell, T. (2007). Particle swarm optimization. *Swarm Intelligence*, *1*(1), 33–57.

Prakash, M. K., Barducci, A., & Parrinello, M. (2010). Probing the mechanism of ph-induced large-scale conformational changes in dengue virus envelope protein using atomistic simulations. *Biophysical Journal*, *99*(2), 588–594.

Rinner, O., Seebacher, J., Walzthoeni, T., Mueller, L., Beck, M., Schmidt, A., et al. (2008). Identification of cross-linked peptides from large sequence databases. *Nature Methods*, *5*(4), 315–318.

Rojas, A. V., Liwo, A., & Scheraga, H. A. (2011). A study of the α-helical intermediate preceding the aggregation of the amino-terminal fragment of the β amyloid peptide (aβ1–28). *The Journal of Physical Chemistry B*, *115*, 12978–12983.

Rubinstein, A., & Sherman, S. (2004). Influence of the solvent structure on the electrostatic interactions in proteins. *Biophysical Journal*, *878*(3), 15441557. http://dx.doi.org/10.1529/biophysj.103.038620.

Rühle, V., Junghans, C., Lukyanov, A., Kremer, K., & Andrienko, D. (2009). Versatile objectoriented toolkit for coarse-graining applications. *Journal of Chemical Theory and Computation*, *5*(12), 3211–3223.

Russel, D., Lasker, K., Phillips, J., Schneidman-Duhovny, D., Velazquez-Muriel, J. A., & Sali, A. (2009). The structural dynamics of macromolecular processes. *Current Opinion in Cell Biology*, *21*, 97–108.

Russel, D., Lasker, K., Webb, B., Velázquez-Muriel, J., Tjioe, E., Schneidman-Duhovny, D., et al. (2012). Putting the pieces together: Integrative modeling platform software for structure determination of macromolecular assemblies. *PLoS Biology*, *10*(1), e1001244.

Santini, S., Mousseau, N., & Derreumaux, P. (2004). In silico assembly of alzheimer's $a\beta_{6-}^{122}$ peptide into *β*-sheets. *Journal of the American Chemical Society*, *126*(37), 11509–11516.

Schneidman-Duhovny, D., Inbar, Y., Nussinov, R., & Wolfson, H. (2005). Patchdock and symmdock: Servers for rigid and symmetric docking. *Nucleic Acids Research*, *33*(Suppl. 2), W363–W367.

Scott, K. A., Bond, P. J., Ivetac, A., Chetwynd, A. P., Khalid, S., & Sansom, M. S. P. (2008). Coarse-grained md simulations of membrane protein-bilayer self-assembly. *Structure*, *16*, 621–630.

Slovic, A. M., Stayrook, S. E., North, B., & DeGrado, W. F. (2005). X-ray structure of a water soluble analog of the membrane protein phospholamban: Sequence determinants defining the topology of tetrameric and pentameric coiled coils. *Journal of Molecular Biology*, *348*, 777–787. http://dx.doi.org/10.1006/jmbi.2005.02.040.

Slovic, A. M., Summa, C. M., Lear, J. D., & DeGrado, W. F. (2003). Computational design of a water-soluble analog of phospholamban. *Protein Science*, *12*, 337–348. http://dx.doi.org/10.1110/ps.0226603.

Smith, G. R., Sternberg, M. J. E., & Bates, P. A. (2005). The relationship between the flexibility of proteins and their conformational states on forming protein-protein complexes with an application to protein-protein docking. *Journal of Molecular Biology*, *347*, 1077–1101.

Spiga, E., Alemani, D., Degiacomi, M. T., Cascella, M., & Dal Peraro, M. (2013). Electrostaticconsistent coarse-grained potentials for molecular simulations of proteins. *Journal of Chemical Theory and Computation*, *9*(8), 3515–3526.

Stark, A., Andrews, C. T., & Elcock, A. H. (2013). Toward optimized potential functions for protein–protein interactions in aqueous solutions: Osmotic second virial coeffcient calculations using the martini coarse-grained force field. *Journal of Chemical Theory and Computation*, *9*, 4176–4185.

Sugita, Y., & Okamoto, Y. (1999). Replica-exchange molecular dynamics method for protein folding. *Chemical Physics Letters*, *314*(27), 141–151.

Takada, S. (2012). Coarse-grained molecular simulations of large biomolecules. *Current Opinion in Structural Biology*, *22*, 130–137.

Thorpe, I. F., Zhou, J., & Voth, G. A. (2008). Peptide folding using multiscale coarse-grained models. *The Journal of Physical Chemistry B*, *112*(41), 13079–13090.

Tozzini, V. (2010). Minimalist models for proteins: A comparative analysis. *Quarterly Reviews of Biophysics*, *43*(3), 333–371. http://dx.doi.org/10.1017/s0033583510000132.

Tozzini, V., Rocchia, W., & McCammon, J. A. (2006). Mapping all-atom models onto one-bead coarse-grained models: General properties and applications to a minimal polypeptide model. *Journal of Chemical Theory and Computation*, *2*(3), 667–673. http://dx.doi.org/10.1021/ct050294k.

Trabuco, L. G., Schreiner, E., Gumbart, J., Hsin, J., Villa, E., & Schulten, K. (2011). Applications of the molecular dynamics flexible fitting method. *Journal of Structural Biology*, (173), 420–427.

Trabuco, L. G., Villa, E., Mitra, K., Frank, J., & Schulten, K. (2008, May). Flexible fitting of atomic structures into electron microscopy maps using molecular dynamics. *Structure*, *16*(5), 673–683. http://dx.doi.org/10.1016/j.str.2008.03.005. Retrieved from, http://dx.doi.org/10.1016/j.str.2008.03.005.

Tristan Bereau, T., Wang, Z. J., & Deserno, M. (2014). More than the sum of its parts: Coarsegrained peptide-lipid interactions from a simple cross-parametrization. *The Journal of Chemical Physics*, *140*, 115101.

Tschöp, W., Kremer, K., Batoulis, J., Bürger, T., & Hahn, O. (1998). Simulation of polymer melts. 1. coarse-graining procedure for polycarbonates. *Acta Polymerica*, *49*, 61–74.

van Gunsteren, W. F., Bakowies, D., Baron, R., Chandrasekhar, I., Christen, M., Daura, X., et al. (2006). Biomolecular modeling: Goals, problems, perspectives. *Angewandte Chemie International Edition*, *45*, 4064–4092.

Visel, A., Blow, M. J., Li, Z., Zhang, T., Akiyama, J. A., Holt, A., et al. (2009). Chip-seq accurately predicts tissue-specific activity of enhancers. *Nature*, *457*(7231), 854–858.

Voelz, V. A., Bowman, G. R., Beauchamp, K., & Pande, V. S. (2010). Molecular simulation of ab initio protein folding for a millisecond folder ntl9 (1–39). *Journal of the American Chemical Society*, *132*(5), 1526–1528.

Wagner, S., Sorg, I., Degiacomi, M., Journet, L., Peraro, M. D., & Cornelis, G. R. (2009). The helical content of the yscp molecular ruler determines the length of the yersinia injectisome. *Molecular Microbiology*, *71*(3), 692–701.

Wagner, S., Stenta, M., Metzger, L. C., Dal Peraro, M., & Cornelis, G. R. (2010). Length control of the injectisome needle requires only one molecule of yop secretion protein p (yscp). *Proceedings of the National Academy of Sciences*, *107*(31), 13860–13865.

Ward, A. B., Sali, A., & Wilson, I. A. (2013). Integrative structural biology. *Science*, *339*, 443–477.

Watson, J. D., & Crick, F. H. (1953). Molecular structure of nucleic acids. *Nature*, *171*(4356), 737–738.

Zacharias, M. (2003). Protein-protein docking with a reduced protein model accounting for side-chain flexibility. *Protein Science*, *12*, 1271–1282.

Zacharias, M. (2013). Combining coarse-grained nonbonded and atomistic bonded interactions for protein modeling. *Proteins*, *81*(1), 81–92.

Zhao, G., Perilla, J. R., Yufenyuy, E. L., Meng, X., Chen, B., Ning, J., et al. (2013). Mature hiv-1 capsid structure by cryo-electron microscopy and all-atom molecular dynamics. *Nature*, *497*, 643–646.

CHAPTER FOUR

Stability of Amyloid Oligomers

Workalemahu M. Berhanu, Ulrich H.E. Hansmann[1]
Department of Chemistry & Biochemistry, University of Oklahoma, Norman, Oklahoma, USA
[1]Corresponding author: e-mail address: uhansmann@ou.edu

Contents

Abstract

Molecular simulations are now commonly used to complement experimental techniques in investigating amyloids and their role in human diseases. In this chapter, we will summarize techniques and approaches often used in amyloid simulations and will present recent success stories. Our examples will be focused on lessons learned from molecular dynamics simulations in aqueous environments that start from preformed aggregates. These studies explore the limitations that arise from the choice of force field, the role of mutations in the growth of amyloid aggregates, segmental polymorphism, and the importance of cross-seeding. Furthermore, they give evidence for potential toxicity mechanisms. We finally discuss the role of molecular simulations in the search for aggregation inhibitors.

1. INTRODUCTION

Within organisms, many proteins are found not only in their native, functional form determined by their sequence of amino acids but also in a sequence-independent amyloid state, characterized by elongated fibers with spines built out of β-sheets strands (Eisenberg & Jucker, 2012). The amyloid fibrils are held together by a dense hydrogen-bond network and steric zipper-like interactions (see Fig. 1) that result in β-sheets staged along

Advances in Protein Chemistry and Structural Biology, Volume 96
ISSN 1876-1623
http://dx.doi.org/10.1016/bs.apcsb.2014.06.006

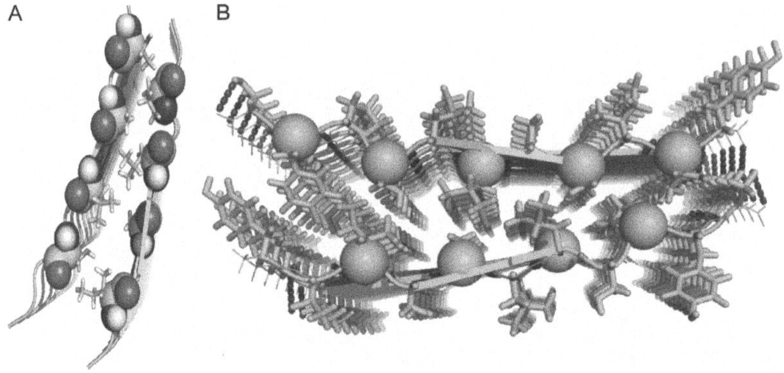

Figure 1 Fibril model of the protein transthyretin, TTR ($_{105}$YTIAALLSPYS $_{115}$) based on a solid-state NMR derived model, illustrating schematically the cross-β-sheets of amyloid fibril: (PDB ID: 2M5N (Fitzpatrick et al., 2013)). (A) A view of the peptide structure down the fiber axis; and (B) viewed from the side, perpendicular to the fiber axis. Hydrogen bonds are shown in red (black in the print version).

the fibril axis (Chiti & Dobson, 2006; Eisenberg & Jucker, 2012; Fitzpatrick et al., 2013; Nelson et al., 2005). The amyloid fibrils are further stabilized by hydrophobic surfaces with good shape complementarity bound by van der Waals and hydrophobic forces (Cheng, Pham, & Nowick, 2013; Sawaya et al., 2007). Deposits of amyloids are associated with a growing number of human illnesses (Chiti & Dobson, 2006; Eisenberg & Jucker, 2012); however, it appears that the most cell-toxic aggregates are not the fibrils themselves but the transient, pre-fibrillar oligomer species (Baglioni et al., 2006; Benilova, Karran, & De Strooper, 2012). Hydrophobic patches of the later can damage cells by disrupting cell membranes or through dislocation of other proteins in the cell (Baglioni et al., 2006; Baral et al., 2014a; Kayed & Lasagna-Reeves, 2013; Olzscha et al., 2011).

Given the common structural and physiochemical characteristics of amyloids, it is important to characterize the mechanisms by which they form and become toxic. This would allow predictions of compounds which can interfere with these processes potentially leading to effective therapeutic agents (Baral et al., 2014a; Bemporad & Chiti, 2012; Merlini, Seldin, & Gertz, 2011). However, despite the highly ordered nature of amyloids, it is still a challenge to resolve their structure with sufficient resolution. Only recently, with advances in experimental techniques such as hydrogen deuterium exchange, X-ray crystallography (on short amyloid segments), solid-state nuclear magnetic resonance (SSNMR), cryo-electron microscopy, and Fourier transform infrared analyses, has it become possible to derive high-resolution models (Comellas & Rienstra, 2013; Middleton et al., 2012; Sawaya et al., 2007; Toyama & Weissman, 2011). Yet, even these experimentally derived amyloid fibril and oligomer models lack the necessary molecular detail (Bemporad & Chiti, 2012) required for the development of drugs that could prevent aggregation or would solubilize aggregates (Jiang et al., 2013). Similarly, multiple experimental methods are available to characterize the protein or amyloid dynamics (Bruggink, Muller, Kuiperij, & Verbeek, 2012), but it is still a challenge to delineate motions of proteins at an atomic scale. For example, experimental techniques do not permit a direct examination of complete, continuous folding pathways at an atomistic level of detail (Klepeis & Shaw, 2014).

One way to extract this missing information is through molecular dynamics (Adcock & McCammon, 2006; Karplus & McCammon, 2002; Shea & Urbanc, 2012) and Monte Carlo (Jonsson, Mohanty, & Irback, 2012; Li, Mohanty, Irback, & Huo, 2008) simulations. While Monte Carlo simulations generate a statistical ensemble of conformations (Rizzuti & Daggett, 2013), molecular dynamics leads to realistic trajectories by integrating Newton's equations of motion for all atoms in a molecule or system of molecules (Dror, Dirks, Grossman, Xu, & Shaw, 2012; Durrant & McCammon, 2011). It follows that molecular dynamics can, in principle, depict the structural features of proteins and relate them to their chemical functions, thereby complementing experiments in probing protein folding and aggregation (Klepeis & Shaw, 2014; Shaw et al., 2010; Straub & Thirumalai, 2011) by making predictions that can guide further experiments (Adcock & McCammon, 2006; Mousseau & Derreumaux, 2005).

The reliability of simulations is based on the accuracy of the underlying empirical energy function or force field (Huang & MacKerell, 2013; Lopes et al., 2013; Vymetal & Vondrasek, 2013), with the danger that predicted mechanisms of protein folding and amyloid formation depend on the selected force field. Hence, proper choice of the force field is critical in all folding and aggregation studies (Freddolino, Park, Roux, & Schulten, 2009; Kia & Darve, 2013; Lindorff-Larsen et al., 2012; Nguyen, Li, & Derreumaux, 2011; Piana, Lindorff-Larsen, & Shaw, 2011; Prigozhin & Gruebele, 2013).

An equally severe problem is the lack of efficiency by molecular dynamics in navigating the rugged hyper surface of protein folding and aggregation. Numerical simulations cover nanoseconds to microseconds, while folding and aggregation happen on a time scale of milliseconds to hours/days. As a consequence, computational results easily depend on the start conformation instead of describing an equilibrium state. The difference in time scales appears because the numerical effort in fixed-temperature molecular dynamics increases exponentially with the size of the system, with even the simplest systems such as a single solvated protein could be made out of several thousand atoms (Piana, Lindorff-Larsen, & Shaw, 2012). This size dependence limits the scope of computer simulations in protein folding and aggregation studies (Kinnear, Jarrold, & Hansmann, 2004; Zimmermann & Hansmann, 2008) and restricts, for instance, the computational design of amyloid inhibitors which requires knowledge of the dynamics and structural details of the toxic oligomers (Caughey & Lansbury, 2003). While a number of authors have reported simulations that start from disordered amyloid peptide building blocks and model the distribution of configurations at early steps of oligomer formation (Anand & Hansmann, 2011; Anand, Nandel, & Hansmann, 2008a, 2008b; Bellesia & Shea, 2009; Coskuner, Wise-Scira, Perry, & Kitahara, 2013; Han & Hansmann, 2011; Kim, Takeda, & Klimov, 2010; Lin & Pande, 2012; Meinke & Hansmann, 2007; Meral & Urbanc, 2013; Reddy, Straubb, & Thirumalai, 2009; Schor, Vreede, & Bolhuis, 2012; Shea & Urbanc, 2012; Viet, Nguyen, Ngo, Li, & Derreumaux, 2013; Wu & Shea, 2013), these studies are computationally very expensive. As a consequence, sampling the entire conformational space is yet not possible for amyloid oligomers (Schor et al., 2012).

One possibility to lessen the computational costs is the use of coarse-grained (Sieradzan, Liwo, & Hansmann, 2012; Wu & Shea, 2011) and discrete molecular dynamics simulations (Urbanc, Betnel, Cruz, Bitan, & Teplow, 2010); however, such models only poorly describe the details of peptide interactions (Saunders & Voth, 2013). Other strategies involve enhanced sampling techniques such as replica exchange (Hansmann, 1997), generalized-ensemble approach (Hansmann, 2002; Hansmann & Okamoto, 1993), meta-dynamics (Laio & Parrinello, 2002), and bias-exchange meta-dynamics (Buchanan et al., 2013; Piana & Laio, 2007). These techniques have been used successfully in simulations of protein folding and amyloid aggregation using that they accelerate sampling of rare events and reduce the size dependence of numerical studies to a power law while keeping the resolution of the biomolecules model (Sinitskiy & Voth, 2013). However, while these techniques have been used in some of the above-cited studies, they all rely on nonphysical dynamics and therefore do not allow a direct evaluation of the kinetics of a system under investigation.

Similar problems appear also in another technique sometimes used in folding studies, the so-called high-temperature unfolding molecular dynamics. Its key assumption is that unfolding is the reverse of the folding process (Daggett & Levitt, 1993) and can be accelerated by raising the temperature (Scalley & Baker, 1997). However, experimental evidence is conflicting on whether the temperature dependence of amyloid fibril nucleation follows Arrhenius law (Arora, Ha, & Park, 2004; Sabate, Villar-Pique, Espargaro, & Ventura, 2012) as assumed in the above approach. In addition, under nonequilibrium conditions, reversibility does not hold. Thermal or chemical denaturants can alter the free-energy landscape and especially the ensemble of transition states, and high-temperature landscapes can differ significantly from physiological energy landscapes (Scheraga, Khalili, & Liwo, 2007). Hence, while the Dagget group has used elevated temperature "unfolding" molecular dynamic simulations that start from crystal or NMR structures (Armen & Daggett, 2005; Armen, DeMarco, Alonso, & Daggett, 2004) to study conformational transition in amyloidosis, it is not clear how reliable these simulations are.

While starting from a formed fibril structure (derived from experimental data), computational stability studies avoid the above problems by not assuming the equivalence of folding and time-reversed unfolding but only aiming to assess their stability. The underlying assumption is that if the model aggregates do not disintegrate during a molecular dynamics run of sufficient length, then the aggregates and proto-fibrils can be considered stable (Berhanu & Hansmann, 2012a). Even though such studies do not involve amyloid assembly, they provide indirect input into the various factors that contribute to or modulate fibril formation. Examples are the nucleation and growth mechanisms of fibril-like oligomers (Horn & Sticht, 2010; Kahler, Sticht, & Horn, 2013), polymorphism (alternative interfaces) (Berhanu & Hansmann, 2012b; Miller, Ma, & Nussinov, 2011a, 2011b; Wu, Bowers, & Shea, 2010), the effects of mutations (Blinov, Dorosh, Wishart, & Kovalenko, 2010; Kassler, Horn, & Sticht, 2010), thermodynamic stability (Blinov et al., 2010), and interactions with membranes (Jang et al., 2013). We will therefore focus in this review on stability studies of preformed amyloid aggregates in explicit water and show that such simulations can not only explain experimental observations but also guide future experiments (Dror et al., 2012; Mousseau & Derreumaux, 2005). A striking example may be recent work by Baral et al. (2014b), who by combining NMR and molecular dynamics simulations of the prion molecule bound to two anti-prion phenothiazines (i.e., promazine and chlorpromazine) explore how these two small molecules stabilize several PrP^C motifs implicated in the transition to the neurotoxic PrP^{sc} state. In this way, they can explain the ability of phenothiazine to inhibit prions and provide a road map for future efforts in computer-aided structure-guided search for anti-prion molecules with better activity and drug properties. In the following, we will describe some of the topics currently under investigation and the conditions required for successful stability investigations.

2. EFFECT OF FORCE FIELD ON THE PREFORMED OLIGOMER

The choice of force field is a challenge when setting up molecular dynamics simulations of amyloid aggregates, as different force fields employ different parameterization strategies and calibration that in turn may influence the outcome of the simulation. The determination of the "best" force field that agrees with experimental structure is therefore critical for the simulation of performed amyloid fibrils (Berhanu & Hansmann, 2012a). By comparing trajectories of simulations of the performed $A\beta_{16-22}$ aggregates that use various common force fields, we found that two recent versions of AMBER force fields have consistently smaller root mean square deviations to the initial configuration than those that rely on one of the other force fields (Berhanu & Hansmann, 2012a). In the simulations that use AMBER force fields, the peptide changes little from the initial conformation exemplifying the stability of the $A\beta_{16-22}$ aggregates in simulations with these two force fields. Simulations that used the GROMOS96-53a6 force field have the largest root mean square deviation, suggesting instability of the aggregates which is probably an artifact of the united-atom approximation in this force field. The force field-induced bias becomes smaller with increasing size of the double sheet aggregates, suggesting that force-field induced differences in stability decrease with size of the system. Other authors have noted also the need for carefully choosing an adequate force field in simulations of amyloids. For instance, Xiao, Zhu, Ji, and Zhang (2013) discuss the impact of an electrostatic polarization term in all-atom force fields on the stability of preformed β-sheet oligomers. Nguyen et al. (2011) observed that the initial steps of oligomer formation and the conformation space accessible to Aβ building blocks may depend on the specific force field.

3. ROLE OF MUTATIONS

As probably the first concrete step in a structure-based drug design of inhibitors to one of the most complex and difficult-to-drug targets in human diseases, Eisenberg and coworkers (Jiang et al., 2013) have designed several new compounds that reduce amyloid toxicity. They have shown that by stabilizing mature fibrils against dissociation small molecules can reduce or nullify their toxicity. These studies rely on a high-resolution structure of an Alzheimer's peptide in an amyloid-like conformation bound to an inhibitor molecule (Landau et al., 2011) and demonstrate the importance of understanding the various conditions that modulate the stability of fibrils and their toxic precursors. These conditions are either internally (variations in sequence) or externally (environment) dependent. The atomic-resolution structures of fibrils formed by short segments of amyloid-related peptides that have been determined by the Eisenberg group (Sawaya et al., 2007) and the various SSNMR amyloid fibril models (Tycko & Wickner, 2013) make it possible to use atomistic molecular dynamics to test how variations in the sequence can change the pathway and the mechanism by that fibrils form. Computational studies by Kassler et al. (2010), starting from preformed Aβ aggregates built from either wild type peptides or one of four pathologic mutants, have demonstrated that the four mutant oligomers are more stable than the oligomers built from the wild type. Their results explain the experimentally observed differences in aggregation propensity between wild type and various pathologic mutants of Aβ. Blinov et al. (2010), using all-atom molecular dynamics on SSNMR models of the $A\beta_{17-42}$ oligomers and pathological mutants, explain the stability and association of aggregates of the wild type and pathological mutants of Aβ from an analysis of their microscopic solvation structure and local physical interactions. Ndlovu, Ashcroft, Radford, and Harris (2012), using steered atomistic molecular dynamics, studied the mechanical stability of the amyloid-forming $hIAPP_{20-29}$ fragment of human amylin for wild type and mutants generating results that could relate it to the degree of structural order in the β-sheet region (which is determined by the amino acids sequence) and the direction of the applied force. In another study, the Buehler group (Paparcone, Pires, & Buehler, 2010), using an all-atom explicit solvent simulation starting from SSNMR Alzheimer's β-amyloid fibrils and considering single point mutants, found that mutations that alter the stability of the D_{23}–K_{28} salt bridge significantly reduce the mechanical strength of the fibrils.

Using all-atom molecular dynamics with an explicit solvent (Berhanu & Hansmann, 2012a), we have investigated in a similar effort how L, I, and A mutations of the aromatic F residues at positions 19 and 20 change the stability of the double-layer hexametric chains of $A\beta_{16-22}$ preformed fibril aggregates. This stability analysis of single and double point mutants located in the hydrophobic core region of Aβ_{16-22} confirms that size and hydrophobicity determine the aggregation process. Our results are in agreement with pervious experiments (Armstrong, Chen, McKoy, & Hecht, 2011; Senguen, Doran, Anderson, & Nilsson, 2011) and indicate that rather than aromatic interactions, the hydrophobicity and β-sheet propensity of residues at positions 19 and 20 are responsible for the aggregation of Aβ. Hence, the substitution of residues in the hydrophobic region Aβ_{16-22} (especially, Phe 19 and 20) with natural and nonnatural amino acids of similar size and hydrophobicity is a promising venue for designing Aβ aggregation inhibitors.

4. STABILITY OF CYLINDRIN β-BARREL AMYLOID OLIGOMER MODELS

Because of the potential health applications, there is considerable interest in the structure of these toxic oligomers, and the process by which they form and/or convert to fibrils. However, these transient structures and processes are difficult to probe in experiments (Bemporad & Chiti, 2012), and the molecular structure of the toxic amyloids and amyloid intermediates is therefore only poorly understood. Known is that they are unstable, rich in β-sheet structure, polymorphic and react with oligomer-recognizing antibodies, i.e., antibodies that bind only to the oligomers but not to fibrils (Benilova et al., 2012; Kayed et al., 2007; Lee, Culyba, Powers, & Kelly, 2011). There seems to be a correlation between toxicity and deviation from the energetically favorable and stable parallel in-register β-sheet amyloid state (Liu et al., 2012; Tycko & Wickner, 2013).

A major breakthrough has been the atomic-resolution structure of "cylindrin," a motif seen for various amyloid-forming proteins and the first reliable example of an oligomer architecture (Apostol, Perry, & Surewicz, 2013; Laganowsky et al., 2012; Liu et al., 2012; Pham, Chim, Goulding, & Nowick, 2013). A similar β-barrel-like structure was also found in computer simulations of amyloid-forming peptide segments from Aβ (Barz, Wales, & Strodel, 2014; De Simone & Derreumaux, 2010; Xie, Luo, & Wei, 2013). The cylindrin oligomers might represent off-pathway particles in fibril formation (see Fig. 2). As the elucidation of pathways for oligomer formation and a structural explanation for their higher toxicity is critical for the identification of therapeutics, the cylindrin β-barrel oligomer conformation (Caughey & Lansbury, 2003) appears to be an excellent target system for computational studies.

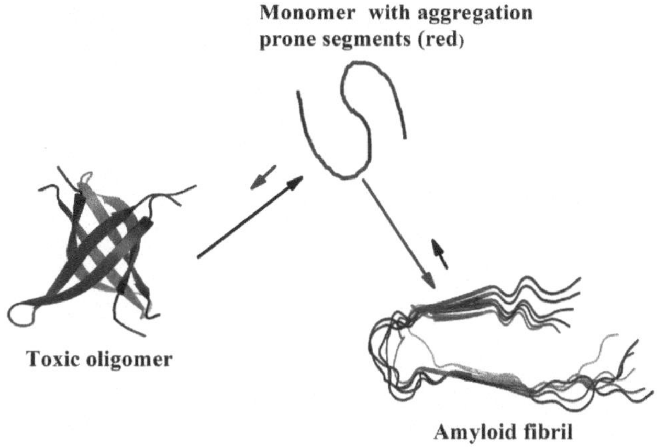

Figure 2 Schematic model of the formation of amyloid fibril and toxic oligomers. Proteins initially populate a native conformation (A) Native protein with 2 amyloid-forming sequences (colored in red; dark gray in the print version) in which amyloid-prone segments (in red; dark gray in the print version) are structured/buried and unable to polymerize. Under certain conditions, native states can convert into aggregation-prone states that may lead to cylindrin-like toxic oligomers. The oligomers could be off-path (B) modeled from the cylindrical barrel (PDB ID: 3SGR) of a segment of the amyloid-forming protein αB crystallin (Laganowsky et al., 2012), or on-path during the fibril formation (C) (PDB ID: 2LMN (Petkova, Yau, & Tycko, 2006)).

We have systematically explored how mutations that replace valine by residues with different hydrophobicity and side chains' size modulate the stability of the cylindrin β-barrel tandem repeat, single-chain cylindrin, and the steric zipper fibril. This has allowed us to identify factors which determine the stability of β-barrel amyloid oligomers (Berhanu & Hansmann, 2013). These multi-copy molecular dynamics simulations demonstrate that substitution of Val-4 and Val-8 by shorter, less hydrophobic Ala residues or flexible Gly residues disrupts the dry interior steric zipper between two neighboring strands. As a result, the Gly and Ala mutants become visibly disordered. In contrast, the single-point mutation of Val2 to Leu, which slightly increases hydrophobicity and size of side chain, leaves the stability of the cylindrin structure unchanged. In order to avoid steric clashes, the pore grows in the double Val to Leu mutants. In turn, this expansion of the pore reduces the number of hydrophobic contacts, hydrogen bonds, and salt bridge, thus lowering the stability of these mutants. Hence, the interactions between hydrophobic residues that fit into the dry inner core are the main factors in stabilizing the cylindrin models. This observation is because the cylindrin geometry leads to a larger number of hydrophobic contacts that, in turn, stabilize the side-chain and main-chain hydrogen bonds as well as the salt bridges on the outer surface along the weak out-off-register interface.

During our simulations, we also observe the presence of water molecules within the β-barrel. This is consistent with a toxicity mechanism where the oligomers act as pores in the cell membrane through which water molecules and ions can penetrate into the cell in a cytotoxic manner.

The differences in the free energy between the fibril models and the corresponding β-barrel cylindrin models indicate that the fibril models are more stable than the β-barrel. The simulation also supports Liu et al. (2012) studies, who found that the β-barrel oligomer model and their related out-of-registered β-sheets are off the pathway to the fibril formation, as the transition from cylindrin to steric zipper involves breaking of hydrogen bonds and re-registration of the strands into an in-register structure.

5. AMYLOID POLYMORPHISM

Amyloid fibrils appear in a large variety of polymorphic structures that result from the same protein precursors and are modulated by the physico-chemical environment (Kodali & Wetzel, 2007; Sarell, Stockley, & Radford, 2013). This polymorphism can take the form of fibril variability through simple side-chain rotations, or that of segmental polymorphism, where different amino acid sequences are incorporated into the β-strand of the core (Wiltzius et al., 2009). It has been proposed (Wiltzius et al., 2009) that structural polymorphism tunes the functional or toxic properties of the amyloid aggregate, either through direct changes in physicochemical properties (e.g., fibril hydrophobicity) or through interactions with other factors (e.g., disaggregating molecular chaperones). For instance, Skeby, Sorensen, and Schiott (2013), using molecular dynamics simulations, have investigated the interactions of certain amyloid imaging agents with an amyloid fibril model. While all these amyloid imaging agents can bind in surface grooves, the particular binding position and affinity of a ligand for an amyloid fibril depend on ligand differences as well as the fibril morphology. Hence, studies of polymorphism are important for understanding the range of potential fibril conformations, for understanding the self-assembly mechanism by which these ordered fibrils assemble and for the design of inhibitor or imaging agents that target these amyloids.

For instance, the Aβ fibrils have complex polymorphism that leads to differences in fibril morphology (Meinhardt, Sachse, Hortschansky, Grigorieff, & Fandrich, 2009) and variations in the degree of toxicity

(Petkova et al., 2005; Seilheimer et al., 1997; Yoshiike, Akagi, & Takashima, 2007). Combining microcrystal structures from various segments of Aβ with previous nuclear magnetic resonance (Petkova et al., 2006) derived structures, Colletier et al. (2011) have proposed several segmental polymorphic fiber models of Aβ that differ in the packing of β-sheets. These differences are caused by differences in length and nature of side chains involved in steric zipper interactions and the hydrophobicity of residues at the β-sheet–β-sheet interface (see Fig. 3). We have investigated the packing and stability of the different models in a series of molecular dynamics simulations (Berhanu & Hansmann, 2012b) and have shown that the stability of these amyloid aggregates depends strongly on the length of the steric zipper, hydrophobicity, and the number of side chains involved at a β-sheet–β-sheet interface. However, all our models retain their original U shape which is consistence with experimental observation of various polymorphic forms *in vitro* (Meinhardt et al., 2009). Our study is another example for the recent molecular dynamic studies on various short segments of pathologic human amyloid polymorphs that have led to fundamental principles for adjusting the characteristics of amyloids as needed for developing molecular therapeutics, biomimetic

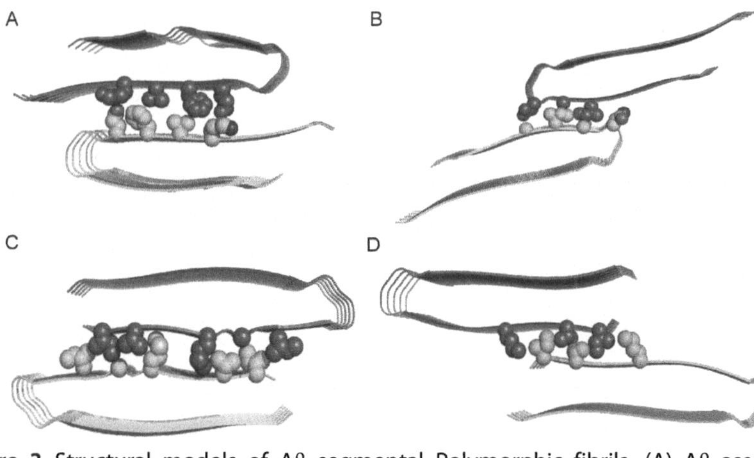

Figure 3 Structural models of Aβ segmental Polymorphic fibrils. (A) Aβ associated through steric zipper residues 16–21 (PDB ID: 3OW9). (B) Aβ associated through steric zipper residues 27–32 (PDB ID: 3Q2X). (C) Aβ associated through steric zipper residues 35–42 (PDB ID: 2Y3K). (D) Aβ associated through long steric zipper of residues 30–35 and 35–42 (PDB ID: 2Y3J and 2Y3K, respectively). *Figure based on amyloid fibril models (http://people.mbi.ucla.edu/sawaya/jmol/fibrilmodels/) from Eisenberg group (Colletier et al., 2011).*

material, and nanotechnology (Berryman, Radford, & Harris, 2011; Yoon, Lee, Kim, Na, & Eom, 2014).

6. AMYLOID AGGREGATION AND CROSS SEEDING

Amyloid fibril formation appears to be a multistep process during which a number of intermediate aggregates are formed. The aggregation starts with the coalescence of peptide monomers to form small oligomers such as dimers, trimmers, and similar small oligomers. These small oligomers then grow further in size and complexity, evolving into proto-fibrils and finally mature amyloid fibrils (Zerovnik et al., 2011). The full mechanism of amyloid aggregation is still unclear, but the availability of atomic-resolution structures allows the modeling of atomistic mechanism of amyloid formation using molecular dynamics methods (Zheng, Jang, Ma, Tsai, & Nussinov, 2007). For instance, using all-atom molecular dynamics simulations, the stability of amyloid oligomers was studied by comparing different candidate β-sheet oligomers, designed to mimic possible nucleus seeds at the very early stage of fibril formation (Horn & Sticht, 2010; Tsai et al., 2005; Zheng et al., 2007). Such computational studies can determine the most stable conformation for a seed nucleus, but do not simulate the actual aggregation scenario of amyloid fibril growth. This aggregation happens on the time scale of minutes to days, far beyond the time scale of microseconds achievable in conventional or even accelerated MD simulations.

Kahler et al. (2013) have proposed that the growth mechanism is initiated by elongation of short proto-filaments; followed by breakage of large proto-filaments that leads to formation of short proto-filament pairs; and as the final step the growth of these proto-filament pairs. Starting from published microcrystal structures (Ivanova, Sievers, Sawaya, Wall, & Eisenberg, 2009; Nelson et al., 2005; Sawaya et al., 2007; Wiltzius et al., 2009, 2008), we have performed molecular dynamics simulations of short, self-assembling segments of pathological amyloids (Sup35, insulin, Aβ, tau, and amylin) in order to gain an insight into their stability and aggregation propensity. Oligomers of $n=1$–5 strands and $m=1$–2 β-sheets (SHm–STn) have been investigated to predict the structural stability of nucleation intermediates and the free energy of aggregation for the oligomers (Berhanu & Masunov, 2011a, 2011b, 2012). We find that the structural stability of the oligomers depends strongly on the number of strands. The aggregate stability depends mainly on the van der Waals and hydrophobic forces. A free energy decomposition shows that the residues situated at the interface contribute favorably to the peptide–peptide association (Berhanu & Masunov, 2011a, 2011b, 2012). A negative total binding free energy indicates that complex formation is favorable for both wild type and mutant. However, the mutants complex is less favorable than the wild type's (Berhanu & Masunov, 2011a, 2011b, 2012). We propose that, depending on the nature of steric zipper interfaces polar/aromatic or small hydrophobic residues, the size of the possible nucleus seed for aggregates may be as small as SH2–ST3 for the polar and as large as SH2–ST6 for the nonpolar peptides (Berhanu & Masunov, 2011a, 2011b, 2012). The stable conformations of the trimer, in contrast to the monomer and the dimer, imply that the trimer may be effective seed for nucleation as has been observed in Aβ oligomer aggregation (Ono, Condron, & Teplow, 2009). Horn et al. (Horn & Sticht, 2010) have also investigated the stability of Aβ(9–42) oligomers (monomer through pentamer) starting from the fibril conformation. They found that the trimmers to pentamer preserved the initial conformation, while the dimer undergoes larger conformational changes. The observed degree of structural similarity with the fibril conformation explains the experimental finding that trimers and tetramers are more potent seeds in fibril formation than dimers. Several similar studies staring with microcrystal structures have provided additional information on the stability of particular fibril structures (Chang et al., 2009; De Simone, Pedone, & Vitagliano, 2008; Lin et al., 2010; Vitagliano, Stanzione, De Simone, & Esposito, 2009).

The formation of amyloids involves at least three steps: a lag phase, a growth (or elongation) phase, and an equilibration phase (Bhak, Choe, & Paik, 2009). The early stage of the amyloid fibril formation, prior to any detection of fibrils, is called the lag phase. Later, the initially soluble fibrils precipitate as they grow to reach their solubility limits. This is called the growth phase. Finally, the system approaches a steady state in what is called the equilibrium phase. Although the first step is under kinetic control, fibril growth is controlled thermodynamically and can be evaluated quantitatively in terms of equilibrium properties such as association constants (Harper, Wong, Lieber, & Lansbury, 1997). As with other nucleated processes, the lag phase in amyloid formation can be accelerated through seeding. Seeding is the addition of preformed fibrils to a monomer solution, thus raising the rate of conversion to amyloid fibrils (Lasagna-Reeves, Castillo-Carranza, Guerrero-Munoz, Jackson, & Kayed, 2010). The seeds can be either homogenous (same peptide) or heterogeneous (related or unrelated peptide), but have to provide a growth face that acts as a template for the polymerization of the complement monomers (see Fig. 4).

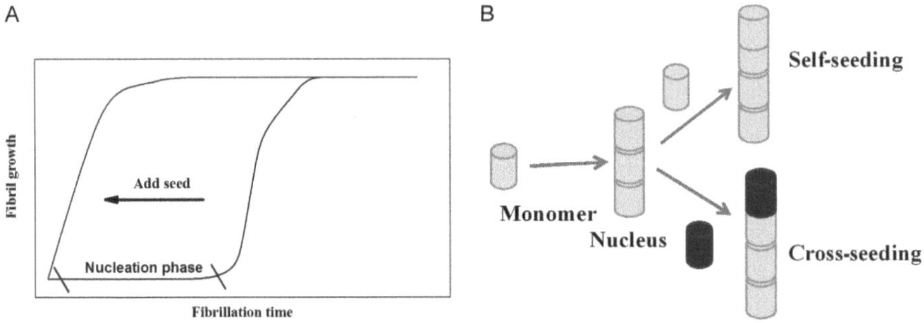

Figure 4 Amyloid fibril formation processes: self-nucleation and the cross-seeding of protein aggregations. (A) The three phases of amyloid aggregation: an initial lag phase followed by an elongation step and a final stationary phase. (B) Preformed amyloid fibrils can act as "seeds" that accelerate self-assembly by reducing the lag time. These seeds can be made of the same protein, or in the case of cross-seeding, made from a different protein.

Seeds composed of one kind of proteins can also enhance aggregation of other proteins. This process is called cross-seeding (Soto, Estrada, & Castilla, 2006). Recent studies of amyloid growth indicate that, in addition to pathogenic self-association, cross seeding plays a critical role in amyloid diseases. Examples are the Aβ–tau, Aβ–amylin, the Aβ–α-synuclein, the Aβ–transthyretin, and the amylin–insulin interactions (Andreetto et al., 2010; Buxbaum et al., 2008; Giasson et al., 2003; Guo, Arai, Miklossy, & McGeer, 2006; Nicolls, 2004).

The possible molecular cross talk between different misfolded protein aggregates may explain the diseases coexistence that has been described in patients afflicted by more than one protein misfolding disorder. This observed clustering of amyloid diseases has been suggested to be a result of direct interaction between misfolded proteins that are similar in their structural features (Morales, Moreno-Gonzalez, & Soto, 2013). Apostol et al. recently have determined the X-ray atomic structures of three, six-residue segments of human, mouse, and hamster prion proteins (Apostol, Wiltzius, Sawaya, Cascio, & Eisenberg, 2011). The sequences encode disparate steric zippers, suggesting that the mismatch between these zipper structures causes interspecies transmission barriers in prion diseases. The functional amyloid Prion-forming domain (PFD) from fungus, on the other hand, is found to be efficient in cross-seeding between different species despite their larger sequence differences (70%): HET-s (218–289) versus FgHET-s (218–289). The model of the FgHET-s (218–289) fibrils based on SSNMR and hydrogen exchange experiments (Wasmer et al., 2010) has a β-solenoid conformation that is similar to HET-s (218–289) structure which may explain the efficient cross-seeding between HET-s (218–289) and FgHET-s (218–289). In a similar way, Miller et al. (2011a, 2011b) using molecular dynamics simulations have investigated the interaction between the Aβ and tau complex. The results from their simulation indicate that the interaction between Aβ oligomers and tau fragments is stronger for the $tau_{275-305}$ segment than for $tau_{306-336}$ and $tau_{337-368}$.

The 40- and 42-residue Aβ-amyloid peptides (Aβ40 and Aβ42) and the 37-residue islet amylin are key amyloid polypeptides in Alzheimer's disease (AD) and in type-2 diabetes (T2D), respectively (Westermark, 2005). Even though human amylin differs significantly in sequence from Aβ peptides, the two peptides are able to cross seed efficiently (Andreetto et al., 2010). Amylin and Aβ amyloid have a similar secondary structure in the fibril state, and amylin has about 65% sequence similarity with Aβ peptide when overlapped in the central core region (residues 15–37) of Aβ peptide. Increasing amounts of evidence suggest that AD and tT2D are linked to each other. A possible explanation is an interaction between Aβ and amylin. Because the resulting amyloid oligomers are difficult to characterize in experiments, we have investigated numerically the stability and conformational changes of mixed Aβ-amylin oligomers (Berhanu, Yasar, & Hansmann, 2013). Our results demonstrate that Aβ is a good template for the growth of amylin and vice versa. During our simulations, water molecules permeate the β-strand–turn–β-strand motif pore of the oligomers, which suggests water leakage as mechanism for toxicity of β-rich amyloid oligomers. Our result on the stability of the heterofibrils between Aβ and amylin and the presence of water channel in the heteroassociation suggests a potential cytotoxic mechanism which connects the pathogenesis of AD and T2D and is in agreement with experimental data (Gal et al., 2013; Jackson et al., 2013).

We have also performed a molecular dynamic study to probe how mutations alter the seeding of amylin fibrils and what is the dependence of seeding on structure similarity between seed and seeded proteins. Investigating the structural stability of decamers of wild-type amylin peptides and their Y37L mutants, and of mixed assemblies of wild-type and mutant amylin molecules, we find that the experimentally observed cross seeding can be explained with the similarity in fibril structure. Amyloids with similar side chains packing at the β-sheet interface are structurally compatible and therefore good templates for homologues peptides. In the Y37L mutants, the lack of tyrosine-specific interactions causes significant larger flexibility of the C terminal region than in the wild-type fibril. This hinders elongation of the Y37L mutant fibril leading to the longer lag times during aggregation that are observed in experiments for this mutant (Bernhardt, Berhanu, & Hansmann, 2013).

In another project, we have used molecular dynamics simulations (Berhanu & Hansmann, 2014) to explain the weak inhibition of human amylin aggregation by rat amylin that was observed in experiments. For this purpose, we have compared the stability of fibril aggregates built from human amylin, rat amylin, and mixtures of both. The atomic models of the human fibril by Wiltzius et al. (2008), a U-shaped model with beta-strands at the N- and C- terminals, are used as a starting point to build rat-amylin and various human/rat hybrids. The structural instability of rat amylin fibrils (due to proline residues in the C terminus) makes N-terminal head-to-head interface between the β-sheets of rat/human amylin mixed fibril more stable than fibrils that are organized through CC terminal interfaces. The simulation points to a possible mechanism for the weak inhibitor activity of rat amylin and for the formation of rat amylin amyloids when provided with human amylin fibrils as template (Berhanu & Hansmann, 2014) that is in agreement with experimental data (Middleton et al., 2012; Young, Cao, Raleigh, Ashcroft, & Radford, 2014).

7. TOXICITY MECHANISM OF AMYLOID FROM MOLECULAR DYNAMIC SIMULATIONS

Interestingly, we also find in the above simulations water in the interior of the fibril structures, which again points to water leakage as the mechanism of cytotoxicity (see Fig. 5). The identification of toxicity mechanism is one of the examples where the increasing importance of molecular dynamics simulations as a tool that complements experiments can be seen best. Understanding the interaction of a protein with cell membranes requires characterization of the position, orientation, structure, and dynamics of a peptide in the lipid bilayer and its effects on surrounding lipids. Recent growth in computational power has now made possible atomistic simulations of lipid or mixed lipid systems with long time and length scales. This has not only fasten efforts to improve the existing force fields for both lipids and proteins (Brooks et al., 2009; Klauda et al., 2010), but also led to computational studies of processes such as membrane assembly, fusion, domain formation, and protein/peptide/DNA (Siu, Vacha, Jungwirth, & Bockmann, 2008). Amyloid peptides have been found to induce ionic conductance in both artificial membranes and native cell plasma membrane (Arispe, Pollard, & Rojas, 1993; Etcheberrigaray, Ito, Kim, & Alkon, 1994; Kawahara, Kuroda, Arispe, & Rojas, 2000; Lin, Bhatia, & Lal,

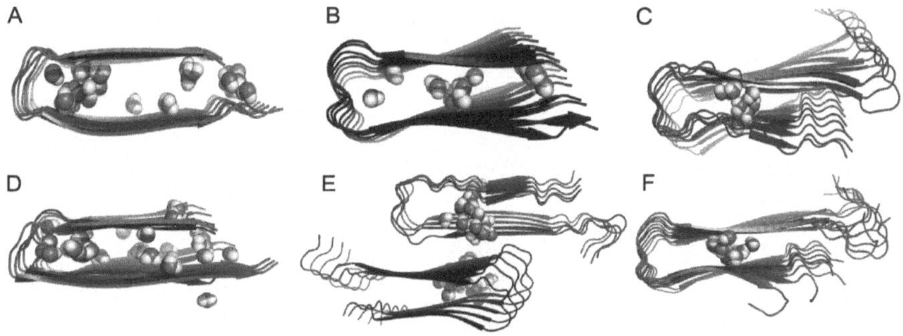

Figure 5 The amyloid aggregates of Aβ, amylin, and their complexes as found in our molecular dynamics simulations can accommodate water molecules in an ambiphilic pore. However, none of the experimental fibril structures of Aβ (PDB ID: 2M4J and 2LMN) and human amylin (Wiltzius et al., 2008) contain water. Water residing in the hydrophilic cavities of homo-oligomers and hetero-oligomers (Berhanu et al., 2013; Berhanu & Hansmann, 2014). Snapshots for (A) wild-type Aβ = brown (dark gray in the print version); (B) wild-type human amylin = blue (black in the print version); (C) wild-type rat amylin = magenta (dark gray in the print version); (D) wild-type Aβ–amylin aggregates complex; Aβ = brown (dark gray in the print version) and amylin = blue (gray in the print version); (E) wild-type human amylin in complex with rat amylin aggregates; human amylin = blue (black in the print version) and rat amylin = magenta (dark gray in the print version); (F) wild-type human amylin in complex with Y37L mutant; human amylin = blue (light gray in the print version) and Y37L mutant = brown (dark gray in the print version). The red (dark gray in the print version) and white (light gray in the print version) spheres represent oxygen and hydrogen atoms in water molecules, respectively.

2001; Lin, Zhu, & Lal, 1999; Quist et al., 2005; Rhee, Quist, & Lal, 1998). However, very little is known about the structures of these aggregates in the membrane. Lashuel, Hartley, Petre, Walz, and Lansbury (2002) have shown a "pore-like" annular structure for amyloidogenic proto-fibrils. Quist et al. (2005), using atomic force microscopy, circular dichroism, gel electrophoresis, and electrophysiological recordings, observed supramolecular conformational changes in amyloid-β(1–40), amylin, and other amyloidic peptides. They form ion-channel-like structures with observable single ion-channel currents in reconstituted membranes. It has been proposed that in amyloid diseases, these ion channels destabilize cellular ionic homeostasis leading ultimately to cell death. In agreement with this idea, molecular dynamics simulations of Aβ oligomers consisting of an SSNMR-based β-arch motif allow the observation of channels with inner diameters in the 1.5–2.6 nm range that are consistent with shape and morphology from electron microscopy/atomic force microscopy (Jang et al., 2009; Jang, Zheng, & Nussinov,

2007; Shafrir, Durell, Anishkin, & Guy, 2010). The results from these studies suggest that other amyloid oligomers, such as amylin and α-synuclein, that form pores in cell membranes (Quist et al., 2005), may have channel structures similar to Aβ oligomers. Sepulveda, Parodi, Peoples, Opazo, and Aguayo (2010) have shown that Aβ perforates neuronal membranes increasing membrane conductance, intracellular calcium, and ethidium bromide influx. Jang et al. (2010) using a mix of computational and experimental techniques showed that non-amyloid-forming Aβ (9–42) and Aβ(17–42) peptides assemble into ion channels that evoke single-channel conductance.

8. CONCLUSIONS AND OUTLOOK

We have shown in this review how molecular dynamic simulations that start from preformed aggregates can complement experiments in investigations of amyloids and their role in human diseases. By restricting themselves on testing the stability of aggregates, these simulations circumvent the sampling problem that otherwise limits the use of computer simulations as virtual microscopes. Our examples demonstrate that while the underlying protein models and energy functions need to be selected carefully, the present generation of force fields is already accurate enough to provide us with meaningful and quantitative predictions that can guide experiments. With the exponentially growing availability of computational resources, we can expect that the role of such computational aggregation studies will even more increase in the near future.

We expect this to be especially true in the search for aggregation inhibitors. For instance, while the early solvable and most toxic amyloid oligomers have not yet been fully uncovered, it is generally presumed that these amyloid fibril precursors are characterized by a β-barrel structure. Cohen et al. (2013) have shown that fibrils promote the aggregation of such toxic oligomers already at low concentration. Lu et al. (2013) in recent study on the structure of β-amyloid fibrils from AD patients did find that not only oligomers but also the fibrils themselves are a likely cause or a contributing agent in AD. These results suggest that future work on the development of inhibitors should focus not only on the toxic oligomers but also on preventing fibril formation, as fibrils can catalyze the aggregation of the toxic oligomers. Developing such inhibitors that both suppress oligomer toxicity and prevent fibril growth is likely necessary for developing effective therapies against amyloid diseases (Dessalew &

Mikre, 2008). A promising start point is cyclic peptides. Designed as sequence and conformational mimics, they have been shown to be effective inhibitors against amyloid formation (Cheng, Liu, Zhao, Eisenberg, & Nowick, 2012; Liu et al., 2011; Richman et al., 2013; Zheng, Baghkhanian, & Nowick, 2013).

Future studies that try to aid the development of inhibitors will also have to account for the inherent polymorphism in amyloids. Lu and colleagues have recently reported the structures of Aβ amyloid fibrils seeded from the brain extracts of two AD patients (Lu et al., 2013). The structure of these amyloid-β fibrils differs from patients with divergent clinical histories (Lu et al., 2013) and is significantly different from previous models of *in vitro* fibrils. While the *in vitro* fibrils show polymorphism, the fibril from patients is not polymorphic (Lu et al., 2013). These novel structural data will open new avenues for a structure-based design of diagnostic imaging agents and aggregation inhibiting drugs (Aguzzi & Gitler, 2013; Lu et al., 2013) and will certainly initiate future computer-aided design of drugs by combining techniques such as QSAR, docking with molecular dynamic simulations (Alonso, Bliznyuk, & Gready, 2006; Berhanu, Pillai, Oliferenko, & Katritzky, 2012). Such combinations have already been proved to reduce the costs and to increase the speed of drug discovery, and their importance will only grow further in future (Sliwoski, Kothiwale, Meiler, & Lowe, 2014).

Finally, we expect that another focal point of activity in computational aggregation studies will be the investigation and design of functional amyloids. Many organisms, ranging from bacteria to human, utilize the unique properties of amyloids (Fowler, Koulov, Balch, & Kelly, 2007). Structural studies of functional and pathological amyloids have revealed cross-β-sheet as a common structural features (Shewmaker, McGlinchey, & Wickner, 2011; Tycko & Wickner, 2013; Van Melckebeke et al., 2010), and while amyloids are usually connected to toxicity and diseases, their high structural stability, insolubility, and stiffness make amyloids fibers especially useful in designing artificial biomaterials for applications in medicine, nanotechnology, and engineering (Gazit, 2007; Knowles & Buehler, 2011; Volpatti & Knowles, 2014). For instance, David Eisenberg's group recently could show that a powder form of amyloid fibers is able to capture selectively carbon dioxide in an efficient and environmentally friendly process (Li et al., 2014). We expect that such application of functional amyloids will increase in importance and will further raise the interest in amyloid fibrils (Adamcik & Mezzenga, 2012).

ACKNOWLEDGMENTS

The work by our group that is described in this review is supported by the National Institutes of Health under Grant No. GM62838 and the National Science Foundation under Grant CHE-1266256, and used resources of the National Energy Research Scientific Computing Center, which is supported by the Office of Science of the U.S. Department of Energy under contract no. DE-AC02-05CH1123. Other parts of the simulations were done on the BOOMER cluster of the University of Oklahoma. Any opinions, findings and conclusions or recommendations expressed in this review are those of the authors and do not necessarily reflect the views of the National Institutes of Health, the National Science Foundation, the Department of Energy, or the University of Oklahoma.

REFERENCES

Adamcik, J., & Mezzenga, R. (2012). Proteins fibrils from a polymer physics perspective. *Macromolecules*, *45*(3), 1137–1150.

Adcock, S. A., & McCammon, J. A. (2006). Molecular dynamics: Survey of methods for simulating the activity of proteins. *Chemical Reviews*, *106*(5), 1589–1615.

Aguzzi, A., & Gitler, A. D. (2013). A template for new drugs against Alzheimer's disease. *Cell*, *154*(6), 1182–1184.

Alonso, H., Bliznyuk, A. A., & Gready, J. E. (2006). Combining docking and molecular dynamic simulations in drug design. *Medicinal Research Reviews*, *26*(5), 531–568.

Anand, P., & Hansmann, U. H. E. (2011). Internal and environmental effects on folding and dimerisation of Alzheimer's beta-amyloid peptide. *Molecular Simulation*, *37*(6), 440–448.

Anand, P., Nandel, F. S., & Hansmann, U. H. E. (2008a). The Alzheimer's beta amyloid (A beta(1-39)) monomer in an implicit solvent. *Journal of Chemical Physics*, *128*(16), 165102.

Anand, P., Nandel, F. S., & Hansmann, U. H. E. (2008b). The Alzheimer beta-amyloid (A beta(1-39)) dimer in an implicit solvent. *Journal of Chemical Physics*, *129*(19), 195102.

Andreetto, E., Yan, L. M., Tatarek-Nossol, M., Velkova, A., Frank, R., & Kapurniotu, A. (2010). Identification of hot regions of the A beta-IAPP interaction interface as high-affinity binding sites in both cross- and self-association. *Angewandte Chemie International Edition in English*, *49*(17), 3081–3085.

Apostol, M. I., Perry, K., & Surewicz, W. K. (2013). Crystal structure of a human prion protein fragment reveals a motif for oligomer formation. *Journal of the American Chemical Society*, *135*(28), 10202–10205.

Apostol, M. I., Wiltzius, J. J. W., Sawaya, M. R., Cascio, D., & Eisenberg, D. (2011). Atomic structures suggest determinants of transmission barriers in mammalian prion disease. *Biochemistry*, *50*(13), 2456–2463.

Arispe, N., Pollard, H. B., & Rojas, E. (1993). Giant multilevel cation channels formed by Alzheimer-disease amyloid beta-protein [a-beta-p-(1-40)] in bilayer-membranes. *Proceedings of the National Academy of Sciences of the United States of America*, *90*(22), 10573–10577.

Armen, R. S., & Daggett, V. (2005). Characterization of two distinct beta(2)-microglobulin unfolding intermediates that may lead to amyloid fibrils of different morphology. *Biochemistry*, *44*(49), 16098–16107.

Armen, R. S., DeMarco, M. L., Alonso, D. O. V., & Daggett, V. (2004). Pauling and Corey's alpha-pleated sheet structure may define the prefibrillar amyloidogenic intermediate in amyloid disease. *Proceedings of the National Academy of Sciences of the United States of America*, *101*(32), 11622–11627.

Armstrong, A. H., Chen, J., McKoy, A. F., & Hecht, M. H. (2011). Mutations that replace aromatic side chains promote aggregation of the Alzheimer's A beta peptide. *Biochemistry*, *50*(19), 4058–4067.

Arora, A., Ha, C., & Park, C. B. (2004). Insulin amyloid fibrillation at above 100 degrees C: New insights into protein folding under extreme temperatures. *Protein Science*, *13*(9), 2429–2436.

Baglioni, S., Casamenti, F., Bucciantini, M., Luheshi, L. M., Taddei, N., Chiti, F., et al. (2006). Prefibrillar amyloid aggregates could be generic toxins in higher organisms. *Journal of Neuroscience*, *26*(31), 8160–8167.

Baral, P. K., Swayampakula, M., Rout, M. K., Kav, N. N., Spyracopoulos, L., Aguzzi, A., et al. (2014a). Structural basis of prion inhibition by phenothiazine compounds. *Structure*, *22*, 1–13.

Baral, P. K., Swayampakula, Mridula, Rout, Manoj Kumar, Kav, N. N., Spyracopoulos, Leo, Aguzzi, Adriano, et al. (2014b). Structural basis of prion inhibition by phenothiazine compounds. *Structure*, *22*, 291–303.

Barz, Bogdan, Wales, David J., & Strodel, Birgit. (2014). A kinetic approach to the sequence-aggregation relationship in disease-related protein assembly. *The Journal of Physical Chemistry. B*, *118*, 1003–1011.

Bellesia, G., & Shea, J. E. (2009). Diversity of kinetic pathways in amyloid fibril formation. *Journal of Chemical Physics*, *131*(11).

Bemporad, F., & Chiti, F. (2012). Protein misfolded oligomers: Experimental approaches, mechanism of formation, and structure-toxicity relationships. *Chemistry & Biology*, *19*(3), 315–327.

Benilova, I., Karran, E., & De Strooper, B. (2012). The toxic A beta oligomer and Alzheimer's disease: An emperor in need of clothes. *Nature Neuroscience*, *15*(3), 349–357.

Berhanu, W. M., & Hansmann, U. H. E. (2012a). Side-chain hydrophobicity and the stability of A beta(16-22) aggregates. *Protein Science*, *21*(12), 1837–1848.

Berhanu, W. M., & Hansmann, U. H. E. (2012b). Structure and dynamics of amyloid-beta segmental polymorphisms. *PLoS One*, 7(7), e41479.

Berhanu, W. M., & Hansmann, U. H. E. (2013). The stability of cylindrin-barrel amyloid oligomer models—A molecular dynamics study. *Proteins*, *81*(9), 1542–1555.

Berhanu, W. M., & Hansmann, U. H. E. (2014). Inter-species cross-seeding: stability and assembly of rat - human amylin aggregates. *PLoS One*, *9*, e97051.

Berhanu, W. M., & Masunov, A. E. (2011a). Can molecular dynamics simulations assist in design of specific inhibitors and imaging agents of amyloid aggregation? Structure, stability and free energy predictions for amyloid oligomers of VQIVYK, MVGGVV and LYQLEN. *Journal of Molecular Modeling*, *17*(10), 2423–2442.

Berhanu, W. M., & Masunov, A. E. (2011b). Molecular dynamic simulation of wild type and mutants of the polymorphic amyloid NNQNTF segments of elk prion: Structural stability and thermodynamic of association. *Biopolymers*, *95*(9), 573–590.

Berhanu, W. M., & Masunov, A. E. (2012). Unique example of amyloid aggregates stabilized by main chain H-bond instead of the steric zipper: Molecular dynamics study of the amyloidogenic segment of amylin wild-type and mutants. *Journal of Molecular Modeling*, *18*(3), 891–903.

Berhanu, W. M., Pillai, G. G., Oliferenko, A. A., & Katritzky, A. R. (2012). Quantitative structure-activity/property relationships: The ubiquitous links between cause and effect. *ChemPlusChem*, 77(7), 507–517.

Berhanu, W. M., Yasar, F., & Hansmann, U. H. E. (2013). In silico cross seeding of A beta and amylin fibril-like oligomers. *ACS Chemical Neuroscience*, *4*(11), 1488–1500.

Bernhardt, N. A., Berhanu, W. M., & Hansmann, U. H. E. (2013). Mutations and seeding of amylin fibril-like oligomers. *Journal of Physical Chemistry B*, *117*(50), 16076–16085.

Berryman, J. T., Radford, S. E., & Harris, S. A. (2011). Systematic examination of polymorphism in amyloid fibrils by molecular-dynamics simulation. *Biophysical Journal*, *100*(9), 2234–2242.

Bhak, G., Choe, Y. J., & Paik, S. R. (2009). Mechanism of amyloidogenesis: Nucleation-dependent fibrillation versus double-concerted fibrillation. *BMB Reports*, *42*(9), 541–551.

Blinov, N., Dorosh, L., Wishart, D., & Kovalenko, A. (2010). Association thermodynamics and conformational stability of beta-sheet amyloid beta(17-42) oligomers: Effects of E22Q (Dutch) mutation and charge neutralization. *Biophysical Journal*, *98*(2), 282–296.

Brooks, B. R., Brooks, C. L., Mackerell, A. D., Nilsson, L., Petrella, R. J., Roux, B., et al. (2009). CHARMM: The biomolecular simulation program. *Journal of Computational Chemistry*, *30*(10), 1545–1614.

Bruggink, K. A., Muller, M., Kuiperij, H. B., & Verbeek, M. M. (2012). Methods for analysis of amyloid-beta aggregates. *Journal of Alzheimer's Disease*, *28*(4), 735–758.

Buchanan, L. E., Dunkelberger, E. B., Tran, H. Q., Cheng, P. N., Chiu, C. C., Cao, P., et al. (2013). Mechanism of IAPP amyloid fibril formation involves an intermediate with a transient beta-sheet. *Proceedings of the National Academy of Sciences of the United States of America*, *110*(48), 19285–19290.

Buxbaum, J. N., Ye, Z., Reixach, N., Friske, L., Levy, C., Das, P., et al. (2008). Transthyretin protects Alzheimer's mice from the behavioral and biochemical effects of A beta toxicity. *Proceedings of the National Academy of Sciences of the United States of America*, *105*(7), 2681–2686.

Caughey, B., & Lansbury, P. T. (2003). Protofibrils, pores, fibrils, and neurodegeneration: Separating the responsible protein aggregates from the innocent bystanders. *Annual Review of Neuroscience*, *26*, 267–298.

Chang, L. K., Zhao, J. H., Liu, H. L., Liu, K. T., Chen, J. T., Tsai, W. B., et al. (2009). Molecular dynamics simulations to investigate the structural stability and aggregation behavior of the GGVVIA oligomers derived from amyloid beta peptide. *Journal of Biomolecular Structure & Dynamics*, *26*(6), 731–740.

Cheng, P. N., Liu, C., Zhao, M. L., Eisenberg, D., & Nowick, J. S. (2012). Amyloid beta-sheet mimics that antagonize protein aggregation and reduce amyloid toxicity. *Nature Chemistry*, *4*(11), 927–933.

Cheng, P. N., Pham, J. D., & Nowick, J. S. (2013). The supramolecular chemistry of beta-sheets. *Journal of the American Chemical Society*, *135*(15), 5477–5492.

Chiti, F., & Dobson, C. M. (2006). Protein misfolding, functional amyloid, and human disease. *Annual Review of Biochemistry*, *75*, 333–366.

Cohen, S. I. A., Linse, S., Luheshi, L. M., Hellstrand, E., White, D. A., Rajah, L., et al. (2013). Proliferation of amyloid-beta 42 aggregates occurs through a secondary nucleation mechanism. *Proceedings of the National Academy of Sciences of the United States of America*, *110*(24), 9758–9763.

Colletier, J. P., Laganowsky, A., Landau, M., Zhao, M. L., Soriaga, A. B., Goldschmidt, L., et al. (2011). Molecular basis for amyloid-beta polymorphism. *Proceedings of the National Academy of Sciences of the United States of America*, *108*(41), 16938–16943.

Comellas, G., & Rienstra, C. M. (2013). Protein structure determination by magic-angle spinning solid-state NMR, and insights into the formation, structure, and stability of amyloid fibrils. *Annual Review of Biophysics*, *42*, 515–536.

Coskuner, O., Wise-Scira, O., Perry, G., & Kitahara, T. (2013). The structures of the E22 delta mutant-type amyloid-beta alloforms and the impact of E22 delta mutation on the structures of the wild-type amyloid-beta alloforms. *ACS Chemical Neuroscience*, *4*(2), 310–320.

Daggett, V., & Levitt, M. (1993). Protein unfolding pathways explored through molecular-dynamics simulations. *Journal of Molecular Biology*, *232*(2), 600–619.

De Simone, A., & Derreumaux, P. (2010). Low molecular weight oligomers of amyloid peptides display beta-barrel conformations: A replica exchange molecular dynamics study in explicit solvent. *Journal of Chemical Physics*, *132*(16).

De Simone, A., Pedone, C., & Vitagliano, L. (2008). Structure, dynamics, and stability of assemblies of the human prion fragment SNQNNF. *Biochemical and Biophysical Research Communications*, *366*(3), 800–806.

Dessalew, N., & Mikre, W. (2008). On the paradigm shift towards multitarget selective drug design. *Current Computer-Aided Drug Design, 4*(2), 76–90.
Dror, R. O., Dirks, R. M., Grossman, J. P., Xu, H. F., & Shaw, D. E. (2012). Biomolecular simulation: A computational microscope for molecular biology. In D. C. Rees (Ed.), *Annual Review of Biophysics: 41.* (pp. 429–452).
Durrant, J. D., & McCammon, J. A. (2011). Molecular dynamics simulations and drug discovery. *BMC Biology, 9,* 71.
Eisenberg, D., & Jucker, M. (2012). The amyloid state of proteins in human diseases. *Cell, 148*(6), 1188–1203.
Etcheberrigaray, R., Ito, E., Kim, C. S., & Alkon, D. L. (1994). Soluble beta-amyloid induction of Alzheimers phenotype for human fibroblast k+ channels. *Science, 264*(5156), 276–279.
Fitzpatrick, A. W. P., Debelouchina, G. T., Bayro, M. J., Clare, D. K., Caporini, M. A., Bajaj, V. S., et al. (2013). Atomic structure and hierarchical assembly of a cross-beta amyloid fibril. *Proceedings of the National Academy of Sciences of the United States of America, 110*(14), 5468–5473.
Fowler, D. M., Koulov, A. V., Balch, W. E., & Kelly, J. W. (2007). Functional amyloid—From bacteria to humans. *Trends in Biochemical Sciences, 32*(5), 217–224.
Freddolino, P. L., Park, S., Roux, B., & Schulten, K. (2009). Force field bias in protein folding simulations. *Biophysical Journal, 96*(9), 3772–3780.
Gal, N., Morag, A., Kolusheva, S., Winter, R., Landau, M., & Jelinek, R. (2013). Lipid bilayers significantly modulate cross-fibrillation of two distinct amyloidogenic peptides. *Journal of the American Chemical Society, 135*(36), 13582–13589.
Gazit, E. (2007). Self-assembled peptide nanostructures: The design of molecular building blocks and their technological utilization [Review]. *Chemical Society Reviews, 36*(8), 1263–1269. http://dx.doi.org/10.1039/b605536m.
Giasson, B. I., Forman, M. S., Higuchi, M., Golbe, L. I., Graves, C. L., Kotzbauer, P. T., et al. (2003). Initiation and synergistic fibrillization of tau and alpha-synuclein. *Science, 300*(5619), 636–640.
Guo, J. P., Arai, T., Miklossy, J., & McGeer, P. L. (2006). A beta and tau form soluble complexes that may promote self aggregation of both into the insoluble forms observed in Alzheimer's disease. *Proceedings of the National Academy of Sciences of the United States of America, 103*(6), 1953–1958.
Han, M., & Hansmann, U. H. E. (2011). Replica exchange molecular dynamics of the thermodynamics of fibril growth of Alzheimer's A beta(42) peptide. *Journal of Chemical Physics, 135*(6).
Hansmann, U. H. E. (1997). Parallel tempering algorithm for conformational studies of biological molecules. *Chemical Physics Letters, 281*(1–3), 140–150.
Hansmann, U. H. E. (2002). Protein-folding simulations in generalized ensembles. *International Journal of Quantum Chemistry, 90*(4–5), 1515–1521.
Hansmann, U. H. E., & Okamoto, Y. (1993). Prediction of peptide conformation by multicanonical algorithm—New approach to the multiple-minima problem. *Journal of Computational Chemistry, 14*(11), 1333–1338.
Harper, J. D., Wong, S. S., Lieber, C. M., & Lansbury, P. T. (1997). Observation of metastable A beta amyloid protofibrils by atomic force microscopy. *Chemistry & Biology, 4*(2), 119–125.
Horn, A. H. C., & Sticht, H. (2010). Amyloid-beta 42 oligomer structures from fibrils: A systematic molecular dynamics study. *Journal of Physical Chemistry B, 114*(6), 2219–2226.
Huang, J., & MacKerell, A. D. (2013). CHARMM36 all-atom additive protein force field: Validation based on comparison to NMR data. *Journal of Computational Chemistry, 34*(25), 2135–2145.

Ivanova, M. I., Sievers, S. A., Sawaya, M. R., Wall, J. S., & Eisenberg, D. (2009). Molecular basis for insulin fibril assembly. *Proceedings of the National Academy of Sciences of the United States of America*, *106*(45), 18990–18995.

Jackson, K., Barisone, G. A., Diaz, E., Jin, L. W., DeCarli, C., & Despa, F. (2013). Amylin deposition in the brain: A second amyloid in Alzheimer disease? *Annals of Neurology*, *74*(4), 517–526.

Jang, H., Arce, F. T., Capone, R., Ramachandran, S., Lal, R., & Nussinov, R. (2009). Misfolded amyloid ion channels present mobile beta-sheet subunits in contrast to conventional ion channels. *Biophysical Journal*, *97*(11), 3029–3037.

Jang, H., Arce, F. T., Ramachandran, S., Capone, R., Azimova, R., Kagan, B. L., et al. (2010). Truncated beta-amyloid peptide channels provide an alternative mechanism for Alzheimer's disease and Down syndrome. *Proceedings of the National Academy of Sciences of the United States of America*, *107*(14), 6538–6543.

Jang, H., Connelly, L., Arce, F. T., Ramachandran, S., Kagan, B. L., Lal, R., et al. (2013). Mechanisms for the insertion of toxic, fibril-like beta-amyloid oligomers into the membrane. *Journal of Chemical Theory and Computation*, *9*(1), 822–833.

Jang, H., Zheng, J., & Nussinov, R. (2007). Models of beta-amyloid ion channels in the membrane suggest that channel formation in the bilayer is a dynamic process. *Biophysical Journal*, *93*(6), 1938–1949.

Jiang, L., Liu, C., Leibly, D., Landau, M., Zhao, M. L., Hughes, M. P., et al. (2013). Structure-based discovery of fiber-binding compounds that reduce the cytotoxicity of amyloid beta. *elife*, *2*, e00857.

Jonsson, S. A., Mohanty, S., & Irback, A. (2012). Distinct phases of free alpha-synuclein—A Monte Carlo study. *Proteins*, *80*(9), 2169–2177.

Kahler, A., Sticht, H., & Horn, A. H. C. (2013). Conformational stability of fibrillar amyloid-beta oligomers via protofilament pair formation—A systematic computational study. *PLoS One*, *8*(7), e0070521.

Karplus, M., & McCammon, J. A. (2002). Molecular dynamics simulations of biomolecules. *Nature Structural Biology*, *9*(9), 646–652.

Kassler, K., Horn, A. H. C., & Sticht, H. (2010). Effect of pathogenic mutations on the structure and dynamics of Alzheimer's A beta(42)-amyloid oligomers. *Journal of Molecular Modeling*, *16*(5), 1011–1020.

Kawahara, M., Kuroda, Y., Arispe, N., & Rojas, E. (2000). Alzheimer's beta-amyloid, human islet amylin, and prion protein fragment evoke intracellular free calcium elevations by a common mechanism in a hypothalamic GnRH neuronal cell line. *Journal of Biological Chemistry*, *275*(19), 14077–14083.

Kayed, R., Head, E., Sarsoza, F., Saing, T., Cotman, C. W., Necula, M., et al. (2007). Fibril specific, conformation dependent antibodies recognize a generic epitope common to amyloid fibrils and fibrillar oligomers that is absent in prefibrillar oligomers. *Molecular Neurodegeneration*, *2*, 18.

Kayed, R., & Lasagna-Reeves, C. A. (2013). Molecular mechanisms of amyloid oligomers toxicity. *Journal of Alzheimer's Disease*, *33*, S67–S78.

Kia, A., & Darve, E. (2013). The accuracy of the CHARMM22/CMAP and AMBER ff99SB force fields for modelling the antimicrobial peptide cecropin P1. *Molecular Simulation*, *39*(11), 922–936.

Kim, S., Takeda, T., & Klimov, D. K. (2010). Mapping conformational ensembles of A beta oligomers in molecular dynamics simulations. *Biophysical Journal*, *99*(6), 1949–1958.

Kinnear, B. S., Jarrold, M. F., & Hansmann, U. H. E. (2004). All-atom generalized-ensemble simulations of small proteins. *Journal of Molecular Graphics & Modelling*, *22*(5), 397–403.

Klauda, J. B., Venable, R. M., Freites, J. A., O'Connor, J. W., Tobias, D. J., Mondragon-Ramirez, C., et al. (2010). Update of the CHARMM all-atom additive force field for lipids: Validation on six lipid types. *Journal of Physical Chemistry B*, *114*(23), 7830–7843.

Klepeis, J. L., & Shaw, D. E. (2014). Assessing the accuracy of physical models used in protein-folding simulations: Quantitative evidence from long molecular dynamics simulations. *Current Opinion in Structural Biology*, *24*, 98–105.

Knowles, T. P. J., & Buehler, M. J. (2011). Nanomechanics of functional and pathological amyloid materials. *Nature Nanotechnology*, *6*(8), 469–479.

Kodali, R., & Wetzel, R. (2007). Polymorphism in the intermediates and products of amyloid assembly. *Current Opinion in Structural Biology*, *17*(1), 48–57.

Laganowsky, A., Liu, C., Sawaya, M. R., Whitelegge, J. P., Park, J., Zhao, M. L., et al. (2012). Atomic view of a toxic amyloid small oligomer. *Science*, *335*(6073), 1228–1231.

Laio, A., & Parrinello, M. (2002). Escaping free-energy minima. *Proceedings of the National Academy of Sciences of the United States of America*, *99*(20), 12562–12566.

Landau, M., Sawaya, M. R., Faull, K. F., Laganowsky, A., Jiang, L., Sievers, S. A., et al. (2011). Towards a pharmacophore for amyloid. *PLoS Biology*, *9*(6), e1001080.

Lasagna-Reeves, C. A., Castillo-Carranza, D. L., Guerrero-Munoz, M. J., Jackson, G. R., & Kayed, R. (2010). Preparation and characterization of neurotoxic tau oligomers. *Biochemistry*, *49*(47), 10039–10041.

Lashuel, H. A., Hartley, D., Petre, B. M., Walz, T., & Lansbury, P. T. (2002). Neurodegenerative disease—Amyloid pores from pathogenic mutations. *Nature*, *418*(6895), 291.

Lee, J., Culyba, E. K., Powers, E. T., & Kelly, J. W. (2011). Amyloid-beta forms fibrils by nucleated conformational conversion of oligomers. *Nature Chemical Biology*, 7(9), 602–609.

Li, D., Furukawa, H., Deng, H., Liu, C., Yaghi, O. M., & Eisenberg, D. S. (2014). Designed amyloid fibers as materials for selective carbon dioxide capture. *Proceedings of the National Academy of Sciences of the United States of America*, *111*, 191–196.

Li, D. W., Mohanty, S., Irback, A., & Huo, S. H. (2008). Formation and growth of oligomers: A Monte Carlo study of an amyloid tau fragment. *PLoS Computational Biology*, *4*(12), e1000238.

Lin, H., Bhatia, R., & Lal, R. (2001). Amyloid beta protein forms ion channels: Implications for Alzheimer's disease pathophysiology. *FASEB Journal*, *15*(13), 2433–2444.

Lin, Y. S., & Pande, V. S. (2012). Effects of familial mutations on the monomer structure of A beta(42). *Biophysical Journal*, *103*(12), L47–L49.

Lin, Y. F., Zhao, J. H., Liu, H. L., Liu, K. T., Chen, J. T., Tsai, W. B., et al. (2010). Structural stability and aggregation behavior of the VEALYL peptide derived from human insulin: A molecular dynamics simulation study. *Biopolymers*, *94*(3), 269–278.

Lin, H., Zhu, Y. W. J., & Lal, R. (1999). Amyloid beta protein (1-40) forms calcium-permeable, Zn^{2+}-sensitive channel in reconstituted lipid vesicles. *Biochemistry*, *38*(34), 11189–11196.

Lindorff-Larsen, K., Maragakis, P., Piana, S., Eastwood, M. P., Dror, R. O., & Shaw, D. E. (2012). Systematic validation of protein force fields against experimental data. *PLoS One*, 7(2), e0032131.

Liu, C., Sawaya, M. R., Cheng, P. N., Zheng, J., Nowick, J. S., & Eisenberg, D. (2011). Characteristics of amyloid-related oligomers revealed by crystal structures of macrocyclic beta-sheet mimics. *Journal of the American Chemical Society*, *133*(17), 6736–6744.

Liu, C., Zhao, M. L., Jiang, L., Cheng, P. N., Park, J., Sawaya, M. R., et al. (2012). Out-of-register beta-sheets suggest a pathway to toxic amyloid aggregates. *Proceedings of the National Academy of Sciences of the United States of America*, *109*(51), 20913–20918.

Lopes, P. E. M., Huang, J., Shim, J., Luo, Y., Li, H., Roux, B., et al. (2013). Polarizable force field for peptides and proteins based on the classical drude oscillator. *Journal of Chemical Theory and Computation*, *9*(12), 5430–5449.

Lu, J. X., Qiang, W., Yau, W. M., Schwieters, C. D., Meredith, S. C., & Tycko, R. (2013). Molecular structure of beta-amyloid fibrils in Alzheimer's disease brain tissue. *Cell*, *154*(6), 1257–1268.

Meinhardt, J., Sachse, C., Hortschansky, P., Grigorieff, N., & Fandrich, M. (2009). A beta(1-40) fibril polymorphism implies diverse interaction patterns in amyloid fibrils. *Journal of Molecular Biology, 386*(3), 869–877.

Meinke, J. H., & Hansmann, U. H. E. (2007). Aggregation of beta-amyloid fragments. *Journal of Chemical Physics, 126*(1), 014706.

Meral, D., & Urbanc, B. (2013). Discrete molecular dynamics study of oligomer formation by n-terminally truncated amyloid beta-protein. *Journal of Molecular Biology, 425*(12), 2260–2275.

Merlini, G., Seldin, D. C., & Gertz, M. A. (2011). Amyloidosis: Pathogenesis and new therapeutic options. *Journal of Clinical Oncology, 29*(14), 1924–1933.

Middleton, C. T., Marek, P., Cao, P., Chiu, C. C., Singh, S., Woys, A. M., et al. (2012). Two-dimensional infrared spectroscopy reveals the complex behaviour of an amyloid fibril inhibitor. *Nature Chemistry, 4*(5), 355–360.

Miller, Y., Ma, B. Y., & Nussinov, R. (2011a). The unique Alzheimer's beta-amyloid triangular fibril has a cavity along the fibril axis under physiological conditions. *Journal of the American Chemical Society, 133*(8), 2742–2748.

Miller, Y., Ma, B., & Nussinov, R. (2011b). Synergistic interactions between repeats in tau protein and Aβ amyloids may be responsible for accelerated aggregation via polymorphic states. *Biochemistry, 50*, 5172–5181.

Morales, R., Moreno-Gonzalez, I., & Soto, C. (2013). Cross-seeding of misfolded proteins: Implications for etiology and pathogenesis of protein misfolding diseases. *PLoS Pathogens, 9*(9), e1003537.

Mousseau, N., & Derreumaux, P. (2005). Exploring the early steps of amyloid peptide aggregation by computers (Review). *Accounts of Chemical Research, 38*(11), 885–891.

Ndlovu, H., Ashcroft, A. E., Radford, S. E., & Harris, S. A. (2012). Effect of sequence variation on the mechanical response of amyloid fibrils probed by steered molecular dynamics simulation. *Biophysical Journal, 102*(3), 587–596.

Nelson, R., Sawaya, M. R., Balbirnie, M., Madsen, A. O., Riekel, C., Grothe, R., et al. (2005). Structure of the cross-beta spine of amyloid-like fibrils. *Nature, 435*(7043), 773–778.

Nguyen, P. H., Li, M. S., & Derreumaux, P. (2011). Effects of all-atom force fields on amyloid oligomerization: Replica exchange molecular dynamics simulations of the A beta(16-22) dimer and trimer. *Physical Chemistry Chemical Physics, 13*(20), 9778–9788.

Nicolls, M. R. (2004). The clinical and biological relationship between type ii diabetes mellitus and Alzheimer's disease. *Current Alzheimer Research, 1*(1), 47–54.

Olzscha, H., Schermann, S. M., Woerner, A. C., Pinkert, S., Hecht, M. H., Tartaglia, G. G., et al. (2011). Amyloid-like aggregates sequester numerous metastable proteins with essential cellular functions. *Cell, 144*(1), 67–78.

Ono, K., Condron, M. M., & Teplow, D. B. (2009). Structure-neurotoxicity relationships of amyloid beta-protein oligomers. *Proceedings of the National Academy of Sciences of the United States of America, 106*(35), 14745–14750.

Paparcone, R., Pires, M. A., & Buehler, M. J. (2010). Mutations alter the geometry and mechanical properties of Alzheimer's A beta(1-40) amyloid fibrils. *Biochemistry, 49*(41), 8967–8977.

Petkova, A. T., Leapman, R. D., Guo, Z. H., Yau, W. M., Mattson, M. P., & Tycko, R. (2005). Self-propagating, molecular-level polymorphism in Alzheimer's beta-amyloid fibrils. *Science, 307*(5707), 262–265.

Petkova, A. T., Yau, W. M., & Tycko, R. (2006). Experimental constraints on quaternary structure in Alzheimer's beta-amyloid fibrils. *Biochemistry, 45*(2), 498–512.

Pham, J. D., Chim, N., Goulding, C. W., & Nowick, J. S. (2013). Structures of oligomers of a peptide from beta-amyloid. *Journal of the American Chemical Society, 135*(33), 12460–12467.

Piana, S., & Laio, A. (2007). A bias-exchange approach to protein folding. *Journal of Physical Chemistry B*, *111*(17), 4553–4559.
Piana, S., Lindorff-Larsen, K., & Shaw, D. E. (2011). How robust are protein folding simulations with respect to force field parameterization? *Biophysical Journal*, *100*, L47–L49.
Piana, S., Lindorff-Larsen, K., & Shaw, D. E. (2012). Protein folding kinetics and thermodynamics from atomistic simulation. *Proceedings of the National Academy of Sciences of the United States of America*, *109*(44), 17845–17850.
Prigozhin, M. B., & Gruebele, M. (2013). Microsecond folding experiments and simulations: A match is made. *Physical Chemistry Chemical Physics*, *15*(10), 3372–3388.
Quist, A., Doudevski, L., Lin, H., Azimova, R., Ng, D., Frangione, B., et al. (2005). Amyloid ion channels: A common structural link for protein-misfolding disease. *Proceedings of the National Academy of Sciences of the United States of America*, *102*(30), 10427–10432.
Reddy, G., Straubb, J. E., & Thirumalai, D. (2009). Dynamics of locking of peptides onto growing amyloid fibrils. *Proceedings of the National Academy of Sciences of the United States of America*, *106*(29), 11948–11953.
Rhee, S. K., Quist, A. P., & Lal, R. (1998). Amyloid beta protein-(1-42) forms calcium-permeable, Zn2+-sensitive channel. *Journal of Biological Chemistry*, *273*(22), 13379–13382.
Richman, M., Wilk, S., Chemerovski, M., Warmlander, Skts, Wahlstrom, A., Graslund, A., et al. (2013). In vitro and mechanistic studies of an antiamyloidogenic self-assembled cyclic D,L-alpha-peptide architecture. *Journal of the American Chemical Society*, *135*(9), 3474–3484.
Rizzuti, B., & Daggett, V. (2013). Using simulations to provide the framework for experimental protein folding studies. *Archives of Biochemistry and Biophysics*, *531*(1–2), 128–135.
Sabate, R., Villar-Pique, A., Espargaro, A., & Ventura, S. (2012). Temperature dependence of the aggregation kinetics of Sup35 and Ure2p yeast prions. *Biomacromolecules*, *13*(2), 474–483.
Sarell, C. J., Stockley, P. G., & Radford, S. E. (2013). Assessing the causes and consequences of co-polymerization in amyloid formation. *Prion*, 7(5), 359–368.
Saunders, M. G., & Voth, G. A. (2013). Coarse-graining methods for computational biology. *Annual Review of Biophysics*, *42*, 73–93.
Sawaya, M. R., Sambashivan, S., Nelson, R., Ivanova, M. I., Sievers, S. A., Apostol, M. I., et al. (2007). Atomic structures of amyloid cross-beta spines reveal varied steric zippers. *Nature*, *447*(7143), 453–457.
Scalley, M. L., & Baker, D. (1997). Protein folding kinetics exhibit an Arrhenius temperature dependence when corrected for the temperature dependence of protein stability. *Proceedings of the National Academy of Sciences of the United States of America*, *94*(20), 10636–10640.
Scheraga, H. A., Khalili, M., & Liwo, A. (2007). Protein-folding dynamics: Overview of molecular simulation techniques. *Annual Review of Physical Chemistry*, *58*, 57–83.
Schor, M., Vreede, J., & Bolhuis, P. G. (2012). Elucidating the locking mechanism of peptides onto growing amyloid fibrils through transition path sampling. *Biophysical Journal*, *103*(6), 1296–1304.
Seilheimer, B., Bohrmann, B., Bondolfi, L., Muller, F., Stuber, D., & Dobeli, H. (1997). The toxicity of the Alzheimer's beta-amyloid peptide correlates with a distinct fiber morphology. *Journal of Structural Biology*, *119*(1), 59–71.
Senguen, F. T., Doran, T. M., Anderson, E. A., & Nilsson, B. L. (2011). Clarifying the influence of core amino acid hydrophobicity, secondary structure propensity, and molecular volume on amyloid-beta 16-22 self-assembly. *Molecular Biosystems*, 7(2), 497–510.
Sepulveda, F. J., Parodi, J., Peoples, R. W., Opazo, C., & Aguayo, L. G. (2010). Synaptotoxicity of Alzheimer beta amyloid can be explained by its membrane perforating property. *PLoS One*, *5*(7), e11820.
Shafrir, Y., Durell, S. R., Anishkin, A., & Guy, H. R. (2010). Beta-barrel models of soluble amyloid beta oligomers and annular protofibrils. *Proteins*, *78*(16), 3458–3472.

Shaw, D. E., Maragakis, P., Lindorff-Larsen, K., Piana, S., Dror, R. O., Eastwood, M. P., et al. (2010). Atomic-level characterization of the structural dynamics of proteins. *Science, 330*(6002), 341–346.

Shea, J. E., & Urbanc, B. (2012). Insights into A beta aggregation: A molecular dynamics perspective. *Current Topics in Medicinal Chemistry, 12*(22), 2596–2610.

Shewmaker, F., McGlinchey, R. P., & Wickner, R. B. (2011). Structural insights into functional and pathological amyloid. *Journal of Biological Chemistry, 286*(19), 16533–16540.

Sieradzan, A. K., Liwo, A., & Hansmann, U. H. E. (2012). Folding and self-assembly of a small protein complex. *Journal of Chemical Theory and Computation, 8*(9), 3416–3422.

Sinitskiy, A. V., & Voth, G. A. (2013). Coarse-graining of proteins based on elastic network models. *Chemical Physics, 422*, 165–174.

Siu, Shirley W. I., Vacha, Robert, Jungwirth, Pavel, & Bockmann, Rainer A. (2008). Biomolecular simulations of membranes: Physical properties from different force fields. *Journal of Chemical Physics, 128*(12), 125103–125112.

Skeby, K. K., Sorensen, J., & Schiott, B. (2013). Identification of a common binding mode for imaging agents to amyloid fibrils from molecular dynamics simulations. *Journal of the American Chemical Society, 135*(40), 15114–15128.

Sliwoski, Gregory, Kothiwale, Sandeepkumar, Meiler, Jens, & Lowe, Edward W., Jr. (2014). Computational methods in drug discovery. *Pharmacological Reviews, 66*, 334–395.

Soto, C., Estrada, L., & Castilla, J. (2006). Amyloids, prions and the inherent infectious nature of misfolded protein aggregates. *Trends in Biochemical Sciences, 31*(3), 150–155.

Straub, J. E., & Thirumalai, D. (2011). Toward a molecular theory of early and late events in monomer to amyloid fibril formation. In S. R. Leone, P. S. Cremer, J. T. Groves, & M. A. Johnson (Eds.), *Annual Review of Physical Chemistry: 62*. (pp. 437–463).

Toyama, B. H., & Weissman, J. S. (2011). Amyloid structure: Conformational diversity and consequences. In R. D. Kornberg, C. R. H. Raetz, J. E. Rothman, & J. W. Thorner (Eds.), *Annual Review of Biochemistry: 80*. (pp. 557–585).

Tsai, H. H., Reches, M., Tsai, C. J., Gunasekaran, K., Gazit, E., & Nussinov, R. (2005). Energy landscape of amyloidogenic peptide oligomerization by parallel-tempering molecular dynamics simulation: Significant role of Asn ladder. *Proceedings of the National Academy of Sciences of the United States of America, 102*(23), 8174–8179.

Tycko, R., & Wickner, R. B. (2013). Molecular structures of amyloid and prion fibrils: Consensus versus controversy. *Accounts of Chemical Research, 46*(7), 1487–1496.

Urbanc, B., Betnel, M., Cruz, L., Bitan, G., & Teplow, D. B. (2010). Elucidation of amyloid beta-protein oligomerization mechanisms: Discrete molecular dynamics study. *Journal of the American Chemical Society, 132*(12), 4266–4280.

Van Melckebeke, H., Wasmer, C., Lange, A., Ab, E., Loquet, A., Bockmann, A., et al. (2010). Atomic-resolution three-dimensional structure of het-s(218-289) amyloid fibrils by solid-state NMR spectroscopy. *Journal of the American Chemical Society, 132*(39), 13765–13775.

Viet, M. H., Nguyen, P. H., Ngo, S. T., Li, M. S., & Derreumaux, P. (2013). Effect of the Tottori familial disease mutation (D7N) on the monomers and dimers of A beta(40) and A beta(42). *ACS Chemical Neuroscience, 4*(11), 1446–1457.

Vitagliano, L., Stanzione, F., De Simone, A., & Esposito, L. (2009). Dynamics and stability of amyloid-like steric zipper assemblies with hydrophobic dry interfaces. *Biopolymers, 91*(12), 1161–1171.

Volpatti, L. R., & Knowles, T. P. J. (2014). Polymer physics inspired approaches for the study of the mechanical properties of amyloid fibrils. *Journal of Polymer Science Part B: Polymer Physics, 52*(4), 281–292.

Vymetal, J., & Vondrasek, J. (2013). Critical assessment of current force fields. Short peptide test case. *Journal of Chemical Theory and Computation, 9*(1), 441–451.

Wasmer, C., Zimmer, A., Sabate, R., Soragni, A., Saupe, S. J., Ritter, C., et al. (2010). Structural similarity between the prion domain of HET-s and a homologue can explain amyloid cross-seeding in spite of limited sequence identity. *Journal of Molecular Biology*, *402*(2), 311–325.

Westermark, P. (2005). Aspects on human amyloid forms and their fibril polypeptides. *The FEBS Journal*, *272*(23), 5942–5949.

Wiltzius, J. J. W., Landau, M., Nelson, R., Sawaya, M. R., Apostol, M. I., Goldschmidt, L., et al. (2009). Molecular mechanisms for protein-encoded inheritance. *Nature Structural & Molecular Biology*, *16*(9), 973–998.

Wiltzius, J. J. W., Sievers, S. A., Sawaya, M. R., Cascio, D., Popov, D., Riekel, C., et al. (2008). Atomic structure of the cross-beta spine of islet amyloid polypeptide (amylin). *Protein Science*, *17*(9), 1467–1474.

Wu, C., Bowers, M. T., & Shea, J. E. (2010). Molecular structures of quiescently grown and brain-derived polymorphic fibrils of the Alzheimer amyloid a beta(9-40) peptide: A comparison to agitated fibrils. *PLoS Computational Biology*, *6*(3), e1000693.

Wu, C., & Shea, J. E. (2011). Coarse-grained models for protein aggregation. *Current Opinion in Structural Biology*, *21*(2), 209–220.

Wu, C., & Shea, J. E. (2013). Structural similarities and differences between amyloidogenic and non-amyloidogenic islet amyloid polypeptide (IAPP) sequences and implications for the dual physiological and pathological activities of these peptides. *PLoS Computational Biology*, *9*(8), e1003211.

Xiao, X. D., Zhu, T., Ji, C. G., & Zhang, J. Z. H. (2013). Development of an effective polarizable bond method for biomolecular simulation. *Journal of Physical Chemistry B*, *117*(48), 14885–14893.

Xie, L. G., Luo, Y., & Wei, G. H. (2013). A beta(16-22) peptides can assemble into ordered beta-barrels and bilayer beta-sheets, while substitution of phenylalanine 19 by tryptophan increases the population of disordered aggregates. *Journal of Physical Chemistry B*, *117*(35), 10149–10160.

Yoon, G., Lee, M., Kim, J. I., Na, S., & Eom, K. (2014). Role of sequence and structural polymorphism on the mechanical properties of amyloid fibrils. *PLoS One*, *9*, e88502.

Yoshiike, Y., Akagi, T., & Takashima, A. (2007). Surface structure of amyloid-beta fibrils contributes to cytotoxicity. *Biochemistry*, *46*(34), 9805–9812.

Young, L. M., Cao, P., Raleigh, D. P., Ashcroft, A. E., & Radford, S. E. (2014). Ion mobility spectrometry-mass spectrometry defines the oligomeric intermediates in amylin amyloid formation and the mode of action of inhibitors. *Journal of the American Chemical Society*, *136*(2), 660–670.

Zerovnik, E., Stoka, V., Mirtic, A., Guncar, G., Grdadolnik, J., Staniforth, R. A., et al. (2011). Mechanisms of amyloid fibril formation—Focus on domain-swapping. *The FEBS Journal*, *278*(13), 2263–2282.

Zheng, J., Baghkhanian, A. M., & Nowick, J. S. (2013). A hydrophobic surface is essential to inhibit the aggregation of a tau-protein-derived hexapeptide. *Journal of the American Chemical Society*, *135*(18), 6846–6852.

Zheng, J., Jang, H., Ma, B., Tsai, C. J., & Nussinov, R. (2007). Modeling the Alzheimer A beta(17-42) fibril architecture: Tight intermolecular sheet-sheet association and intramolecular hydrated cavities. *Biophysical Journal*, *93*(9), 3046–3057.

Zimmermann, O., & Hansmann, U. H. E. (2008). Understanding protein folding: Small proteins in silico. *Biochimica et Biophysica Acta*, *1784*(1), 252–258.

CHAPTER FIVE

Recent Advances in Transferable Coarse-Grained Modeling of Proteins

Parimal Kar*, Michael Feig*,†,1
*Department of Biochemistry and Molecular Biology, Michigan State University, East Lansing, Michigan, USA
†Department of Chemistry, Michigan State University, East Lansing, Michigan, USA
[1]Corresponding author: e-mail address: feig@msu.edu

Contents

Abstract

Computer simulations are indispensable tools for studying the structure and dynamics of biological macromolecules. Biochemical processes occur on different scales of length and time. Atomistic simulations cannot cover the relevant spatiotemporal scales at which the cellular processes occur. To address this challenge, coarse-grained (CG) modeling of the biological systems is employed. Over the last few years, many CG models for proteins continue to be developed. However, many of them are not transferable with respect to different systems and different environments. In this review, we

Advances in Protein Chemistry and Structural Biology, Volume 96
ISSN 1876-1623
http://dx.doi.org/10.1016/bs.apcsb.2014.06.005

discuss those CG protein models that are transferable and that retain chemical specificity. We restrict ourselves to CG models of soluble proteins only. We also briefly review recent progress made in the multiscale hybrid all-atom/CG simulations of proteins.

1. INTRODUCTION

Classical all-atom molecular dynamics simulations with explicit solvent and ions are routinely used to investigate the dynamics of biological macromolecules. Such simulations can reveal the mechanism of biochemical processes occurring on different scales of length and time (Karplus & McCammon, 2002). With the availability of highly optimized atomistic force fields, specialized supercomputers, scalable codes, and enhanced sampling methodologies, one can study the dynamics of biomolecules in a realistic fashion and explain and predict experimental results. Most notable recent highlights are folding studies of small proteins by Freddolino, Liu, Gruebele, and Schulten (2008) and Shaw et al. (2010). The Schulten group has obtained a continuous trajectory of 10 μs while the Shaw group has performed 1 ms long simulation that would have been unthinkable not long ago. Other examples include simulations of proteins in cellular environments (Harada, Sugita, & Feig, 2012; Harada, Tochio, Kigawa, Sugita, & Feig, 2013) to investigate the effect of cellular-like environments on the stability and diffusive properties of biological macromolecules and simulations of very large complexes such as RNA polymerase II (Wang, Predeus, Burton, & Feig, 2013) and MutSα/MutSβ (Law & Feig, 2011; Sharma, Predeus, Kovacs, & Feig, 2014). Despite such successes, atomistic simulations are still several magnitudes away from being able to effectively cover the spatiotemporal scales of most cellular processes.

To address this challenge, the most promising strategy is a reductionist approach, i.e., a simplification of the underlying model that is used to describe the biological system (see Fig. 1). This can be achieved by coarse-graining of the system. In coarse-grained (CG) models, several atoms are grouped into a single particle (CG bead) and the interactions between these CG beads are then described by effective potentials that incorporate both energetic and entropic contributions resulting from integrating out (or averaging over) the neglected atomistic details. The coarse-graining of the system can greatly reduce the number of degrees of freedom of the system compared to a complete all-atom description of the system. As a result,

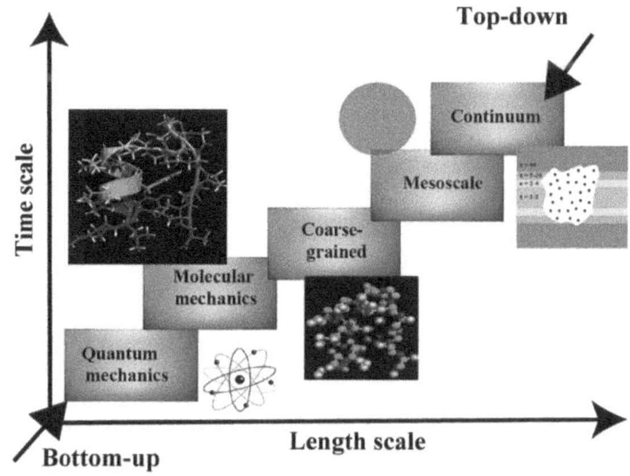

Figure 1 Schematic representation of various computational approaches that is used to cover different scales of length and time associated with different biophysical processes. These methods range from highly accurate but also the most computational demanding *ab initio* molecular dynamics (quantum) in the lower left corner to the highly efficient but very low-detail continuum models. In between, there exist atomistic and coarse-grained models followed by mesoscale models.

less computational power is required to simulate the system. Furthermore, the interactions between CG sites are usually "softer" which allows for longer integration time steps in molecular dynamics simulations. Typical time steps that can be used with CG models are tens of femtoseconds compared to 1–2 fs used in all-atom MD simulations. CG models combined with the dissipative particle dynamics allow time steps of 100 fs or above. Interactions in CG models tend to be of shorter range than in all-atom models due to the use of effective energy terms and typically the expensive treatment of long-range electrostatics can be avoided. This results in additional significant speedups in CG simulations compared with atomistic simulations. Finally, CG simulations generate inherently accelerated dynamics since the energy landscapes in CG models are smoother compared with atomistic models. Because of these advantages over fully atomistic simulations, CG models have the potential to cover significantly larger spatial and longer temporal scales compared to all-atom models.

In the past few years there has been a flurry of CG models for proteins with varying complexities (Majek & Elber, 2009; Tozzini, 2005). These CG models are employed for various applications, such as study of protein folding (Maisuradze, Senet, Czaplewski, Liwo, & Scheraga, 2010; Thorpe, Zhou, & Voth, 2008) problems, protein–protein docking (Basdevant, Borgis, & Ha-Duong, 2013), structure predictions (Kolinski, 2004;

Maupetit, Tuffery, & Derreumaux, 2007), scoring protein decoys (Miyazawa & Jernigan, 1985), structural features of viral capsid (Arkhipov, Freddolino, Imada, Namba, & Schulten, 2006; Krishna, Ayton, & Voth, 2010), structure and dynamics of membrane or protein-membrane systems (Ayton & Voth, 2009), and influence of certain amino acid substitutions in the HIV-1 Gag polyprotein and how the resulting physical effects influence HIV-1 virion formation and structure (Ayton & Voth, 2010).

There is no unique way of coarse-graining a protein (Sinitskiy, Saunders, & Voth, 2012). It depends on the levels of granularity and specific applications. The CG mapping schemes of proteins fall into two major categories: shape-based (Arkhipov et al., 2006) and residue-based approaches (Tozzini, 2005). The shape-based CG models emphasize outwards appearance and are thus not suitable for studying protein folding problems or distinguishing the dynamics between the wild-type and mutant variants of a protein. Residues-based models are suited best for studying internal dynamics and effects of chemical specificity of different residues.

Depending on the CG mapping scheme and specific applications, CG force fields may range from simple effective interaction potentials to more realistic all-atom like force fields. Generally, two different approaches, namely bottom-up (structure-based coarse-graining) and top-down (thermodynamic-based coarse-graining), are followed to generate a CG force field. In the bottom-up approach, CG force fields are parameterized based on comparisons to reference atomistic simulations. Several approaches, such as inverse Monte Carlo (Lyubartsev & Laaksonen, 1995), iterative Boltzmann inversion (IBI; Reith, Putz, & Muller-Plathe, 2003), force matching (FM; Izvekov & Voth, 2005a; Noid, Chu, et al., 2008; Noid, Liu, et al., 2008), fluctuation matching (Chu, Izvekov, & Voth, 2006), minimization of relative entropy (Chaimovich & Shell, 2011), or conditional reversible work (Brini, Herbers, Deichmann, & van der Vegt, 2012; Brini & van der Vegt, 2012) are employed in bottom-up approaches to match a set of target distributions from atomistic simulations. The bottom-up approaches are also highly amenable to multiscale approaches since the fine-grained (AA) information is propagated to a lower resolution (CG) model. On the other hand, in top-down approaches, CG force fields may be parameterized against experimental data, in particular thermodynamic data such as oil/water partitioning coefficients and density. Among various bottom-up approaches, IBI and FM methods are most popular and are frequently used to parameterize CG force fields for biomolecular simulations. IBI reproduces the structure of the underlying atomistic model by means of an iterative optimization scheme through which the effective potentials between CG sites are obtained. This method is motivated by Henderson's uniqueness theorem (Henderso, 1974) which states that, for a given pair radial distribution function, there is a unique underlying pair potential that will produce it. On the other hand, FM does not depend on the pair correlation functions. Instead, the idea of FM is to match the average force on CG beads in the CG system to that expected from the atomistic model. We are not going to discuss the underlying theories of different parameterization procedures employed in the bottom-up approaches and for an extended overview of such topics, the reader is referred to a recent review by Noid (Ingolfsson et al., 2014).

A major challenge remains in developing CG models that are highly predictive with respect to (unknown) specific behavior of specific systems rather than simply recapitulating either results from atomistic simulations or the generally expected characteristics of protein biophysics. This generally requires that CG models exhibit a sufficient degree of both transferability and chemical specificity. The range of CG models is very broad given different levels of coarse-graining and various philosophies of the parameterization of the CG force fields. Here, we will focus primarily on recent advances in developing such transferable CG force fields as they show the most promise in broad future applications of CG methodologies.

This review is organized as follows: First, we give a review of popular CG models for proteins. Next, we discuss recent advances in hybrid all-atom/coarse-grained (AA/CG) simulation methodologies. Finally, we conclude with a short outlook.

2. CG MODELS OF PROTEIN

Generally, a CG model involves the following four choices (Saunders & Voth, 2013): (i) resolution of the model appropriate for the size of the system and the types of questions that are going to be addressed; (ii) mapping of the CG beads to AA sites; (iii) type of interaction potential, and (iv) parameterization of the interaction potential to faithfully describe the energetics and dynamics of the biological system in comparison with experiment and AA simulations. CG models of varying complexity have been proposed for proteins since the pioneering work of Levitt and

Warshel (1975). Their resolution may vary from one or a few beads-per-amino-acids to near-atomistic resolutions (Bahar & Jernigan, 1997; Brown, Fawzi, & Head-Gordon, 2003; Buchete, Straub, & Thirumalai, 2004; Fernandez & Colubri, 2002; Fujitsuka, Takada, Luthey-Schulten, & Wolynes, 2004; Ghavami, van der Giessen, & Onck, 2013; Irback, Sjunnesson, & Wallin, 2000; Tozzini, 2005; Tozzini & McCammon, 2004). The coarser the description, the larger the gain in computational speedup. However, coarser models lack transferability and may not be suitable for multiscale simulations.

Early on, one-bead models (Honeycutt & Thirumalai, 1990; Korkuta & Hendrickson, 2009; Tozzini & McCammon, 2004) were most popular. For instance, Tozzini and McCammon have developed a Cα-based single bead CG model to study the dynamics of flap opening in HIV-1 protease (Tozzini & McCammon, 2004). In their model, each amino acid is represented by a single bead placed on the C_α atom. The force field was parameterized based on the Boltzmann inversion procedure. A similar approach was adopted by Korkuta and Hendrickson (2009) to build the virtual atom molecular mechanics (MM) force field. To study the disordered state of proteins, Ghavami and coworkers (Ghavami et al., 2013) have proposed a one-bead-per-amino-acid model. Generally, one-bead models are highly system-specific, such as the widely used Gō models, and they typically lack chemical specificity of amino acid side-chains (SC). Furthermore, one-bead models are typically not well suited for multiscale simulations because of their low resolution.

In contrast to other CG models, a model developed by the Voth group did not rely on a predefined potential but instead utilized the multiscale coarse-graining (MS-CG) method (Izvekov, Chung, & Rice, 2010; Izvekov & Voth, 2005a, 2005b). In this approach, the CG model and interaction potential was derived via a force-matching method in which atomistic force information from reference simulations is matched to forces from CG potentials within a variational framework (Noid, Chu, et al., 2008; Noid, Liu, et al., 2008). One of the advantages of the MS-CG approach is that it can be applied at arbitrary resolutions. The MS-CG method has been applied to the study of protein folding and dynamics (Hills, Lu, & Voth, 2010; Thorpe et al., 2008). However, MS-CG potentials are commonly not transferable. These models will not be described here. The reader is referred to earlier reviews on such models (Head-Gordon & Brown, 2003; Hills et al., 2010; Takada, 2012; Tozzini, 2010a).

Side-chain chemical specificity is essential for transferability and is the key for describing packing interactions within and between proteins. Adding one or several beads for the side-chain in addition to a single bead for the backbone can greatly improve the specificity of local interactions (Bahar & Jernigan, 1997; Mukherjee & Bagchi, 2004). A detailed description of the backbone, on the other hand, allows a faithful reproduction of secondary structure propensities. In the last few years, a number of CG models were developed that have relatively high resolution, with several beads per residue, and that take into account the chemical specificity of SC. These models employ different types of interaction potentials but generally offer at least some degree of transferability. In the following, the main features of the most prominent examples of such models are described. Table 1 summarizes mapping schemes and different force field terms for these models.

2.1. UNRES model

The UNited RESidue (UNRES) model was initially developed by the Scheraga group almost two decades ago as a knowledge-based CG model (Liwo, Oldziej, Kazmierkiewicz, Groth, & Czaplewski, 1997; Liwo, Oldziej, Pincus, et al., 1997; Liwo, Pincus, Wawak, et al., 1997). However, over the years, it has undergone several refinement stages and finally evolved into a fully physics-based model. In the UNRES model (Khalili, Liwo, Rakowski, Grochowski, & Scheraga, 2005; Liwo, Oldziej, Kazmierkiewicz, et al., 1997; Liwo, Oldziej, Pincus, et al., 1997; Liwo, Pincus, Wawak, et al., 1997), a polypeptide chain is represented as a sequence of α-carbon atoms with attached united SC, whose sizes depend on the nature of the amino acid residue and united peptide groups (P), each of the latter being positioned in the middle between two consecutive C_α atoms (Fig. 2). The α-carbon atoms are linked by virtual C_α...C_α bonds of length 3.8 Å corresponding to *trans* peptide groups. It should be noted here that these C_α atoms are not interaction sites *per se* but rather they assist only in the definition of the geometry. Each SC is represented as an ellipsoid and is attached to the respective C_α atom by a virtual C_α...SC bond.

The force field is derived as a restricted free energy function, which corresponds to averaging the energy over the degrees of freedom that are neglected in the UNRES model. The UNRES effective energy function is given by Eq. (1) (Maisuradze, Liwo, & Scheraga, 2010).

Table 1 Key features of different popular CG models for proteins

Model	UNRES	MARTINI	OPEP	SCORPION	PaLaCe	Bereau and Deserno	PRIMO	GBEMP	PRIME
BB	1	1	5	1	3	3	3	2	3
SC (non-Gly)	1	1–5	1	1–2	1–2	1	1–5	2–3	1
Virtual bead	X	–	–	–	X	–	X	–	–
V^{bond}	X	X	X	–	X	X	X	X	X
V^{angle}	X	X	X	–	X	X	X	X	X
$V^{torsion}$	X	X	X	–	X	X	X	X	X
$V^{double\text{-}torsion}$	X	–	–	–	–	–	–	–	–
V^{vdW}	GB[a] 12-6	LJ 12-6	r^{-8} + AF[b]	r^{-8} + G[c]	LJ 8-6	WCA	LJ 12-6	GB[a] 12-6	Hard-sphere
V^{ele}	X	X	–	X	X	–	X	–	–
$V^{H\text{-}bond}$	–	–	X	–	X	X	X	–	X
Solvent	Effective[d]	CG	Effective[d]	CG	CV[e]	Effective[d]	GBMV[f]	GK[g]	Effective[d]
Membrane	–	X	–	–	–	–	X	–	–
Restraint	–	ENM	–	ENM	–	–	–	–	–
$V^{multibody}$	X	–	–	–	–	–	–	–	–

Dipole	–	–	–	–	–	X	–	–	–
Multipole	–	–	–	–	–	–	–	X	–
Speedup	~1000	75–100	30–40	100–130	~1000	NA[h]	10–15	~50	~1000

[a]Gay–Bern.
[b]r^{-8} combine with analytical formulation.
[c]r^{-8} combine with Gaussian attractive.
[d]Implicitly included within nonbonded parameters.
[e]Circular variance.
[f]Generalized Born/molecular volume.
[g]Generalized Kirkwood.
[h]Not available.
BB denotes backbone while SC denotes side-chain. Here, V^{bond}, V^{aangle}, $V^{torsion}$, $V^{double\text{-}torsion}$, V^{vdW}, V^{ele}, $V^{H\text{-}bond}$, and $V^{multibody}$ represent bond, angle, torsion, double-torsion, van der Waals, electrostatic, hydrogen bonding, and multibody potentials, respectively

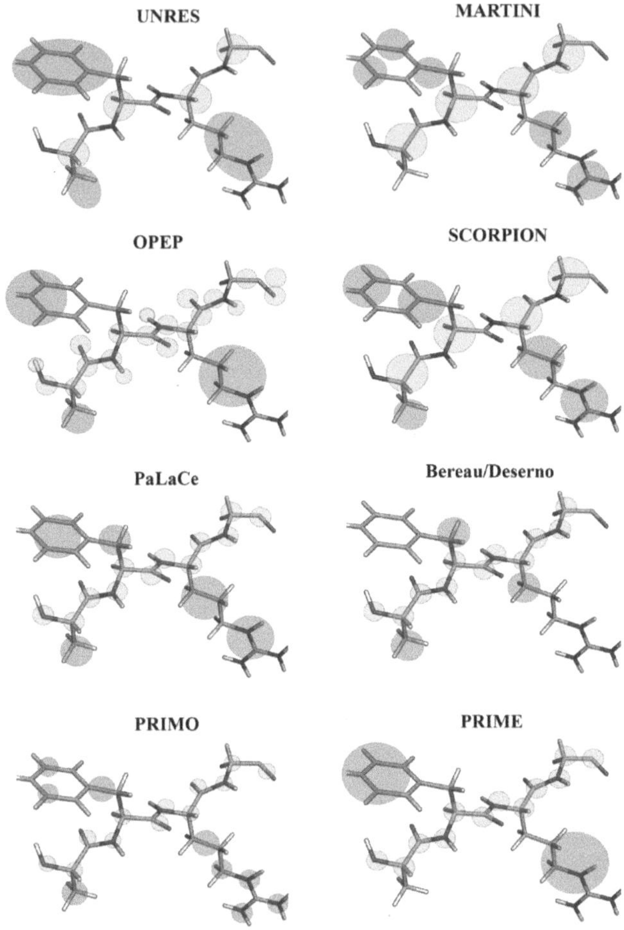

Figure 2 Mapping strategies for various CG models for an AlaPheArgGly peptide. Backbone beads are shown in yellow (light gray in the print version) color while side-chain beads are shown in red (dark gray in the print version) color.

$$
\begin{aligned}
U^{\mathrm{UNRES}} = {} & w_{\mathrm{SC}} \sum_{i<j} U_{\mathrm{SC}_i\mathrm{SC}_j} + w_{\mathrm{SCp}} \sum_{i \neq j} U_{\mathrm{SC}_i\mathrm{p}_j} \\
& + w_{\mathrm{pp}} f_2(T) \sum_{i<j-1} U_{\mathrm{p}_i\mathrm{p}_j} + w_{\mathrm{tor}} f_2(T) \sum_i U_{\mathrm{tor}}(\gamma_i) \\
& + w_{\mathrm{tord}} f_3(T) \sum_i U_{\mathrm{tord}}(\gamma_i, \gamma_{i+1}) + w_{\mathrm{b}} \sum_i U_{\mathrm{b}}(\theta_i) \\
& + w_{\mathrm{rot}} \sum_i U_{\mathrm{rot}}\left(\alpha_{\mathrm{SC}_i}, \beta_{\mathrm{SC}_i}, \theta_i\right) + w_{\mathrm{bond}} \sum_i U_{\mathrm{bond}}(d_i) \\
& + \sum_{m=3}^{6} w_{\mathrm{corr}}^{m} f_m(T) U_{\mathrm{corr}}^{(m)} + w_{\mathrm{SS}} \sum_i U_{\mathrm{SS},i}
\end{aligned}
\tag{1}
$$

where *w* represents the relative weights of each term. The initial three terms represent nonbonded interaction energies between SC–SC, SC–P, and P–P beads. The next five terms describe the bonded interaction energies corresponding to torsion, double-torsion, angle, SC rotamers, and bonds, respectively. The last two remaining terms model multibody (correlation) interaction energies and the energy of forming disulfide bonds, respectively. Notably, the correlation term is temperature dependent. The force field is optimized based on the concept of a hierarchical protein–energy landscape (Liwo et al., 2002).

The potential function of UNRES is parameterized against free energies from atomistic polypeptide simulations in explicit water using a functional form that resembles atomistic force fields but with additional terms. In UNRES, solvent effects are represented implicitly. Applications of UNRES to date have been largely limited to *de novo* protein folding studies (Lee, Liwo, & Scheraga, 1999; Liwo, Khalili, & Scheraga, 2005; Maisuradze, Senet, et al., 2010) and protein structure prediction (Nanias, Chinchio, Oldziej, Czaplewski, & Scheraga, 2005; Oldziej et al., 2005). Recently, Solernou and Fernandez-Recio (2011) have developed a CG potential by extending the CG potential UNRES with additional terms accounting for a solvation energy contribution and an explicit electrostatic interactions to predict the protein–protein interactions.

Among all of the physics-based CG models, UNRES performs exceptionally well in folding proteins to their native-like conformations. This model has been used successfully in CASP structure prediction competition. For instance, the target T0215 was fold to a C_α RMSD of 3.5 Å from the native state. UNRES has been used to fold multichain proteins also. For example, UNRES was used to *ab initio* folding of the multichain protein with Protein Data Bank (PDB) ID 1G6U and the lowest RMSD obtained was 2.4 Å from the native state (Rojas, Liwo, & Scheraga, 2007). Their simulations reveal that individual chain folded independently to their native structure and then assembled into a complex structure.

The UNRES model is physically accurate and highly transferable but it suffers from a number of drawbacks. First, the model requires a very complex reverse-mapping scheme to convert from UNRES to all-atom models and is thus difficult to use within the context of multiscale AA/CG simulations. Second, UNRES is neither transferable to the lipid environments nor is it easy to extend the model for other biomolecules such as nucleic acids because the necessary model reparameterization would likely be very time-consuming.

2.2. MARTINI model

The most popular and most widely used CG model for lipids as well as proteins is MARTINI (Marrink, de Vries, & Mark, 2004; Marrink, Risselada, Yefimov, Tieleman, & de Vries, 2007; Monticelli et al., 2008). This model has been employed in a wide range of applications (Marrink, de Vries, Harroun, Katsaras, & Wassall, 2008; Marrink, de Vries, & Tieleman, 2009; Marrink & Tieleman, 2013; Periole, Huber, Marrink, & Sakmar, 2007; Periole, Knepp, Sakmar, Marrink, & Huber, 2012; Risselada & Marrink, 2008; Sengupta, Leontiadou, Mark, & Marrink, 2007; Smirnova, Marrink, Lipowsky, & Knecht, 2010; Treptow, Marrink, & Tarek, 2008). The model was initially developed for lipid molecules and later was extended to proteins and several other biomolecules. The MARTINI model is based on a four-to-one mapping, i.e., on average four heavy atoms plus associated hydrogens are represented by a single CG site (Fig. 2). However, ring-like molecules (e.g., benzene, cholesterol, and several of the amino acids) are mapped with a higher resolution of up to two nonhydrogen atoms to one MARTINI particle. The backbone of each amino acid is modeled with a single bead placed at the center of mass of the backbone, while up to four beads are used for SC. In contrast to UNRES, solvent is treated at the CG level in MARTINI. Four real water molecules are mapped to a single CG water bead. Recently, polarizable CG water models (Wu, Cui, & Yethiraj, 2010; Yesylevskyy, Schafer, Sengupta, & Marrink, 2010) were also developed to be used with the MARTINI model. Ions are represented by a single CG bead, which represents both the ion itself and its first hydration shell. Based on the nature of different chemical groups, four main types of particle are assigned in the MARTINI model, namely polar (P), nonpolar (N), apolar (C), and charged (Q). Within a main type, subtypes are assigned according to hydrogen bonding capabilities or the degree of polarity. In total, MARTINI has 18 particle types.

The MARTINI CG force field consists of generic bonded (bond, angle, dihedral, and improper dihedral) and nonbonded (van der Waals and electrostatics) interaction potentials. Notably, the electrostatic interaction is described by the Coulomb potential with a relative dielectric constant ($\varepsilon_{rel}=15$) for explicit screening. However, a lower relative dielectric constant of 2.5 is suggested when combining MARTINI with a polarizable water model. The parameterization of this model relies on a combination of bottom-up and top-down approaches. For instance, a bottom-up approach was adopted to derive the bonded terms and was obtained from the underlying atomistic geometry or by comparing to atomistic simulations. On the other hand, the nonbonded interaction potentials were optimized based on a systematic comparison to experimental theromodynamic quantities, such as free energy of hydration and vaporization, and the partitioning free energies between oil and aqueous phases. Recently, the nonbonded parameters were further optimized (de Jong et al., 2013) based the binding studies of Wimley–White pentapeptides to a 2-oleoyl-1-pamlitoyl-*sn*-glyecro-3-phosphocholine bilayer (Singh & Tieleman, 2011) and SC analog dimerization free energies (de Jong, Periole, & Marrink, 2012). This led to the addition of an explicit polarization term to polar SCs such as glutamine and asparagine. The MARTINI CG model can achieve about a factor of 75–100 speedup with respect to fully atomistic simulations (Gu, Bai, Li, & Wang, 2012).

MARTINI suffers from not being able to accurately represent amino-acid specific secondary structure propensities. As a consequence, a weak bias has to be employed to maintain secondary structures based on what is known from native structures. Therefore, MARTINI is not suitable for modeling processes in which secondary structure changes and, more generally, folding and unfolding play a substantial role. However, movements of secondary structures with respect to each other can be studied with the MARTINI model. Furthermore, although MARTINI is very successful in modeling interactions with lipid bilayers and lipid–water interfaces, it is not applicable to systems involving solid/fluid or gas/fluid interfaces because of the fluid-phase based parameterization.

It is worth mentioning here that Peraro and coworkers (Spiga, Alemani, Degiacomi, Cascella, & Dal Peraro, 2013) have adopted the MARTINI mapping scheme and have developed an electrostatic-consistent CG potential for proteins in which they have introduced an explicit dipole defined by three consecutive C_α beads along with the treatment of nonradial dipole–dipole interactions in dynamics. As a result, in contrast to MARTINI, this model can maintain stable secondary elements of unspecific proteins and sample conformational transitions.

2.3. OPEP model

The optimized potential for efficient structure prediction (OPEP; Barducci, Bonomi, & Derreumaux, 2011; Derreumaux, 1999; Maupetit et al., 2007; Sterpone et al., 2014) is a transferable CG force field for proteins developed by Derreumaux and coworkers. OPEP is an intermediate resolution model where each amino acid is represented with six beads. The backbone is retained in full atomic detail while SCs are represented by a single bead or centroid (Fig. 2). However, the amino acid proline (Pro) is represented by all heavy atoms. The centroids are placed at the center of mass of the heavy atoms and calculated employing the rotamer distribution of each SC. Each SC bead is described by an appropriate van der Waals radius, and their positions are defined with respect to the backbone heavy atoms N, C_α, and C'. The OPEP energy function consists of bonded and non-bonded terms. Bonded terms include bonds, angles, dihedrals, and improper torsions (to maintain the *cis* or *trans* conformation of the peptide bond and the chirality of the amino acid). Additional terms confine the dihedrals to the most populated regions in the Ramachandran plot via harmonic barriers. The force constants and equilibrium values associated with these terms were derived from the AMBER all-atom force field (Cornell et al., 1995). Non-bonded interactions in OPEP consist of van der Waals interactions and explicit hydrogen bonding potentials. The hydrogen-bonding potential consists of two-body and four-body terms, and the later term was used to capture the strong cooperative nature of the amide hydrogen bonds. The van der Waals interactions are knowledge-based as van der Waals radii were calculated using a dataset of thousands of PDB structures. The van der Waals interactions combine backbone–backbone, backbone–SC terms with 210 SC pair interactions. The aqueous solvent is treated implicitly in OPEP. A genetic algorithm using a scoring function, which requires that the native structure has the lowest energy and that native-like structures have an energy higher than the native structure but lower than non-native conformations, was used to optimize the OPEP force field. In OPEPv4, the van der Waals term was composed of an r^{-8} term combined with an analytical formulation that quickly vanishes as a function of distance (Chebaro, Pasquali, & Derreumaux, 2012). This analytical function has a steeper profile with compared to a 12-6 Lennard–Jones potential. In OPEPv4, authors have changed the ε_{ij} coupling constants of some interaction pairs present in α-helices based on a statistical analysis of PDB structures. To further improve OPEP, Sterpone and coworkers (Sterpone, Nguyen, Kalimeri, & Derreumaux, 2013) recently derived new effective potentials for describing the ion pair interactions from all-atom PMF (potential of mean force), that feature a single minimum for Lys–Asp and Lys–Glu pairs and two minima for Arg–Asp and Arg–Glu. The OPEP force field has been used successfully for predicting protein stabilities as well as folding and aggregation studies of small peptides

(Chebaro, Mousseau, & Derreumaux, 2009; Chebaro et al., 2012) and the structure prediction of small proteins (Maupetit, Derreumaux, & Tuffery, 2009).

OPEP is highly transferable among different proteins, but it is limited to aqueous solvent environments. OPEP requires an integration time step of 1.5 fs due to the detailed backbone and hydrogen bonds. This limits the amount of sampling that can be accomplished. Furthermore, a single bead representation of the SC puts a limit on capturing the chemical specificity of SCs specificity and creates ambiguity in when reconstructing all-atom models from the CG model. Therefore, applications within multiscale frameworks may be limited. Furthermore, due to the lack of chemical SC specificity, this model may not be suitable for investigating protein–protein interactions.

2.4. SCORPION model

A major challenge in computational modeling of protein–protein interactions is the *de novo* prediction of protein–protein interfaces. To access the large sampling space, CG simulations are often employed. Basdevant and coworkers (Basdevant et al., 2013) have developed the physics-based simplified CG model Solvated COaRse-grained Protein interactION (SCORPION) which targets applications in protein–protein recognition. In SCORPION, the backbone of each amino acid is represented by a single bead while one or two beads are used to describe SCs depending on their size (Fig. 2). The CG beads are located at the geometric center of the heavy atoms. This mapping scheme yields 29 CG types. An elastic network model (ENM) model was introduced in SCORPION to account for the internal protein flexibility. Because the ENM model maintains the internal protein structure, no bonded terms are necessary in SCORPION. Following classical all-atom force fields, nonpolar and electrostatic interactions were expressed as sums of pair wise van der Waals and Coulombic interaction potentials and were optimized in a consistent bottom-up approach. Van der Waals interactions were extracted by fitting van der Waals SC pair interactions to PMFs extracted from all-atom MD simulations that were performed in vacuum with vanishing charges (i.e., no Coulomb interactions) in order to account for the purely nonelectrostatic interaction. The resulting van der Waals interaction potential is softer than its atomistic counterpart and is composed of a very smooth repulsive part in $1/r$ and a Gaussian attractive part (Basdevant et al., 2013). The CG point charges for a protein were obtained by reproducing the electrostatic potentials obtained with an all-atom model in vacuum. The motivation for carrying out the parameterization in vacuum was to obtain a model that can then be coupled to any implicit or explicit solvent model. While this makes sense in principle, this approach neglects solute polarization that is induced upon transfer from vacuum to solvent environments and that may require a polarizable version of SCORPION. As a step in that direction, the authors of SCORPION have recently developed a polarizable coarse-grained solvent model, which can capture such polarization effects in part and was validated successfully against solvation free energies of peptides. The resulting CG model was found to be about 100–130 times faster than the atomistic simulation.

The combined protein and water model was used successfully in mechanistic studies of protein–protein recognition. For instance, they have conducted seven long molecular dynamics simulations starting from several initial configurations to predict the structure of barnase/barstar complex. Notably, the simulations were carried out at a very high temperature (500 K) to accelerate the phase-space sampling. Within 1–500 ns, they observed that five of seven simulations adopted a conformation very similar to the native complex structure with RMSDs from the crystal structure varying between 0.5 and 2 Å. Furthermore, their simulations reveal that this protein–protein recognition process is mainly driven by the van der Waals and electrostatic interactions between two partner proteins. These results agree with the previous all-atom simulations (Hoefling & Gottschalk, 2010).

The current model suffers from two main drawbacks: lack of bonded interactions and the use of high temperature that affects the balance between the enthalpic and entropic contributions.

2.5. PaLaCe model

Recently, Pasi, Lavery, and Ceres (PaLaCe; Pasi, Lavery, & Ceres, 2013) developed a new CG model for proteins for investigating the mechanical properties of proteins. This model uses a two-tier representation of the polypeptide chains: one representation is used for bonded interactions and for backbone hydrogen bonds and involves atomic beads while a different level is used for nonbonded terms, other than main chain hydrogen bonds, and involves a pseudo atom (or CG) representation of the amino acids. One or two CG beads are used for SCs while three beads (N, C_α, and C′) are used to represent the backbone so that an explicit description of hydrogen bonds becomes possible (Fig. 2). The potential energy function of PaLaCe consists of physics-based bonded and nonbonded terms with an additional term for the backbone hydrogen bonding potential, which is introduced to allow secondary structures changes. The van der Waals interactions in PaLaCe are characterized by a Lennard–Jones-like interaction but with a softer repulsive term than atomistic force fields. The hydrogen bonding potential is composed of 12-6 Lennard–Jones and electrostatic contributions. The electrostatic component of the hydrogen bonding potential is modeled via Coulomb potential damped by a distance-dependent dielectric constant. The solvent is treated implicitly with a one-body solvation term based on circular variance, which is used to describe the occlusion of a CG atom within a protein structure. The PaLaCe potential was parameterized following a bottom-up approach by applying the Boltzmann inversion technique to conformational probability distributions derived from a large data set of protein structures followed by iterative refinement. PaLaCe was shown to maintain stable structures of folded proteins. It can also correctly model large-scale, force-induced conformational changes for the immunoglobulin-like domain of the giant protein titin. PaLaCe is reported to be faster than equivalent all-atom simulations in an explicit solvent by roughly 1000 times. The model is limited to aqueous solvent environments only.

2.6. Bereau and Deserno model

Another transferable CG model for proteins was designed by Bereau and Deserno (2009). It uses an intermediate level of description of the peptide and treats solvent effects implicitly. Each amino acid is described by four beads: three beads for the backbone and one bead for SC. The CG beads for backbones are located at N, C_α, and C' while the CG bead for SCs is located at C_β (Fig. 2). For nonglycine amino acids, all SC beads are assigned the same van der Waals radius. The bonded terms are generic and the corresponding geometric parameters were taken from existing peptide models (Ding, Borreguero, Buldyrey, Stanley, & Dokholyan, 2003; Irback et al., 2000; Takada, Luthey-Schulten, & Wolynes, 1999). The spring constants are set high enough and an approximate 5% flexibility around their reference values was given to account for thermal fluctuations. The peptide bond is modeled by a single minimum centered around the *trans* conformation while two minima centered around *cis* and *trans* conformations are employed to model the isomerization.

The SC nonbonded interactions were described by the venerable knowledge-based potential from Miyazawa and Jernigan (1985) that resulted from a statistical analysis of SC contacts in a protein structure database. No explicit electrostatic interactions were included since they are implicitly accounted for in several of the other terms. The backbone bead interaction potential includes a short-range, purely repulsive Week–Chandler–Anderson (WCA) potential (Weeks, Chandler, & Andersen, 1971) to capture locally excluded volume interactions. Furthermore, a backbone hydrogen bonding potential as a function of the relative distance and orientation of an amide and a carbonyl group was added following Irback et al. (2000), where a radial 12-10 Lennard–Jones potential is combined with an angular term. It has been shown previously (Chen, Su, & Mou, 2006) that carbonyl and amide groups at the peptide bond form dipoles that interact with each other and that the nearest-neighbor interaction is enough to favor β- over α-content (Mu & Gao, 2007). Therefore, an extra term describing dipole–dipole interactions of neighboring residues is also included in this model. A single cosine function centered around φ, $\psi = 0$ with an identical amplitude along both coordinates was employed to model the neighboring dipole potential. These terms were optimized by reproducing dihedral angle distributions of GGG and GAG tripeptides and global folding properties of a three-helix bundle. The resulting CG force field was used for studying protein folding (Bereau, Bachmann, & Deserno, 2010) and aggregation (Bereau & Deserno, 2009). One of the drawbacks of this model is that the stability of β-sheet structures and proteins with mixed α/β-content is apparently underestimated. The model is implemented in the ESPRESSO package (Limbach, Arnold, Mann, & Holm, 2006).

2.7. PRIMO model

Very recently, we developed the intermediate resolution CG model for proteins, PRIMO (Gopal, Mukherjee, Cheng, & Feig, 2010; Kar, Gopal, Cheng, Predeus, & Feig, 2013). The backbone in PRIMO is described by three CG beads (N, C_α, and combined CO) while one to five sites are used for SCs (Fig. 2). A unique feature of PRIMO is that the CG beads are mapped in such a way that atomistic models can be reconstructed from the CG beads using an analytical procedure that is extremely fast and provides near-atomistic accuracy in the reconstructed models. This allows PRIMO to be tied closely to fully atomistic models in order to maximize transferability. It also positions PRIMO as an ideal CG model for multiscale models. Like some of the other CG models, PRIMO uses virtual sites that are constructed on the fly from the CG interaction sites in order to maintain better bonding geometries.

PRIMO is largely a physics-based model with an all-atom like force field as shown in Eq. (2):

$$
\begin{aligned}
U^{\mathrm{PRIMO}} = & \sum_{i=1}^{N_{\mathrm{bond}}} k_i^{\mathrm{bond}} (l_i - l_{i,0})^2 + \sum_{i=1}^{N_{\mathrm{angle}}} k_i^{\mathrm{angle}} (\theta_i - \theta_{i,0})^2 \\
& + \sum_{i=1}^{N_{\mathrm{bond-spline}}} s_i^{1\mathrm{D}}\left(l_i^{1-2}\right) + \sum_{i=1}^{N_{\mathrm{angle-spline}}} s_i^{1D}\left(l_i^{1-3}\right) \\
& + \sum_{i=1}^{N_{\mathrm{virtual-atom}}} k_i^{\mathrm{virtual-bond}} (i_i - l_{i,0})^2 + \sum_{i=1}^{N_{\mathrm{virtual-atom}}} k_i^{\mathrm{virtual-angle}} (\theta_i - \theta_{i,0})^2 \\
& + \sum_{i=1}^{N_{\mathrm{torsion}}} \sum_{j=1}^{N_i^{\mathrm{mult}}} k_{i,j}^{\mathrm{torsion}} (1 + \cos(n_{i,j}\varphi_i - \varphi_{i,j,0})) + \sum_{i=1}^{N_{\mathrm{torsion-spline}}} s_{1\mathrm{D}}^{i}\left(l_i^{1-4}\right) \\
& + \sum_{i=1}^{N_{\mathrm{CMAP}}} s_i^{2\mathrm{D}} (\varphi_i, \psi_i) + \sum_{i=1}^{N_{\mathrm{HBOND}}} s_i^{2\mathrm{D}}\left(\cos\theta, l_i^{\mathrm{N-CO}}\right) \\
& + \sum_{i=1}^{N_{\mathrm{atom}}-1} \sum_{j=i+1}^{N_{\mathrm{atom}}} 4\varepsilon_{ij} \left[\left(\frac{\sigma_{ij}}{r_{ij}}\right)^{12} - \left(\frac{\sigma_{ij}}{r_{ij}}\right)^{6} \right] + \sum_{i=1}^{N_{\mathrm{atom}}-1} \sum_{j=i+1}^{N_{\mathrm{atom}}} \frac{q_i q_j}{4\pi\varepsilon_0 r_{ij}} \\
& + \Delta G_{\mathrm{solv}}^{\mathrm{GB}} + \sum_{i=1}^{N_{\mathrm{atom}}} \gamma_i (\mathrm{SASA})_i
\end{aligned}
\tag{2}
$$

The bonded terms are described by harmonic potentials as well as spline potentials to model nonharmonic bonding functions between CG beads. The interaction potential also includes a distance- and angle-dependent explicit hydrogen-bonding potential term because hydrogen bonding cannot be captured with electrostatics alone in CG models. A CMAP (MacKerell, Feig, & Brooks, 2004) correction term is used to provide a correct description of the backbone (φ,ψ) and SC (χ) distributions. PRIMO uses an implicit solvent potential to capture solvent effects. PRIMO was optimized by direct comparison with the latest all-atom CHARMM36 force field (Best, Mittal, Feig, & MacKerell, 2012; Best, Zhu, et al., 2012) against a diverse set of decoys of peptides and proteins. There is a high energetic correspondence between PRIMO and CHARMM which provides the basis for a seamless mixing of PRIMO and CHARMM in multiscale AA/CG simulations (Predeus, Gul, Gopal, & Feig, 2012). In contrast to most other similar CG models, PRIMO does not require any bias towards known structural information and is suitable for running stable MD simulations of proteins and for folding of small peptides from their extended conformations. PRIMO allows a time step of 4–6 fs and achieves a speedup of about 10–15 times faster compared to atomistic simulations. This speedup is significantly less compared to other CG models in part due to the relatively high resolution but also because of the expensive GBMV implicit solvent model (Lee, Feig, Salsbury, & Brooks, 2003) that is currently used in combination with PRIMO. PRIMO essentially maintains the same level of transferability of all-atom force fields, including transferability to different environments. For example, it is possible to transfer to membrane environments by simply switching the GBMV (Lee et al., 2003) implicit solvent model for the heterogeneous dielectric generalized Born (Tanizaki & Feig, 2005, 2006) implicit membrane model without further optimizations as demonstrated recently (Kar, Gopal, Cheng, Panahi, & Feig, 2014).

2.8. GBEMP model

In general, CG beads are considered to be isotropic. However, in the context of biomolecular systems, CG beads are anisotropic because a single CG bead represents a group of atoms. Motivated by this observation Shen and coworkers (Shen, Li, Ren, Zhang, & Li, 2014) have recently introduced a physics-based CG model, named GBEMP, based on the anisotropic Gay–Berne (GB) and point electric multipole (EMP) potentials that are used to describe the nonbonded interactions. In GBEMP, the bonded terms consist of bond, angle and dihedral potentials. Instead of using simple harmonic potentials, the bond stretching term is described by the fourth-order Taylor expansion of the Morse potential while the angle bending term adopts a sixth-order potential. A Fourier series expansion is used to model the torsional potential. The generalized Kirkwood (GK; Schnieders & Ponder, 2007) implicit solvent model was employed to describe the solvent effect in GBEMP. In an alternative to the traditional leap-frog or Verlet algorithm, Euler's rigid body integrator is used to propagate the equation of motion.

The force field was parameterized based on a combination of experimental and all-atom data. The bonded terms were parameterized by fitting to the PMFs obtained from atomistic molecular dynamics simulations of dipeptides using CHARMM27/CMAP (MacKerell et al., 2004) force field (Brooks et al., 2009; MacKerell et al., 1998). The parameters for torsional potentials were further optimized by iteratively matching to the experimental results for the distributions of backbone and SC torsions. The nonbonded terms were parameterized by reproducing the individual energy components of the atomistic force field. The model was validated by reproducing experimental backbone and SC conformations and stable MD simulations of two proteins. The resulting GBEMP CG force field is general and transferable, suitable for simulating biological macromolecules. The model is also suitable for "serial" or "parallel" multiscale AA/CG simulations. The GBEMP model is about 50 times faster compared to atomistic force fields with point charge models.

2.9. PRIME model

Hall and coworkers (Cheon, Chang, & Hall, 2010; Smith & Hall, 2000) have introduced an implicit solvent intermediate resolution protein model, namely PRIME based on the discontinuous molecular dynamics (Dokholyan, Buldyrev, Stanley, & Shakhnovich, 1998) methodology. Here, each amino acid is described by four beads: three beads for the backbone and a single bead for SC (Fig. 2). The backbone beads are comprised of the united atoms NH, $C_\alpha H$, and CO. Furthermore, SCs are distinguished into 14 groups depending on their size, charge, hydrophobicity, polarity, and potential for SC hydrogen bonding. All forces are modeled either by hard-sphere or by square-well potentials with realistic diameters. For instance, a square-well potential was employed to model the interactions between hydrophobic SCs while the backbone hydrogen bonding potential was represented by a directionally dependent square-well attraction between NH and CO. The resulting force field contains 19 energy parameters. These parameters were obtained by using a perceptron-learning algorithm (Krauth & Mezard, 1987) together with a modified stochastic learning algorithm that optimizes the energy gap between 711 known native states from the PDB and decoys generated by gapless threading.

The SC representation in PRIME is too coarse to correctly capture the SC entropy and the size effect for the packing of SCs that make critical contributions to protein folding. Therefore, Ding and coworkers (Ding, Buldyrev, & Dokholyan, 2005) have extended this model by incorporating one or two more additional effective SC atoms. For the β-branched amino acids (Thr, Ile, and Val), two beads representing the two branches after C_β were introduced while an additional bead was incorporated for bulky amino acids (Arg, Lys, and Trp).

3. HYBRID ALL-ATOM/COARSE-GRAINED (AA/CG) MODELING OF PROTEINS

Generally, AA/CG approaches may be divided into three categories (Ayton, Noid, & Voth, 2007; Kamerlin, Vicatos, Dryga, & Warshel, 2011; Kamerlin & Warshel, 2011): (1) Serial approaches where the information from fine-grained simulations are transferred to a CG model, and subsequently only the CG model is explored; (2) a tightly coupled approach where both levels are present simultaneously; and (3) a loosely coupled approach, where fine-grained and CG levels of resolution are alternated (see Fig. 3). The first type is essentially the idea behind Boltzmann inversion (Reith et al., 2003) and FM methods (Ayton et al., 2007; Izvekov & Voth, 2005a). A typical example of the second type may be alternating simulations at the CG and all-atom levels, where the CG simulations are used to identify an ensemble of possible conformational states that are then connected via fine-grained simulations to obtain accurate conformational PMFs (Fujisaki, Shiga, & Kidera, 2012; Fujisaki, Shiga, Moritsugu, & Kidera, 2013; Moritsugu, Terada, & Kidera, 2012). The third type would involve either resolution exchange simulations, with multiple replicas simulated at different resolutions (Liu & Voth, 2007; Lyman, Ytreberg, & Zuckerman, 2006; Lyman & Zuckerman, 2006), or hybrid models (Izvekov & Voth, 2005a; Noid, Chu, et al., 2008; Noid, Liu, et al., 2008), where part of a system is represented in atomistic detail while other parts remain at the CG level (Kamerlin & Warshel, 2011; Kamerlin et al., 2011; Murtola, Bunker, Vattulainen, Deserno, & Karttunen, 1869–1892; Peter & Kremer, 2009; Praprotnik, Delle Site, & Kremer, 2008). Such hybrid approaches are most attractive for detailed mechanistic studies of specific parts of a large system where an explicit description at the atomistic level

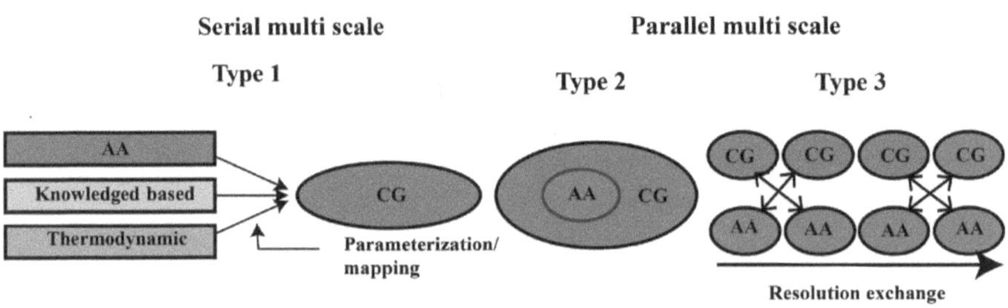

Figure 3 Schematic of the three different types of multiscale AA/CG approaches frequently employed in biomolecular simulations.

at the region of interest is vital but computational costs of simulating the entire complex in full atomistic detail are prohibitive.

In the first two approaches, a reverse mapping (Cheng, Gopal, Law, & Feig, 2012; Wassenaar, Pluhackova, Bockmann, Marrink, & Tieleman, 2014) scheme is employed once at the end (type 1) or infrequently (type 3) to reconstruct the high-resolution model from the corresponding low-resolution model. In the case of tightly coupled AA/CG simulations, the conversion between AA and CG regions is more critical as it has to be done frequently and has to maintain detailed balance when going back and forth between different resolutions. In the case of hybrid AA/CG models, the treatment at the boundary between the different regions becomes a key issue. The boundary may be treated as rigid (fixed-resolution AA/CG method; Michel, Orsi, & Essex, 2008; Neri, Anselmi, Carnevale, Vargiu, & Carloni, 2006; Orsi, Sanderson, & Essex, 2009; Shi, Izvekov, & Voth, 2006) analogous to the conventional hybrid quantum mechanics/MM (QM/MM; Lin & Truhlar, 2007; Warshel, 2003) approach or it may be open (adaptive AA/CG method; Potestio et al., 2013; Praprotnik et al., 2008). In the fixed-resolution AA/CG scheme, the coupling is continuous in time, and the atomistic and CG components of the system directly interact. A limiting case of the adaptive AA/CG scheme are the resolution exchange methods already mentioned above.

In this section, we will focus primarily on tightly coupled hybrid AA/CG models as they promise to be the most effective multiscale scheme in the foreseeable future. For an extended review on the multiscale modeling of proteins, readers are referred to recent reviews (Kamerlin & Warshel, 2011; Nielsen, Bulo, Moore, & Ensing, 2010; Sherwood, Brooks, & Sansom, 2008; Tozzini, 2010b; Zhou, 2014).

3.1. PACE model

Multiscale AA/CG methods are developed to reduce the computational cost while capturing the relevant interactions at the atomic level. Since the solvent degrees of freedom are much larger compared to the solute (protein) degrees of freedom, a significant speedup can be achieved already just by representing the solvent at the CG level while the solute is described at the atomistic level. Motivated by this idea, Schulten and coworkers have proposed protein in atomistic details coupled with CG environment (PACE; Han & Schulten, 2012). More specifically PACE combines a united atom protein model with solvent, water or lipids, represented with the

MARTINI model. The potential energy function of PACE is given by the following equation.

$$U = E_{\text{bond}} + E_{\text{angle}} + E_{\text{dihedral}} + E_{\text{improper}} + E_{\varphi,\psi,\chi_1} + E_{\text{CGW-CGW}} + E_{\text{CGW-UA}} + E_{\text{vdW}} + E_{\text{polar}} \quad (3)$$

The first four terms are generic bonded interaction potentials while the fifth term accounts for backbone and SC rotamers. The remaining terms capture nonbonded interactions that consist of standard Lennard–Jones 12-6 potentials for $E_{\text{CGW-CGW}}$ interactions between CG waters, for $E_{\text{CGW-UA}}$ interactions between CG water and protein sites, and for E_{vdW} interactions between nonpolar protein sites. The interactions between charged groups are modeled with a Coulomb potential using partial charges from the all-atom force field. The cross-resolution parameters were optimized by following the parameterization scheme of MARTINI, i.e., by reproducing experimental thermodynamic properties. Moreover, in the context of the hybrid force field, the interactions of atomic sites in proteins were reparameterized by using different reference data, such as PMF of polar interactions from atomistic simulations and statistical backbone potentials from a PDB coil library (Han & Schulten, 2012). The resulting hybrid model was able to reproduce the statistical backbone and SC potentials of all amino acids accurately. The force field was further validated with *ab initio* folding simulations of a series of peptides and proteins. For instance, PACE (Han & Schulten, 2012) has been employed to fold a 73-residue protein (α3D; Walsh, Cheng, Bryson, Roder, & DeGrado, 1999), which is a *de novo* designed three-helix-bundle protein. Starting from the denatured state, Han and Schulten (Han & Schulten, 2012) folded this protein to an average C_α RMSD of 3.4 Å in three extended (10–30 μs) simulations. This result is very close to the result (3.1 Å) obtained by Shaw and coworkers in fully atomistic simulations (Lindorff-Larsen, Piana, Dror, & Shaw, 2011).

While replacing only the solvent with a CG representation limits the computational gains that can be realized, this approach competes with already established implicit solvent models. Furthermore, the initial tests indicate that the stability of native proteins may be underestimated in PACE, suggesting that the balance between the atomistic protein model and CG solvent may be challenging. Nevertheless, recent studies have extended PACE further towards protein-membrane systems (Wan, Han, & Wu, 2012).

3.2. ATTRACT model

Zacharias et al. recently developed a multiscale AA/CG model (Zacharias, 2013) by combing a modified GROMOS united atom force field with the ATTRACT (Fiorucci & Zacharias, 2010) CG model for studying protein–protein interactions. In this model, bonded interactions are generic and are described by atomistic force field while nonbonded interactions are calculated via CG terms. Backbone atoms were retained in full atomic detail to allow accurate representations of hydrogen-bonded secondary structures and an explicit term for hydrogen bonds of main chain was incorporated to maintain a better secondary structures. In the multiscale scheme, all SC heavy atoms were treated as virtual atoms and nonbonded SC–SC as well as main chain–SC interactions were calculated via the ATTRACT CG model instead of standard atomistic Lennard–Jones and Coulomb terms. In ATTRACT, the backbone is represented by two pseudoatoms (N and O) and SCs are represented by one or two pseudo atoms depending on their size. Effective interactions between pseudo atoms are described by soft distance-dependent Lennard–Jones-type potentials. The attractive and repulsive parameters for each pseudo atom pair were iteratively optimized based on predictions of protein–protein complexes (Fiorucci & Zacharias, 2010). Hence, the parameterization of this model represents the effective interactions between pseudo centers at protein–protein interfaces. The parameters implicitly account for solvation effects, and no explicit solvent molecules were included during simulations. Promising results were obtained for protein–protein interactions with this hybrid methodology. However, the current scheme is of limited use for studies of internal protein dynamics. Although ATTRACT can be used to investigate the dynamics of protein–protein recognition process, computational efficiency is an issue.

3.3. Promising hybrid models

Rzepiela and coworkers (Rzepiela, Louhivuori, Peter, & Marrink, 2011) have proposed a mixed model between all-atom and MARTINI CG force fields where all-atom–CG interactions are replaced by CG interactions through dummy sites. This avoids the need for reparameterization of cross-resolution terms from the CG perspective but it does not fully address the environment seen by the atomistic part. Following this protocol, Sokkar, Choi, and Rhee (2013) found that the MARTINI force field couples too strongly with the atomistic model leading to complete unfolding of a test protein within a very short time.

In a similar effort, the Carloni group (Neri, Anselmi, Cascella, Maritan, & Carloni, 2005) has also developed a method in which the biologically relevant region of the protein is represented at the atomic level, while the rest of the protein is treated at the CG level, by only considering C_α centroids. The atomistic region is treated using a standard molecular force field, the CG region using a simplified Go potential and the interface region is also treated with an AA force field. This AA/CG scheme shows promising results for the HIV-1 protease (HIV-1 PR) and human β-secretase (BACE). As solvent–protein interactions are modeled in terms of viscosity and environment random forces in the framework of stochastic dynamics, this current model is only suitable for simulating globular proteins with buried active sites. Further improvements are necessary for modeling the solvent flow in the AA part.

The Pantano group has recently introduced a CG model for DNA (Dans, Zeida, Machado, & Pantano, 2010) and later combined it with the atomistic AMBER force field (Perez et al., 2007) for multiscale AA/CG simulations of DNA (Machado, Dans, & Pantano, 2011). Villa and coworkers developed a multiscale method for simulating protein–DNA complexes (Villa, Balaeff, Mahadevan, & Schulten, 2004) and conducted a multiscale simulation to investigate a complex between the lac repressor protein (Lacl) and a 107 bp DNA segment (Villa, Balaeff, & Schulten, 2005). The Lacl complex with the operator DNA segment is modeled at the level of atomic details while a continuum model is employed for the 75 bp DNA loop. Using this multiscale strategy, they have investigated the change in structural dynamics of the repressor due to the strained DNA loop.

Finally, PRIMO (Kar et al., 2013) is another promising CG model that can be used to carry out multiscale AA/CG simulations of proteins. Recently, Predeus and coworkers (Predeus et al., 2012) have employed PRIMO (Kar et al., 2013) and CHARMM (Brooks et al., 2009) force fields simultaneously for conformational sampling of trp-cage and melittin peptides in the presence of crowder proteins. A three-component multiscale scheme was adopted in their simulations. The peptides of interest were modeled with the atomistic CHARMM force field (Brooks et al., 2009) while the PRIMO force field was employed to model the crowder proteins and the solvent was treated implicitly via the GBMV (Lee et al., 2003) scheme. Their simulations yielded significant populations of unfolded structures of trp-cage and melittin in the presence of cellular crowders. It should be noted here that there were no direct bonded interactions between AA and CG regions in this multiscale AA/CG hybrid model. Both regions were interacting with each other via nonbonded interactions only. However, PRIMO can also be coupled with the atomistic force field to describe the direct bonded interactions between AA and CG atoms with minimum effort. The coupling can be achieved via two ways as shown in Fig. 4A and B. A crude way of combing PRIMO with CHARMM is to let the CG sites directly interact with the atomistic sites. One such example is shown in Fig. 4C. Here, 1–14 residues of trp-cage are modeled with the CHARMM force filed while residues from 15 to 20 are described by the PRIMO force field. The cross-resolution parameters can be taken from the existing PRIMO or CHARMM model as shown in Fig. 4D. An advantage of such a mixing scheme is that it avoids an extensive parameterization of the interfacial parameters.

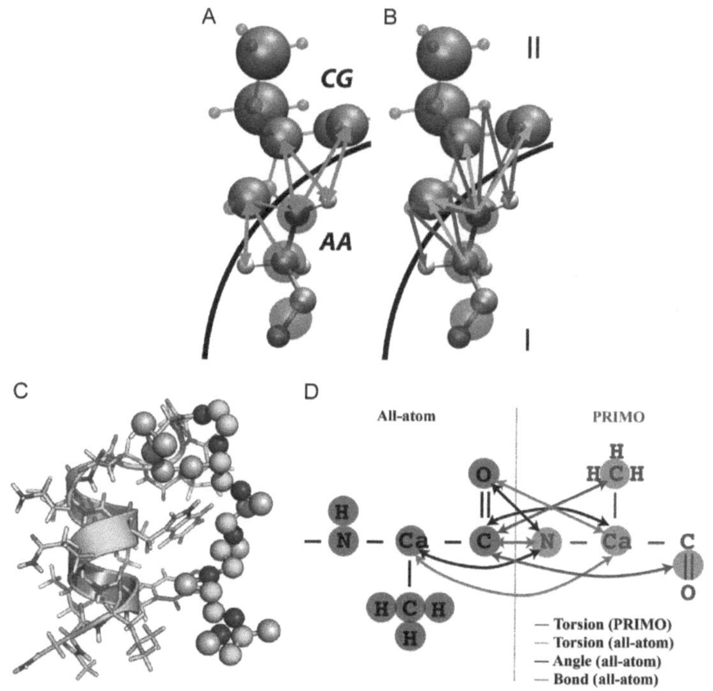

Figure 4 A schematic representation of coupling of PRIMO and CHARMM force fields when both parts interact with each other via bonded interactions: (A and B) Coupling schemes, (C) multiresolution representation of trp-cage where 1–14 residues are described by AA force field (stick), and (D) interfacial bonded parameters are taken from existing atomistic or coarse-grained force field (black, atomistic; red (gray in the print version), atomistic; brown (light gray in the print version), atomistic; and green (dark gray in the print version), PRIMO).

A sophisticated way of achieving multiscale coupling is shown in Fig. 4B. Here, the AA part is converted to CG on the fly and let it interacts with the corresponding CG site. Furthermore, on the CG side, virtual sites are constructed on the fly and let it interacts with the corresponding atomistic bead from the AA part. This is a very robust approach and can be easily implemented as the reverse mapping in PRIMO allows a near-atomistic accuracy of 0.1 Å.

4. OUTLOOK

During the last years we have seen the development of numerous CG models that are developed to study various biophysical processes. In this chapter, we have focused on those CG models that maximize transferability and are thus more universally applicable. We also discuss extensively multiscale hybrid AA/CG methods that rely on transferable CG models and are starting to become useful for practical applications. While we believe, that the use of CG models and hybrid AA/CG schemes will continue to grow in the future many limitations remain as described extensively, e.g., Saunders and Voth (2013). Still, even most 'transferable' CG models currently available are biased towards known data and therefore suffer from limitations when applied to a full range of current problems (protein dynamics, protein–protein interactions, folding, etc.), when considering different environments, when combined with different biomolecules, or under varying thermodynamic conditions. One question how to further address this challenge is how to balance bottom-up and top-down approaches in developing effective CG models as both have strengths and weaknesses. A mixed approach was adopted in the development of MARTINI force field with overall good success. However, there are also fundamental questions of what degree transferability is theoretically possible at a chosen level of resolution (Mullinax & Noid, 2010; Saunders & Voth, 2013).

The coupling of AA and CG models in hybrid multiscale schemes is still in its infancies. While first success stories have been reported, a robust strategy for tightly coupling AA and CG levels across an interface or between replicas at different resolutions is lacking. One of the outstanding issues is the preservation of detailed balance at both the AA and CG levels. Other questions are at what levels CG and AA models can be effectively combined since CG models that are too coarse are most certainly going to lead to problematic behavior when coupled directly to detailed atomistic potential.

Another topic that has remained largely unaddressed is how the dynamics resulting from CG and hybrid AA/CG models reproduces the correct kinetic behavior. For thermodynamic questions this can be neglected but as CG models are increasingly applied to cellular scales (Elcock, 2011; Feig & Sugita, 2013; Predeus et al., 2012) questions related to diffusive properties become more relevant (Thomas & Elcock, 2006, 2011). While it is expected that CG models generally exhibit accelerated kinetics, it is much less clear whether different types of motion are affected similarly. Even more problematic is how dynamic properties may be affected when mixing AA and CG models. Therefore, a future challenge will be to develop a CG model that is dynamically accurate as well as computationally efficient (Saunders & Voth, 2013).

5. CONCLUSIONS

In this chapter, we have reviewed recent advances in developing transferable, more universally applicable CG models of proteins. While there is clear progress in recent years, many challenges remain. Another promising area is the development of multiscale AA/CG hybrid methods that may benefit most from more transferable CG models. It is likely that with further increased accuracy and efficiency, CG and AA/CG schemes will play an even greater role in the future in addressing questions related to biomolecular dynamics and interactions on ever increasing spatiotemporal scales all the way to cellular scales.

ACKNOWLEDGMENT

This work was supported by National Institute of Health Grants GM084953 and GM092949.

REFERENCES

Arkhipov, A., Freddolino, P. L., Imada, K., Namba, K., & Schulten, K. (2006). Coarse-grained molecular dynamics simulations of a rotating bacterial flagellum. *Biophysical Journal, 91*, 4589–4597.

Ayton, G. S., Noid, W. G., & Voth, G. A. (2007). Multiscale modeling of biomolecular systems: In serial and in parallel. *Current Opinion in Structural Biology, 17*, 192–198.

Ayton, G. S., & Voth, G. A. (2009). Systematic multiscale simulation of membrane protein systems. *Current Opinion in Structural Biology, 19*, 138–144.

Ayton, G. S., & Voth, G. A. (2010). Multiscale computer simulation of the immature HIV-1 virion. *Biophysical Journal, 99*, 2757–2765.

Bahar, I., & Jernigan, R. L. (1997). Inter-residue potentials in globular proteins and the dominance of highly specific hydrophilic interactions at close separation. *Journal of Molecular Biology, 266*, 195–214.

Barducci, A., Bonomi, M., & Derreumaux, P. (2011). Assessing the quality of the OPEP coarse-grained force field. *Journal of Chemical Theory and Computation, 7*, 1928–1934.
Basdevant, N., Borgis, D., & Ha-Duong, T. (2013). Modeling protein–protein recognition in solution using the coarse-grained force field SCORPION. *Journal of Chemical Theory and Computation, 9*, 803–813.
Bereau, T., Bachmann, M., & Deserno, M. (2010). Interplay between secondary and tertiary structure formation in protein folding cooperativity. *Journal of the American Chemical Society, 132*, 13129–13131.
Bereau, T., & Deserno, M. (2009). Generic coarse-grained model for protein folding and aggregation. *Journal of Chemical Physics, 130*, 235106.
Best, R. B., Mittal, J., Feig, M., & MacKerell, A. D. (2012). Inclusion of many-body effects in the additive CHARMM protein CMAP potential results in enhanced cooperativity of alpha-Helix and beta-hairpin formation. *Biophysical Journal, 103*, 1045–1051.
Best, R. B., Zhu, X., Shim, J., Lopes, P. E. M., Mittal, J., Feig, M., et al. (2012). Optimization of the additive CHARMM all-atom protein force field targeting improved sampling of the backbone phi, psi and side-chain chi(1) and chi(2) dihedral angles. *Journal of Chemical Theory and Computation, 8*, 3257–3273.
Brini, E., Herbers, C. R., Deichmann, G., & van der Vegt, N. F. A. (2012). Thermodynamic transferability of coarse-grained potentials for polymer-additive systems. *Physical Chemistry Chemical Physics, 14*, 11896–11903.
Brini, E., & van der Vegt, N. F. A. (2012). Chemically transferable coarse-grained potentials from conditional reversible work calculations. *Journal of Chemical Physics, 137*, 154113.
Brooks, B. R., Brooks, C. L., Mackerell, A. D., Nilsson, L., Petrella, R. J., Roux, B., et al. (2009). CHARMM: The biomolecular simulation program. *Journal of Computational Chemistry, 30*, 1545–1614.
Brown, S., Fawzi, N. J., & Head-Gordon, T. (2003). Coarse-grained sequences for protein folding and design. *Proceedings of the National Academy of Sciences of the United States of America, 100*, 10712–10717.
Buchete, N. V., Straub, J. E., & Thirumalai, D. (2004). Continuous anisotropic representation of coarse-grained potentials for proteins by spherical harmonics synthesis. *Journal of Molecular Graphics and Modelling, 22*, 441–450.
Chaimovich, A., & Shell, M. S. (2011). Coarse-graining errors and numerical optimization using a relative entropy framework. *Journal of Chemical Physics, 134*, 094112.
Chebaro, Y., Mousseau, N., & Derreumaux, P. (2009). Structures and thermodynamics of Alzheimer's amyloid-beta A beta(16–35) monomer and dimer by replica exchange molecular dynamics simulations: Implication for full-length A beta fibrillation. *Journal of Physical Chemistry B, 113*, 7668–7675.
Chebaro, Y., Pasquali, S., & Derreumaux, P. (2012). The coarse-grained OPEP force field for non-amyloid and amyloid proteins. *Journal of Physical Chemistry B, 116*, 8741–8752.
Chen, N. Y., Su, Z. Y., & Mou, C. Y. (2006). Effective potentials for folding proteins. *Physical Reviews Letters, 96*, 078103.
Cheng, Y. M., Gopal, S. M., Law, S. M., & Feig, M. (2012). Molecular dynamics trajectory compression with a coarse-grained model. *IEEE ACM Transactions on Computational Biology and Bioinformatics, 9*, 476–486.
Cheon, M., Chang, I., & Hall, C. K. (2010). Extending the PRIME model for protein aggregation to all 20 amino acids. *Proteins, 78*, 2950–2960.
Chu, J. W., Izvekov, S., & Voth, G. A. (2006). The multiscale challenge for biomolecular systems: Coarse-grained modeling. *Molecular Simulation, 32*, 211–218.
Cornell, W. D., Cieplak, P., Bayly, C. I., Gould, I. R., Merz, K. M., Ferguson, D. M., et al. (1995). A 2nd generation force-field for the simulation of proteins, nucleic-acids, and organic-molecules. *Journal of the American Chemical Society, 117*, 5179–5197.

Dans, P. D., Zeida, A., Machado, M. R., & Pantano, S. (2010). A coarse grained model for atomic-detailed DNA simulations with explicit electrostatics. *Journal of Chemical Theory and Computation, 6*, 1711–1725.
de Jong, D. H., Periole, X., & Marrink, S. J. (2012). Dimerization of amino acid side chains: Lessons from the comparison of different force fields. *Journal of Chemical Theory and Computation, 8*, 1003–1014.
de Jong, D. H., Singh, G., Bennett, W. F. D., Arnarez, C., Wassenaar, T. A., Schafer, L. V., et al. (2013). Improved parameters for the martini coarse-grained protein force field. *Journal of Chemical Theory and Computation, 9*, 687–697.
Derreumaux, P. (1999). From polypeptide sequences to structures using Monte Carlo simulations and an optimized potential. *Journal of Chemical Physics, 111*, 2301–2310.
Ding, F., Borreguero, J. M., Buldyrey, S. V., Stanley, H. E., & Dokholyan, N. V. (2003). Mechanism for the alpha-helix to beta-hairpin transition. *Proteins: Structure, Function, and Genetics, 53*, 220–228.
Ding, F., Buldyrev, S. V., & Dokholyan, N. V. (2005). Folding Trp-cage to NMR resolution native structure using a coarse-grained protein model. *Biophysical Journal, 88*, 147–155.
Dokholyan, N. V., Buldyrev, S. V., Stanley, H. E., & Shakhnovich, E. I. (1998). Discrete molecular dynamics studies of the folding of a protein-like model. *Folding and Design, 3*, 577–587.
Elcock, A. H. (2010). Models of macromolecular crowding effects and the need for quantitative comparisons with experiment. *Current Opinion in Structural Biology, 20*, 196–206.
Feig, M., & Sugita, Y. (2013). Reaching new levels of realism in modeling biological macromolecules in cellular environments. *Journal of Molecular Graphics and Modelling, 45*, 144–156.
Fernandez, A., & Colubri, A. (2002). Pathway heterogeneity in protein folding. *Proteins: Structure, Function, and Genetics, 48*, 293–310.
Fiorucci, S., & Zacharias, M. (2010). Binding site prediction and improved scoring during flexible protein–protein docking with ATTRACT. *Proteins, 78*, 3131–3139.
Freddolino, P. L., Liu, F., Gruebele, M., & Schulten, K. (2008). Ten-microsecond molecular dynamics simulation of a fast-folding WW domain. *Biophysical Journal, 94*, L75–L77.
Fujisaki, H., Shiga, M., & Kidera, A. (2012). A multi scale approach for path sampling: Applications to peptides. *Biophysical Journal, 102*, 22.
Fujisaki, H., Shiga, M., Moritsugu, K., & Kidera, A. (2013). Multiscale enhanced path sampling based on the Onsager–Machlup action: Application to a model polymer. *Journal of Chemical Physics, 139*, 054117.
Fujitsuka, Y., Takada, S., Luthey-Schulten, Z. A., & Wolynes, P. G. (2004). Optimizing physical energy functions for protein folding. *Proteins: Structure, Function, and Genetics, 54*, 88–103.
Ghavami, A., van der Giessen, E., & Onck, P. R. (2013). Coarse-grained potentials for local interactions in unfolded proteins. *Journal of Chemical Theory and Computation, 9*, 432–440.
Gopal, S. M., Mukherjee, S., Cheng, Y. M., & Feig, M. (2010). PRIMO/PRIMONA: A coarse-grained model for proteins and nucleic acids that preserves near-atomistic accuracy. *Proteins, 78*, 1266–1281.
Gu, J. F., Bai, F., Li, H. L., & Wang, X. C. (2012). A Generic force field for protein coarse-grained molecular dynamics simulation. *International Journal of Molecular Sciences, 13*, 14451–14469.
Han, W., & Schulten, K. (2012). Further optimization of a hybrid United-atom and coarse-grained force field for folding simulations: Improved backbone hydration and interactions between charged side chains. *Journal of Chemical Theory and Computation, 8*, 4413–4424.

Harada, R., Sugita, Y., & Feig, M. (2012). Protein crowding affects hydration structure and dynamics. *Journal of the American Chemical Society*, *134*, 4842–4849.
Harada, R., Tochio, N., Kigawa, T., Sugita, Y., & Feig, M. (2013). Reduced native state stability in crowded cellular environment due to protein–protein interactions. *Journal of the American Chemical Society*, *135*, 3696–3701.
Head-Gordon, T., & Brown, S. (2003). Minimalist models for protein folding and design. *Current Opinion in Structural Biology*, *13*, 160–167.
Henderso, R. l. (1974). Uniqueness theorem for fluid pair correlation-functions. *Physics Letters A*, *49*, 197–198.
Hills, R. D., Jr., Lu, L., & Voth, G. A. (2010). Multiscale coarse-graining of the protein energy landscape. *PLoS Computational Biology*, *6*, e1000827.
Hoefling, M., & Gottschalk, K. E. (2010). Barnase–Barstar: From first encounter to final complex. *Journal of Structural Biology*, *171*, 52–63.
Honeycutt, J. D., & Thirumalai, D. (1990). Metastability of the folded states of globular-proteins. *Proceedings of the National Academy of Sciences of the United States of America*, *87*, 3526–3529.
Ingolfsson, H. I. L., Lopez, C. A., Uusitalo, J. J., de Jong, D. H., Gopal, S., Periole, X., et al. (2014). The power of coarse-graining in biomolecular simulations. *WIREs Computational Molecular Science*, *4*, 225–248.
Irback, A., Sjunnesson, F., & Wallin, S. (2000). Three-helix-bundle protein in a Ramachandran model. *Proceedings of the National Academy of Sciences of the United States of America*, *97*, 13614–13618.
Izvekov, S., Chung, P. W., & Rice, B. M. (2010). The multiscale coarse-graining method: Assessing its accuracy and introducing density dependent coarse-grain potentials. *Journal of Chemical Physics*, *133*, 064109.
Izvekov, S., & Voth, G. A. (2005a). A multiscale coarse-graining method for biomolecular systems. *Journal of Physical Chemistry B*, *109*, 2469–2473.
Izvekov, S., & Voth, G. A. (2005b). Multiscale coarse graining of liquid-state systems. *Journal of Chemical Physics*, *123*, 134105.
Kamerlin, S. C. L., Vicatos, S., Dryga, A., & Warshel, A. (2011). Coarse-grained (multiscale) simulations in studies of biophysical and chemical systems. *Annual Review of Physical Chemistry*, *62*, 41–64.
Kamerlin, S. C. L., & Warshel, A. (2011). Multiscale modeling of biological functions. *Physical Chemistry Chemical Physics*, *13*, 10401–10411.
Kar, P., Gopal, S. M., Cheng, Y.-M., Panahi, A., & Feig, M. (2014). Transferring the PRIMO coarse-grained force field to the membrane environment: Simulations of membrane proteins and helix–helix association. *Journal of Chemical Theory and Computation*. http://dx.doi.org/10.1021/ct500443v.
Kar, P., Gopal, S. M., Cheng, Y. M., Predeus, A., & Feig, M. (2013). PRIMO: A transferable coarse-grained force field for proteins. *Journal of Chemical Theory and Computation*, *9*, 3769–3788.
Karplus, M., & McCammon, J. A. (2002). Molecular dynamics simulations of biomolecules. *Nature Structural Biology*, *9*, 646–652.
Khalili, M., Liwo, A., Rakowski, F., Grochowski, P., & Scheraga, H. A. (2005). Molecular dynamics with the united-residue model of polypeptide chains. I. Lagrange equations of motion and tests of numerical stability in the microcanonical mode. *Journal of Physical Chemistry B*, *109*, 13785–13797.
Kolinski, A. (2004). Protein modeling and structure prediction with a reduced representation. *Acta Biochimica Polonica*, *51*, 349–371.
Korkuta, A., & Hendrickson, W. A. (2009). A force field for virtual atom molecular mechanics of proteins. *Proceedings of the National Academy of Sciences of the United States of America*, *106*, 15667–15672.

Krauth, W., & Mezard, M. (1987). Learning algorithms with optimal stability in neural networks. *Journal of Physics A: Mathematical and General, 20*, L745–L752.

Krishna, V., Ayton, G. S., & Voth, G. A. (2010). Role of protein interactions in defining HIV-1 viral capsid shape and stability: A coarse-grained analysis. *Biophysical Journal, 98*, 18–26.

Law, S. M., & Feig, M. (2011). Base-flipping mechanism in postmismatch recognition by MutS. *Biophysical Journal, 101*, 2223–2231.

Lee, M. S., Feig, M., Salsbury, F. R., & Brooks, C. L. (2003). New analytic approximation to the standard molecular volume definition and its application to generalized born calculations. *Journal of Computational Chemistry, 24*, 1348, 24, 1821-1821.

Lee, J., Liwo, A., & Scheraga, H. A. (1999). Energy-based de novo protein folding by conformational space annealing and an off-lattice united-residue force field: Application to the 10–55 fragment of staphylococcal protein A and to apo calbindin D9K. *Proceedings of the National Academy of Sciences of the United States of America, 96*, 2025–2030.

Levitt, M., & Warshel, A. (1975). Computer-simulation of protein folding. *Nature, 253*, 694–698.

Limbach, H. J. A., Arnold, A., Mann, B. A., & Holm, C. (2006). ESPRESSO—An extensible simulation package for research on soft matter systems. *Computer Physics Communications, 174*, 704–727.

Lin, H., & Truhlar, D. G. (2007). QM/MM: What have we learned, where are we, and where do we go from here? *Theoretical Chemistry Accounts, 117*, 185–199.

Lindorff-Larsen, K., Piana, S., Dror, R. O., & Shaw, D. E. (2011). How fast-folding proteins fold. *Science, 334*, 517–520.

Liu, P., & Voth, G. A. (2007). Smart resolution replica exchange: An efficient algorithm for exploring complex energy landscapes. *Journal of Chemical Physics, 126*, 045106.

Liwo, A., Arlukowicz, P., Czaplewski, C., Oldziej, S., Pillardy, J., & Scheraga, H. A. (2002). A method for optimizing potential-energy functions by a hierarchical design of the potential-energy landscape: Application to the UNRES force field. *Proceedings of the National Academy of Sciences of the United States of America, 99*, 1937–1942.

Liwo, A., Khalili, M., & Scheraga, H. A. (2005). Ab initio simulations of protein folding pathways by molecular dynamics with the united-residue (UNRES) model of polypeptide chains. *The FEBS Journal, 272*, 359–360.

Liwo, A., Oldziej, S., Kazmierkiewicz, R., Groth, M., & Czaplewski, C. (1997). Design of a knowledge-based force field for off-lattice simulations of protein structure. *Acta Biochimica Polonica, 44*, 527–547.

Liwo, A., Oldziej, S., Pincus, M. R., Wawak, R. J., Rackovsky, S., & Scheraga, H. A. (1997). A united-residue force field for off-lattice protein-structure simulations. 1. Functional forms and parameters of long-range side-chain interaction potentials from protein crystal data. *Journal of Computational Chemistry, 18*, 849–873.

Liwo, A., Pincus, M. R., Wawak, R. J., Rackovsky, S., Oldziej, S., & Scheraga, H. A. (1997). A united-residue force field for off-lattice protein-structure simulations. 2. Parameterization of short-range interactions and determination of weights of energy terms by Z-score optimization. *Journal of Computational Chemistry, 18*, 874–887.

Lyman, E., Ytreberg, F. M., & Zuckerman, D. M. (2006). Resolution exchange simulation. *Physical Reviews Letters, 96*, 028105.

Lyman, E., & Zuckerman, D. M. (2006). Resolution exchange simulation with incremental coarsening. *Journal of Chemical Theory and Computation, 2*, 656–666.

Lyubartsev, A. P., & Laaksonen, A. (1995). Calculation of effective interaction potentials from radial-distribution functions—A reverse Monte-Carlo approach. *Physical Review E, 52*, 3730–3737.

Machado, M. R., Dans, P. D., & Pantano, S. (2011). A hybrid all-atom/coarse grain model for multiscale simulations of DNA. *Physical Chemistry Chemical Physics, 13*, 18134–18144.

MacKerell, A. D., Bashford, D., Bellott, M., Dunbrack, R. L., Evanseck, J. D., Field, M. J., et al. (1998). All-atom empirical potential for molecular modeling and dynamics studies of proteins. *Journal of Physical Chemistry B, 102*, 3586–3616.

MacKerell, A. D., Feig, M., & Brooks, C. L. (2004). Improved treatment of the protein backbone in empirical force fields. *Journal of the American Chemical Society, 126*, 698–699.

Maisuradze, G. G., Liwo, A., & Scheraga, H. A. (2010). Relation between free energy landscapes of proteins and dynamics. *Journal of Chemical Theory and Computation, 6*, 583–595.

Maisuradze, G. G., Senet, P., Czaplewski, C., Liwo, A., & Scheraga, H. A. (2010). Investigation of protein folding by coarse-grained molecular dynamics with the UNRES force field. *The Journal of Physical Chemistry. A, 114*, 4471–4485.

Majek, P., & Elber, R. (2009). A coarse-grained potential for fold recognition and molecular dynamics simulations of proteins. *Proteins, 76*, 822–836.

Marrink, S. J., de Vries, A. H., Harroun, T. A., Katsaras, J., & Wassall, S. R. (2008). Cholesterol shows preference for the interior of polyunsaturated lipid membranes. *Journal of the American Chemical Society, 130*, 10–11.

Marrink, S. J., de Vries, A. H., & Mark, A. E. (2004). Coarse grained model for semiquantitative lipid simulations. *Journal of Physical Chemistry B, 108*, 750–760.

Marrink, S. J., de Vries, A. H., & Tieleman, D. P. (2009). Lipids on the move: Simulations of membrane pores, domains, stalks and curves. *Biochimica et Biophysica Acta, 1788*, 149–168.

Marrink, S. J., Risselada, H. J., Yefimov, S., Tieleman, D. P., & de Vries, A. H. (2007). The MARTINI force field: Coarse grained model for biomolecular simulations. *Journal of Physical Chemistry B, 111*, 7812–7824.

Marrink, S. J., & Tieleman, D. P. (2013). Perspective on the Martini model. *Chemical Society Reviews, 42*, 6801–6822.

Maupetit, J., Derreumaux, P., & Tuffery, P. (2009). PEP-FOLD: An online resource for de novo peptide structure prediction. *Nucleic Acids Research, 37*, W498–W503.

Maupetit, J., Tuffery, P., & Derreumaux, P. (2007). A coarse-grained protein force field for folding and structure prediction. *Proteins, 69*, 394–408.

Michel, J., Orsi, M., & Essex, J. W. (2008). Prediction of partition coefficients by multiscale hybrid atomic-level/coarse-grain simulations. *Journal of Physical Chemistry B, 112*, 657–660.

Miyazawa, S., & Jernigan, R. L. (1985). Estimation of effective interresidue contact energies from protein crystal-structures—Quasi-chemical approximation. *Macromolecules, 18*, 534–552.

Monticelli, L., Kandasamy, S. K., Periole, X., Larson, R. G., Tieleman, D. P., & Marrink, S. J. (2008). The MARTINI coarse-grained force field: Extension to proteins. *Journal of Chemical Theory and Computation, 4*, 819–834.

Moritsugu, K., Terada, T., & Kidera, A. (2012). Disorder-to-order transition of an intrinsically disordered region of sortase revealed by multiscale enhanced sampling. *Journal of the American Chemical Society, 134*, 7094–7101.

Mu, Y., & Gao, Y. Q. (2007). Effects of hydrophobic and dipole–dipole interactions on the conformational transitions of a model polypeptide. *Journal of Chemical Physics, 127*, 105102.

Mukherjee, A., & Bagchi, B. (2004). Contact pair dynamics during folding of two small proteins: Chicken villin head piece and the Alzheimer protein beta-amyloid. *Journal of Chemical Physics, 120*, 1602–1612.

Mullinax, J. W., & Noid, W. G. (2010). Recovering physical potentials from a model protein databank. *Proceedings of the National Academy of Sciences of the United States of America, 107*, 19867–19872.

Murtola, T., Bunker, A., Vattulainen, I., Deserno, M., & Karttunen, M. (1869–1892). Multiscale modeling of emergent materials: Biological and soft matter. *Physical Chemistry Chemical Physics, 2009*, 11.

Nanias, M., Chinchio, M., Oldziej, S., Czaplewski, C., & Scheraga, H. A. (2005). Protein structure prediction with the UNRES force-field using replica-exchange Monte Carlo-with-minimization; comparison with MCM, CSA, and CFMC. *Journal of Computational Chemistry*, *26*, 1472–1486.

Neri, M., Anselmi, C., Carnevale, V., Vargiu, A. V., & Carloni, P. (2006). Molecular dynamics simulations of outer-membrane protease T from E-coli based on a hybrid coarse-grained/atomistic potential. *Journal of Physics. Condensed Matter*, *18*, S347–S355.

Neri, M., Anselmi, C., Cascella, M., Maritan, A., & Carloni, P. (2005). Coarse-grained model of proteins incorporating atomistic detail of the active site. *Physical Reviews Letters*, *95*, 218102.

Nielsen, S. O., Bulo, R. E., Moore, P. B., & Ensing, B. (2010). Recent progress in adaptive multiscale molecular dynamics simulations of soft matter. *Physical Chemistry Chemical Physics*, *12*, 12401–12414.

Noid, W. G., Chu, J. W., Ayton, G. S., Krishna, V., Izvekov, S., Voth, G. A., et al. (2008). The multiscale coarse-graining method. I. A rigorous bridge between atomistic and coarse-grained models. *Journal of Chemical Physics*, *128*, 244114.

Noid, W. G., Liu, P., Wang, Y., Chu, J. W., Ayton, G. S., Izvekov, S., et al. (2008). The multiscale coarse-graining method. II. Numerical implementation for coarse-grained molecular models. *Journal of Chemical Physics*, *128*, 244115.

Oldziej, S., Czaplewski, C., Liwo, A., Chinchio, M., Nanias, M., Vila, J. A., et al. (2005). Physics-based protein-structure prediction using a hierarchical protocol based on the UNRES force field: Assessment in two blind tests. *Proceedings of the National Academy of Sciences of the United States of America*, *102*, 7547–7552.

Orsi, M., Sanderson, W. E., & Essex, J. W. (2009). Permeability of small molecules through a lipid bilayer: A multiscale simulation study. *Journal of Physical Chemistry B*, *113*, 12019–12029.

Pasi, M., Lavery, R., & Ceres, N. (2013). PaLaCe: A coarse-grain protein model for studying mechanical properties. *Journal of Chemical Theory and Computation*, *9*, 785–793.

Perez, A., Marchan, I., Svozil, D., Sponer, J., Cheatham, T. E., Laughton, C. A., et al. (2007). Refinenement of the AMBER force field for nucleic acids: Improving the description of alpha/gamma conformers. *Biophysical Journal*, *92*, 3817–3829.

Periole, X., Huber, T., Marrink, S. J., & Sakmar, T. P. (2007). G protein-coupled receptors self-assemble in dynamics simulations of model bilayers. *Journal of the American Chemical Society*, *129*, 10126–10132.

Periole, X., Knepp, A. M., Sakmar, T. P., Marrink, S. J., & Huber, T. (2012). Structural determinants of the supramolecular organization of G protein-coupled receptors in bilayers. *Journal of the American Chemical Society*, *134*, 10959–10965.

Peter, C., & Kremer, K. (2009). Multiscale simulation of soft matter systems—From the atomistic to the coarse-grained level and back. *Soft Matter*, *5*, 4357–4366.

Potestio, R., Fritsch, S., Espanol, P., Delgado-Buscalioni, R., Kremer, K., Everaers, R., et al. (2013). Hamiltonian adaptive resolution simulation for molecular liquids. *Physical Reviews Letters*, *110*, 108301.

Praprotnik, M., Delle Site, L., & Kremer, K. (2008). Multiscale simulation of soft matter: From scale bridging to adaptive resolution. *Annual Review of Physical Chemistry*, *59*, 545–571.

Predeus, A. V., Gul, S., Gopal, S. M., & Feig, M. (2012). Conformational sampling of peptides in the presence of protein crowders from AA/CG-multiscale simulations. *Journal of Physical Chemistry B*, *116*, 8610–8620.

Reith, D., Putz, M., & Muller-Plathe, F. (2003). Deriving effective mesoscale potentials from atomistic simulations. *Journal of Computational Chemistry*, *24*, 1624–1636.

Risselada, H. J., & Marrink, S. J. (2008). The molecular face of lipid rafts in model membranes. *Proceedings of the National Academy of Sciences of the United States of America*, *105*, 17367–17372.

Rojas, A. V., Liwo, A., & Scheraga, H. A. (2007). Molecular dynamics with the united-residue force field: Ab initio folding simulations of multichain proteins. *Journal of Physical Chemistry B*, *111*, 293–309.
Rzepiela, A. J., Louhivuori, M., Peter, C., & Marrink, S. J. (2011). Hybrid simulations: Combining atomistic and coarse-grained force fields using virtual sites. *Physical Chemistry Chemical Physics*, *13*, 10437–10448.
Saunders, M. G., & Voth, G. A. (2013). Coarse-graining methods for computational biology. *Annual Review of Biophysics*, *42*, 73–93.
Schnieders, M. J., & Ponder, J. W. (2007). Polarizable atomic multipole solutes in a generalized Kirkwood continuum. *Journal of Chemical Theory and Computation*, *3*, 2083–2097.
Sengupta, D., Leontiadou, H., Mark, A. E., & Marrink, S. J. (2008). Toroidal pores formed by antimicrobial peptides show significant disorder. *Biochimica et biophysica Acta (BBA) – Biomembranes*, *178*, 2308–2317.
Sharma, M. P., Predeus, A. V., Kovacs, N., & Feig, M. (2014). MutS differential mismatch recognition specificities of eukaryotic MutS homologs, MutSα and MutSβ. *Biophysical Journal*, *106*, 2483–2492.
Shaw, D. E., Maragakis, P., Lindorff-Larsen, K., Piana, S., Dror, R. O., Eastwood, M. P., et al. (2010). Atomic-level characterization of the structural dynamics of proteins. *Science*, *330*, 341–346.
Shen, H. J., Li, Y., Ren, P. Y., Zhang, D. L., & Li, G. H. (2014). Anisotropic coarse-grained model for proteins based on Gay–Berne and electric multipole potentials. *Journal of Chemical Theory and Computation*, *10*, 731–750.
Sherwood, P., Brooks, B. R., & Sansom, M. S. P. (2008). Multiscale methods for macromolecular simulations. *Current Opinion in Structural Biology*, *18*, 630–640.
Shi, Q., Izvekov, S., & Voth, G. A. (2006). Mixed atomistic and coarse-grained molecular dynamics: Simulation of a membrane-bound ion channel. *Journal of Physical Chemistry B*, *110*, 15045–15048.
Singh, G., & Tieleman, D. P. (2011). Using the Wimley-white hydrophobicity scale as a direct quantitative test of force fields: The MARTINI coarse-grained model. *Journal of Chemical Theory and Computation*, *7*, 2316–2324.
Sinitskiy, A. V., Saunders, M. G., & Voth, G. A. (2012). Optimal number of coarse-grained sites in different components of large biomolecular complexes. *Journal of Physical Chemistry B*, *116*, 8363–8374.
Smirnova, Y. G., Marrink, S. J., Lipowsky, R., & Knecht, V. (2010). Solvent-exposed tails as prestalk transition states for membrane fusion at low hydration. *Journal of the American Chemical Society*, *132*, 6710–6718.
Smith, A. V., & Hall, C. K. (2000). Bridging the gap between homopolymer and protein models: A discontinuous molecular dynamics study. *Journal of Chemical Physics*, *113*, 9331–9342.
Sokkar, P., Choi, S. M., & Rhee, Y. M. (2013). Simple method for simulating the mixture of atomistic and coarse-grained molecular systems. *Journal of Chemical Theory and Computation*, *9*, 3728–3739.
Solernou, A., & Fernandez-Recio, J. (2011). pyDockCG: New coarse-grained potential for protein protein docking. *Journal of Physical Chemistry B*, *115*, 6032–6039.
Spiga, E., Alemani, D., Degiacomi, M. T., Cascella, M., & Dal Peraro, M. (2013). Electrostatic-consistent coarse-grained potentials for molecular simulations of proteins. *Journal of Chemical Theory and Computation*, *9*, 3515–3526.
Sterpone, F., Melchionna, S., Tuffery, P., Pasquali, S., Mousseau, N., Cragnolini, T., et al. (2014). The OPEP protein model: From single molecules, amyloid formation, crowding and hydrodynamics to DNA/RNA systems. *Chemical Society Reviews*, *43*, 4871–4893.
Sterpone, F., Nguyen, P. H., Kalimeri, M., & Derreumaux, P. (2013). Importance of the ion–pair interactions in the OPEP coarse-grained force field: Parameterization and validation. *Journal of Chemical Theory and Computation*, *9*, 4574–4584.

Takada, S. (2012). Coarse-grained molecular simulations of large biomolecules. *Current Opinion in Structural Biology, 22*, 130–137.

Takada, S., Luthey-Schulten, Z., & Wolynes, P. G. (1999). Folding dynamics with nonadditive forces: A simulation study of a designed helical protein and a random heteropolymer. *Journal of Chemical Physics, 110*, 11616–11629.

Tanizaki, S., & Feig, M. (2005). A generalized Born formalism for heterogeneous dielectric environments: Application to the implicit modeling of biological membranes. *Journal of Chemical Physics, 122*, 124706.

Tanizaki, S., & Feig, M. (2006). Molecular dynamics simulations of large integral membrane proteins with an implicit membrane model. *Journal of Physical Chemistry B, 110*, 548–556.

Thomas, A. S., & Elcock, A. H. (2006). Direct observation of salt effects on molecular interactions through explicit-solvent molecular dynamics simulations: Differential effects on electrostatic and hydrophobic interactions and comparisons to Poisson–Boltzmann theory. *Journal of the American Chemical Society, 128*, 7796–7806.

Thomas, A. S., & Elcock, A. H. (2011). Direct measurement of the kinetics and thermodynamics of association of hydrophobic molecules from molecular dynamics simulations. *Journal of Physical Chemistry Letters, 2*, 19–24.

Thorpe, I. F., Zhou, J., & Voth, G. A. (2008). Peptide folding using multiscale coarse-grained models. *Journal of Physical Chemistry B, 112*, 13079–13090.

Tozzini, V. (2005). Coarse-grained models for proteins. *Current Opinion in Structural Biology, 15*, 144–150.

Tozzini, V. (2010a). Minimalist models for proteins: A comparative analysis. *Quarterly Reviews of Biophysics, 43*, 333–371.

Tozzini, V. (2010b). Multiscale modeling of proteins. *Accounts of Chemical Research, 43*, 220–230.

Tozzini, V., & McCammon, J. A. (2005). A coarse grained model for the dynamics of flap opening in HIV-1 protease. *Chemical Physics Letters, 413*, 123–128.

Treptow, W., Marrink, S. J., & Tarek, M. (2008). Gating motions in voltage-gated potassium channels revealed by coarse-grained molecular dynamics simulations. *Journal of Physical Chemistry B, 112*, 3277–3282.

Villa, E., Balaeff, A., Mahadevan, L., & Schulten, K. (2004). Multiscale method for simulating protein–DNA complexes. *Multiscale Modeling and Simulation, 2*, 527–553.

Villa, E., Balaeff, A., & Schulten, K. (2005). Structural dynamics of the lac repressor-DNA complex revealed by a multiscale simulation. *Proceedings of the National Academy of Sciences of the United States of America, 102*, 6783–6788.

Walsh, S. T. R., Cheng, H., Bryson, J. W., Roder, H., & DeGrado, W. F. (1999). Solution structure and dynamics of a de novo designed three-helix bundle protein. *Proceedings of the National Academy of Sciences of the United States of America, 96*, 5486–5491.

Wan, C. K., Han, W., & Wu, Y. D. (2012). Parameterization of PACE force field for membrane environment and simulation of helical peptides and helix–helix association. *Journal of Chemical Theory and Computation, 8*, 300–313.

Wang, B. B., Predeus, A. V., Burton, Z. F., & Feig, M. (2013). Energetic and structural details of the trigger-loop closing transition in RNA polymerase II. *Biophysical Journal, 105*, 767–775.

Warshel, A. (2003). Computer simulations of enzyme catalysis: Methods, progress, and insights. *Annual Review of Biophysics and Biomolecular Structure, 32*, 425–443.

Wassenaar, T. A., Pluhackova, K., Bockmann, R. A., Marrink, S. J., & Tieleman, D. P. (2014). Going backward: A flexible geometric approach to reverse transformation from coarse grained to atomistic models. *Journal of Chemical Theory and Computation, 10*, 676–690.

Weeks, J. D. C., Chandler, D., & Andersen, H. C. (1971). Role of repulsive forces in determining the equilibrium structure of simple liquids. *Journal of Chemical Physics, 54*, 5237–5247.

Wu, Z., Cui, Q. A., & Yethiraj, A. (2010). A new coarse-grained model for water: The importance of electrostatic interactions. *Journal of Physical Chemistry B*, *114*, 10524–10529.

Yesylevskyy, S. O., Schafer, L. V., Sengupta, D., & Marrink, S. J. (2010). Polarizable water model for the coarse-grained MARTINI force field. *PLoS Computational Biology*, *6*, e1000810.

Zacharias, M. (2013). Combining coarse-grained nonbonded and atomistic bonded interactions for protein modeling. *Proteins*, *81*, 81–92.

Zhou, H.-X. (2014). Theoretical frameworks for multiscale modeling and simulation. *Current Opinion in Structural Biology*, *25*, 67–76.

CHAPTER SIX

Studying Allosteric Regulation in Metal Sensor Proteins Using Computational Methods

Dhruva K. Chakravorty*[,1], Kenneth M. Merz Jr.[†,‡]
*Department of Chemistry, University of New Orleans, New Orleans, Louisiana, USA
†Department of Chemistry, Michigan State University, East Lansing, Michigan, USA
‡Department of Biochemistry and Molecular Biology, Michigan State University, East Lansing, Michigan, USA
[1]Corresponding author: e-mail address: dhruva.chakravorty@uno.edu

Contents

Abstract

In this chapter, we describe advances made in understanding the mechanism of allosteric regulation of DNA operator binding in the ArsR/SmtB family of metal-sensing proteins using computational methods. The paradigm, zinc-sensing transcriptional

Advances in Protein Chemistry and Structural Biology, Volume 96
ISSN 1876-1623
http://dx.doi.org/10.1016/bs.apcsb.2014.06.009

repressor *Staphylococcus aureus* CzrA represents an excellent model system to understand how metal sensor proteins maintain cellular metal homeostasis. Here, we discuss studies that helped to characterize a metal ion-mediated hydrogen-bonding pathway (HBP) that plays a dominant role in the allosteric mechanism of DNA operator binding in these proteins. The chapter discusses computational methods used to provide a molecular basis for the large conformational motions and allosteric coupling free energy (~6 kcal/mol) associated with Zn(II) binding in CzrA. We present an accurate and convenient means by which to include metal ions in the nuclear magnetic resonance (NMR) structure determination process using molecular dynamics (MD) constrained by NMR-derived data. The method provides a realistic and physically viable description of the metal-binding site(s) and has potentially broad applicability in the structure determination of metal ion-bound proteins, protein folding, and metal template protein-design studies. Finally, our simulations provide strong support for a proposed HBP that physically connects the metal-binding residue, His97, to the DNA-binding interface through the αR helix that is present only in the Zn(II)-bound state. We find the interprotomer hydrogen bond interaction to be significantly stronger (~8 kcal/mol) at functional allosteric metal-binding sites compared to the *apo* proteins. This interaction works to overcome the considerable disorder at these hydrogen-bonding sites in *apo* protein and functions as a "switch" to lock in a weak DNA-binding conformation once metal is bound. This interaction is found to be considerably weaker in nonresponsive metal-binding sites. These findings suggest a conserved functional role of metal-mediated second-shell coordination hydrogen bonds at allosterically responsive sites in zinc-sensing transcription regulators.

ABBREVIATIONS

AdcR *Mycobacaterium tuberculosis* AdcR
CadC *Mycobacaterium tuberculosis* CadC
CzrA *Staphylococcus aureus* CzrA
DFT density functional theory
ESP electrostatic surface potential
HBP hydrogen-bonding pathway
HSQC spectroscopy heteronuclear single quantum coherence spectroscopy
MD molecular dynamics
MH metal ion homeostasis
MRD-NMR Metal ion restrained dynamics-NMR structure refinement method
NMR nuclear magnetic resonance
NmtR *Mycobacaterium tuberculosis* NmtR
PDB protein data bank
PMF potential of mean force
QM/MM quantum-mechanical molecular-mechanical
QM/MM MD quantum-mechanical molecular-mechanical molecular dynamics
RMSF root mean square fluctuations
SAPT symmetry adapted perturbation theory
SCC-DFTB selfconsistent charge density functional tight-binding
SmtB *Mycobacaterium tuberculosis* SmtB
XAS X-ray absorption spectroscopy

1. INTRODUCTION

Metal ions play a central role in a diverse range of cellular functions (Thomson & Gray, 1998; Tus, Rakipovic, Peretin, Tomic, & Sikic, 2012; Waldron & Robinson, 2009; Waldron, Rutherford, Ford, & Robinson, 2009). These processes in turn depend on the complex machinery of metal ion homeostasis (MH) that helps maintain optimal intracellular concentrations of metal ions and is critical for the survival of living organisms (Giedroc & Arunkumar, 2007). Metal ion-sensing proteins are critical components of the MH machinery. These proteins perform the crucial function of regulating the expression of proteins that are involved in the uptake, transfer and efflux of metal ions, and are consequently referred to as transcriptional regulator proteins. As such, transcriptional regulators may act as activators or repressors of protein expression (Giedroc & Arunkumar, 2007). The ArsR/SmtB family of metal ion-sensing transcriptional repressor proteins is widely distributed in nature and helps provide heavy metal resistance in pathogenic prokaryotic organisms like *Staphylococcus aureus*, *Mycobacterium tuberculosis*, *Listeria monocytogenes*, and *Escherichia coli* (Giedroc & Arunkumar, 2007). ArsR/SmtB family proteins function as metalloregulators by a mechanism of allosteric inhibition (Arunkumar, Campanello, & Giedroc, 2009; Giedroc & Arunkumar, 2007; Grossoehme & Giedroc, 2009). In this mechanism, metal ion binding at a distant metal ion-binding site triggers a conformational change in the protein which in turn reduces the protein's DNA-binding affinity by up to 6 kcal/mol (Arunkumar et al., 2009; Giedroc & Arunkumar, 2007; Grossoehme & Giedroc, 2009; Ma, Jacobsen, & Giedroc, 2009). Members of this protein family have evolved from a common ancestry and are characterized by a helix-turn-helix fold common to transcriptional regulator proteins (Pennella, Shokes, Cosper, Scott, & Giedroc, 2003; Ramos et al., 2005). While insights into the role of allostery in the mechanisms of metal ion and DNA binding to ArsR/SmtB proteins is of fundamental interest, a strong coupling between heavy metal tolerance and antibiotic resistance makes it imperative to understand the structure and function of these proteins. Effectively targeting these metalloregulatory proteins could provide a means to impair heavy metal resistance and associated antibiotic resistance in pathogens, potentially leading to the development of novel therapeutic agents (Chakravorty, Parker, et al., 2013).

The paradigm Zn(II)- and Co(II)-responsive sensor protein, *S. aureus* CzrA is an extensively studied member of the ArsR/SmtB family of proteins and serves as an excellent model system to expand our understanding of the general allosteric mechanism of regulation in transcriptional regulators (Arunkumar et al., 2009; Busenlehner, Pennella, & Giedroc, 2003; Chakravorty, Wang, Lee, Giedroc, & Merz, 2012; Eicken et al., 2003; Giedroc & Arunkumar, 2007; Grossoehme & Giedroc, 2009; Pennella, Arunkumar, & Giedroc, 2006; Reyes-Caballero, Campanello, & Giedroc, 2011). CzrA is a homodimeric protein that houses two metal-recognition sites distant from the DNA-binding interface (Eicken et al., 2003; Fig. 1). The protein binds to DNA sequences containing conserved core sequence (5′-**TGAA**(x)$_6$**TTCA**-3′) with binding affinity of ~15 kcal/mol (Grossoehme & Giedroc, 2009). Zinc is a potent allosteric negative regulator of DNA-binding. Under conditions of excess cellular zinc concentrations CzrA binds a zinc ion at each metal-binding site. Zinc binding to CzrA induces a quaternary structural change that causes the protein to transition from a high-DNA-binding affinity "closed" conformation to an "open" conformation that reduces the protein's DNA-binding affinity by ~6 kcal/mol (Arunkumar et al., 2009; Chakravorty et al., 2012; Guerra & Giedroc, 2012; Reyes-Caballero et al., 2011; Fig. 2). The reduced DNA-binding affinity leads to transcriptional derepression of *czrB*, which encodes a Zn(II)/Co(II) cation diffusion facilitator pump that promotes zinc efflux from the cytoplasm.

Mutagenesis, nuclear magnetic resonance (NMR), isothermal titration calorimetry (ITC), and crystallographic investigations by the Giedroc group have provided considerable insight into the mechanism of negative allostery

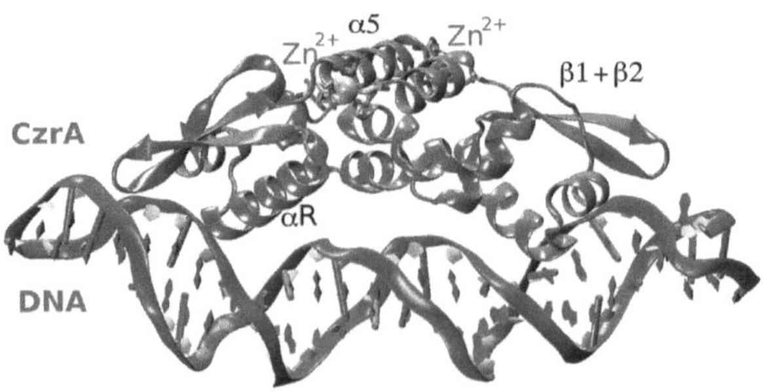

Figure 1 A snapshot from our MD simulations of the Zn_2·CzrA·DNA ternary complex. CzrA and its cognate DNA are shown in cartoon depiction. The zinc ions are depicted as silver spheres (gray in print version) while the zinc-binding residues in the α5 helices are shown in stick notation.

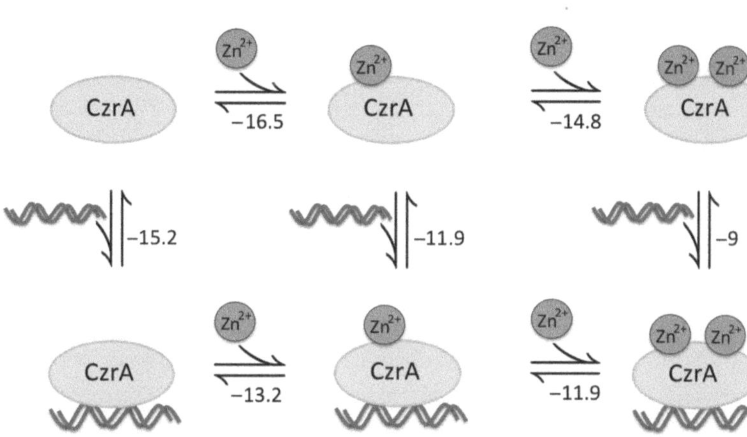

Figure 2 Cartoon depiction of thermodynamic cycle of CzrA binding to zinc and DNA obtained from thermodynamic experiments. Binding affinity values ($\Delta G_{\text{binding}}$) in units of kcal/mol are indicated. Zinc ion binding reduces the CzrA-binding affinity to DNA by 6 kcal/mol.

in CzrA (Arunkumar, Pennella, Kong, & Giedroc, 2007; Campanello et al., 2013; Eicken et al., 2003; Grossoehme & Giedroc, 2009). Protein structures from crystallographic and NMR investigations, combined with heteronuclear single quantum coherence (HSQC) data from NMR experiments have helped to identify the distinct conformational states for the allosteric forms of CzrA (Arunkumar et al., 2009, 2007; Eicken et al., 2003). The allosteric forms of CzrA include the *apo* state (Apo·CzrA), the metal ion-bound holo state with two Zn(II) ions (Zn_2·CzrA), the DNA-bound state (CzrA·DNA) and the ternary state in which the protein is bound to both Zn(II) ions and DNA (Zn_2·CzrA·DNA). X-ray crystal structures from the protein data bank (PDB) of Apo·CzrA (PDB code 1R1U) and Zn_2·CzrA (PDB code 1R1V) are similar to each other (Busenlehner et al., 2003; Eicken et al., 2003; Pennella et al., 2003). In contrast, the solution structure of Zn_2·CzrA (PDB code 2KJC) appears to adopt a "more open" conformation with an average backbone root mean square deviation (RMSD) of 2.53 Å relative to the 1R1V crystal structure of Zn_2·CzrA (Arunkumar et al., 2009). The solution structures of Zn_2·CzrA in conjunction with available $^{1}H,^{15}N$-HSQC data suggest that Zn(II) binding induces a quaternary structural change in Apo·CzrA. Despite providing a structural basis for the mechanism of allosteric regulation, both solution and crystal structures of Zn_2·CzrA suffered from inaccuracies (Chakravorty, Wang, et al., 2013). While the solution structures of Zn_2·CzrA were obtained from heavily deuterated samples and did not model the metal ion during the NMR structure calculation, it is likely that crystal packing effects influenced the crystal structures of *apo*- and Zn(II)-bound CzrA (Arunkumar et al., 2009; Campanello et al., 2013; Chakravorty, Wang, et al., 2013). In the Zn(II)-bound allosteric forms, two Zn(II) ions bind to the dimer interface at the antiparallel α5 helices utilizing two residues from each subunit (Asp84, His86, His97′, and His100′) to form a tetrahedral coordination environment (Busenlehner et al., 2003; Eicken et al., 2003; Pennella et al., 2003; Fig. 3). While these residues are largely pre-arranged for metal ion binding in the crystal structure of *apo* CzrA as well, dynamics experiments and chemical shift perturbations studies suggest that there is motional disorder in solution in this region of the molecule in this allosteric form (Grossoehme & Giedroc, 2009). The solution structure of *apo* CzrA in a DNA-bound conformation (PDB code 2KJB) reveals a large quaternary structural conformational change in *apo* CzrA (Grossoehme & Giedroc, 2009). This DNA molecule was, however, not included in this structural model. The DNA-binding region lies distal to the Zn(II)-binding site and is characterized by the residues in α1, α2, and αR helices and the two β-wings. The ternary structure of CzrA with both Zn(II) ions and DNA bound is a functional intermediate that remains to be solved experimentally.

The available proteins structures of CzrA in conjunction with previous dynamics and thermodynamics measurements from solution NMR and ITC experiments provided a basis for our computational efforts to understand the

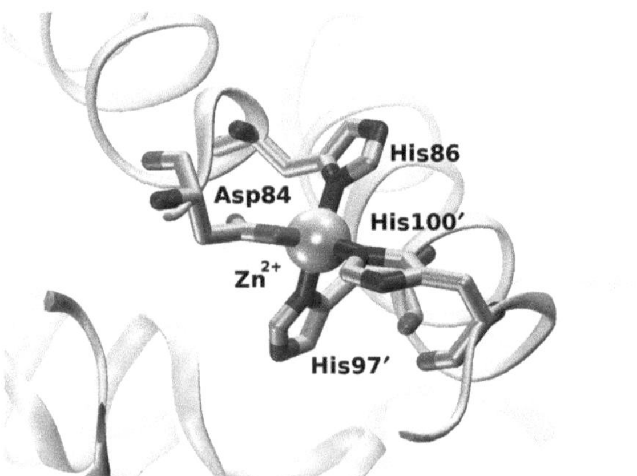

Figure 3 Depiction of the zinc-binding site in CzrA as observed in the 1R1V crystal structure of Zn_2·CzrA. The zinc ion is depicted as a silver sphere (gray in print version) while the zinc-binding residues (Asp84, His86, His97′, and His100′) are shown in stick notation.

pathways dictating the conformational changes in CzrA in response to zinc and DNA binding. In this chapter, we describe our computational efforts that significantly expanded our understanding of the regulatory mechanism in CzrA (Chakravorty et al., 2012). The computational efforts but were accompanied by experiments performed by the Giedroc group. In the proceeding section, we describe the computational methods used to study the conformational dynamics of CzrA. We describe the means to develop accurate metal ion force fields parameters for classical MD simulations and provide details of the accurate metal ion restrained dynamics approach (MRD-NMR) towards refining metal ion sites in NMR structures of metal ion-bound proteins (Chakravorty, Wang, et al., 2013). We show that explicit all atom MD simulations and successfully capture the "closed-to-open" conformational change in CzrA and help identify a metal-mediated hydrogen bond that has been implicated in the long-range allosteric mechanism of a number of metal-sensing proteins (Chakravorty et al., 2012; Chakravorty, Wang, et al., 2013). Using a combination of classical MD, quantum-mechanical molecular/mechanical molecular dynamics (QM/MM MD) simulations, QM/MM calculations, and *ab initio* QM calculations, we analyze the adaptability of CzrA to bind both Zn(II) and DNA in order to provide insights into the mechanism of allosteric regulation that are broadly applicable to metal-sensing transcriptional repressors.

2. METHODS

2.1. Classical MD simulations

2.1.1 Metal ion force fields

Our studies of the allosteric mechanism in CzrA relied upon developing an accurate potential to describe metal ions in classical MD simulations. To date, a variety of "bonded," "semi-bonded," "non-bonded," and "polarizable" approaches to model metal ion in classical MD simulations have been proposed (Chakravorty, Parker, et al., 2013; Chakravorty et al., 2012; Chakravorty, Wang, et al., 2013; Chakravorty, Wang, Ucisik, & Merz, 2011; Hoops, Anderson, & Merz, 1991; Lee et al., 2012; Li & Merz, 2014; Li, Roberts, Chakravorty, & Merz, 2013; Peters et al., 2010; Ponomarev, Click, & Kaminski, 2011; Ponomarev & Kaminski, 2011; Ucisik, Chakravorty, & Merz, 2013; Wu, Piquemal, Chaudret, Reinhardt, & Ren, 2010; Yang, Chakravorty, & Merz, 2010). Owing to its accuracy and relative ease of generation, the bonded model approach has been widely used in MD simulations and was our method of choice to represent zinc and its coordinating ligands in our simulations of Zn_2·CzrA and Zn_2·CzrA·DNA. The bonded model represents a standardized implementation of the "pseudo-bond" or "dummy-cation" approach that is commonly used to model the metal ion-binding sites in NMR structure determination calculations. In the bonded model approach, metal ion parameters used to represent the bond, angle, and dihedral terms for a MD force field are derived from geometry optimized structures of the metal-bound complex (Peters et al., 2010). These calculations can be started from crystal structure geometries or X-ray absorption spectroscopy (XAS) data (Chakravorty et al., 2012, 2011; Lee et al., 2012; Ucisik et al., 2013). As previously mentioned, in CzrA, zinc ions exist in a tetrahedral coordination environment at both metal-binding sites in the α5 helix of each subunit. Asp84 ($O^{\delta1}$) and His86 ($N^{\delta1}$) residues from the first subunit, and His97′ ($N^{\delta1}$) and His100′ ($N\varepsilon^2$) residues from the second protomer together bind to Zn(II) at each metal-binding site. The Zn-ligand force field parameters were calculated using the MTK++ program for the 1R1V Zn-crystal structure geometry (Peters et al., 2010). The details for these calculations have been provided elsewhere (Chakravorty et al., 2012). In brief, our Zn(II)-bonding parameters for CzrA were obtained by performing a frequency calculation on a density functional theory (DFT)-optimized structure (TPSSKCIS/LANL2DZ-6-31G*) of the metal-binding region using the Gaussian09 electronic structure theory program (Cramer & Truhlar, 2009; Frisch et al., 2009; Tao & Perdew, 2003; Wadt & Hay, 1985). The TPSS exchange functional paired with the KCIS correlation functional was chosen as it has been shown to have low average unsigned errors in computational studies that compared predicted heats of formation of Zn(II) complexes to experimental data (Weaver, Yang, & Merz, 2009; Yang, Weaver, & Merz, 2009). We have also successfully calculated metal ion parameters using similar basis sets with the M052X, B3LYP, M06-L, and M06-2X DFT functionals as well (Cramer & Truhlar, 2009; Zhao & Truhlar, 2008). The charges for the metal ion and its binding residues were calculated using the restrained electrostatic potential (RESP) methodology while all remaining amino acids were treated with standard *ff99SB* parameters (Cornell, Cieplak, Bayly, & Kollman, 1993; Hornak et al., 2006). The same set of calculated force field parameters were employed to model Zn(II) ions at both metal-recognition sites in CzrA (Chakravorty et al., 2012).

2.1.2 Preparing the proteins for simulation

Our simulations were propagated from experimentally determined crystal structures or NMR solution structures of the metal ion bound or *apo* protein in their biologically active states (Chakravorty, Parker, et al., 2013; Chakravorty et al., 2012; Lee et al., 2012). The DNA-bound state of CzrA was represented by a structure of a 28-base pair palindromic hairpin DNA molecule docked into the 2KJB structure of *apo* CzrA in the DNA-binding "closed" conformation (Arunkumar et al., 2009). This structure is referred to as GDR in this chapter. The protonation states of charged nonmetal-binding residues of the protein were obtained from the H++ protonation state calculation server (Gordon et al., 2005). The MTK++-derived force field parameters were employed to represent the metal ions and their ligating residues in simulations of $Zn_2 \cdot$ CzrA and $Zn_2 \cdot$ CzrA·DNA using protonation states calculated based on the 1R1V crystal structure geometry of $Zn_2 \cdot$ CzrA. On the other hand, determining the protonation states for metal-binding ligands in *apo* CzrA was a challenging task. Towards this, we performed multiple MD simulations with trial protonation states determined from protonation state calculations to determine the protonation states for the charged metal ion-binding residues in *apo* CzrA. A protonation state of CzrA with deprotonated Asp84 and protonated His97 was used in our simulations. This combination of protonation states was found to maintain the orientation of the metal-binding region for metal ion capture and simultaneously accounts for proton release on Zn(II) binding (Reyes-Caballero et al., 2011).

2.1.3 Simulation protocols

Well-established protocols were used to simulate the protein and protein–DNA complexes in our studies (Chakravorty & Hammes-Schiffer, 2010; Chakravorty, Kumarasiri, Soudackov, & Hammes-Schiffer, 2008; Chakravorty, Soudackov, & Hammes-Schiffer, 2009). All classical MD simulations were performed using the AMBER 11 suite of programs on a potential energy surface described by the AMBER *ff99SB* force field with parambsc0 modifications for DNA (Case et al., 2010). The protein and protein–DNA complexes were solvated in a periodically replicated rectilinear box of explicit SPC/E water molecules (Jorgensen, Chandrasekhar, Madura, Impey, & Klein, 1983). Explicit counter ions were added to net-neutralize the solvated system prior to simulations (Joung & Cheatham, 2008). For simulations of DNA-bound proteins, a 50 mM NaCl buffer solution was introduced in addition to neutralizing counter ions in order to mimic the experimental conditions under which the NMR samples were analyzed (Arunkumar et al., 2009). Each system was then energy minimized over multiple stages to relax the protein structure in its solvated environment. The energy minimized system was then gradually heated to 300 K over 100 ps of MD using a simulated annealing like approach for the canonical ensemble (NVT), during which all heavy backbone atoms were kept restrained with a weak harmonic potential. In the final stage of equilibration, all restraints were removed from protein atoms and 1 ns of MD was propagated for an isothermal and isobaric (NPT) ensemble at 300 K (Allen & Tildesley, 1987). After equilibration, each system was simulated in the NPT ensemble at 300 K with a timestep of 2 fs. A Langevin thermostat with a collision frequency of 1 per ps was employed to maintain the temperature of the system (Allen & Tildesley, 1987). The SHAKE algorithm was utilized to constrain heavy atom bonds to hydrogen atoms and long-range electrostatic interactions were calculated using the particle mesh Ewald method (Allen & Tildesley, 1987; Darden, York, & Pedersen, 1993). All analyses were performed using AMBERTools and the visual MD suite of programs (Case et al., 2010).

2.2. QM/MM and QM/MM MD calculations

Owing to the nature of its implementation in MD simulations, the bonded model approach is unable to account for changes in metal ion coordination. Examples of such changes include changes in the coordination number of the metal ion brought about by the introduction of water molecules, or ligand-exchange brought about by conformational changes near the metal-binding site. In order to account for such possible variations in the metal ion coordination environment, while accurately sampling the protein's conformational ensembles, we employed semi-empirical QM/MM MD calculations performed at the SCC-DFTB3/ff99SB level of theory to supplement our classical MD simulations (Gaus, Cui, & Elstner, 2011; Seabra, Walker, Elstner, Case, & Roitberg, 2007). QM/MM MD calculations offer a means to simulate metal ions in proteins without having to parameterize force field specific parameters (Fig. 4). These calculations were performed on a solvated and well-equilibrated protein structure. The metal ion recognition sites were treated at the QM level, while the remainder of the solvated protein and water were represented by a molecular-mechanical force field (Senn & Thiel, 2009). Side-chain cuts between the α- and β-protein backbone carbon atoms were employed to introduce the metal-binding residues into the QM region. Though similar QM/MM MD calculations have been successfully utilized to study the conformational motions of metalloproteins and investigate enzymatic reactions, they could potentially suffer from inaccuracies in the SCC-DFTB Hamiltonian (Amin, Wood, Johnson, & Young-Dixon, 2008; Sedlak et al., 2013; Settergren,

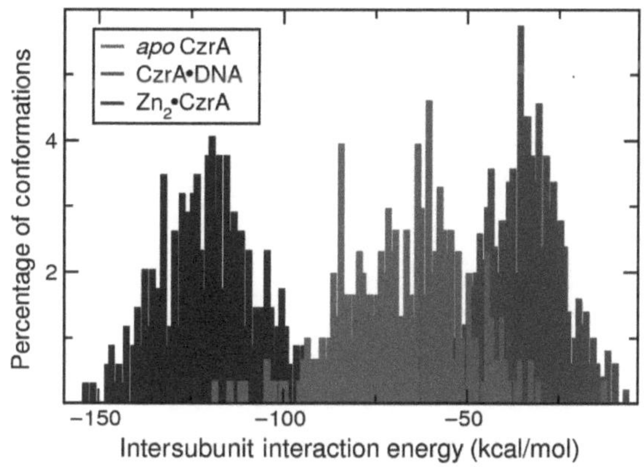

Figure 4 Intersubunit interaction energies between the CzrA monomer units from snapshots obtained from our simulations of *apo* CzrA, Zn_2·CzrA, and CzrA·DNA. These calculations find that zinc binding stabilizes the CzrA dimer while DNA binding destabilizes it.

Buhlmann, & Amin, 2008). In an effort to further validate our bonded metal ion model and rule out any errors arising from the use of SCC-DFTB3, we also performed DFT-based QM/MM energy minimization calculations to study the metal ion-binding sites in CzrA. These calculations were performed on a solvated structure and well-equilibrated structure of Zn_2· CzrA using the Qsite program in the Schrödinger 2011 suite of programs (Qsite, 2011). Much like our QM/MM MD simulations, the metal-binding sites were treated at the QM level (M05-2X/LACVP*) while the remaining system was modeled using the OPLS 2005 force field in these calculations. The LACVP* basis set applies the LANL2DZ psuedopotential-based basis set for Zn(II) and the 6–31G* basis set for nonmetal atoms (Qsite, 2011). Both QM/MM MD simulations and the QM/MM energy minimization calculations found no changes in the zinc coordination environment in CzrA, and validated our choice of the bonded metal ion model (Chakravorty et al., 2012; Chakravorty, Wang, et al., 2013).

3. UNDERSTANDING PROTEIN–DNA INTERACTIONS

In this section, we describe our efforts to understand the influence of zinc binding and DNA binding on the conformational dynamics of CzrA using computational methods. MD simulations performed over biologically relevant timescales provide detailed atomic insights into backbone and side-chain motions that may be readily compared to data from NMR spectroscopy (Chakravorty, Parker, et al., 2013). Starting from the structures of the four allosteric end states of the zinc sensor CzrA, namely Apo·CzrA, Zn_2·CzrA, CzrA·DNA, and Zn_2·CzrA·DNA, we generated over 700 ns of MD data in an effort to understand the underlying mechanism of zinc-induced derepression (Chakravorty et al., 2012). In support of our bonded model description of zinc employed in simulations of Zn_2·CzrA and Zn_2·CzrA·DNA, we found that the metal ions maintained its coordination environment with the ligating residues over the course of 2 ns of QM/MM MD simulations. Our DFT-QM/MM calculations further validated the choice of a bonded model, and showed that the Zn(II) ion remains bonded to its ligating residues. These results, taken in consideration with the high-Zn(II)-binding affinity of CzrA ($\Delta G \sim -16$ kcal/mol), suggested that the crystal structure geometry adopted in the bonded model treatment of Zn(II) ions was a suitable representation of the metal-binding sites in CzrA.

3.1. Conformational changes: open and closed conformations

Zn(II) binding modulates the conformational dynamics of CzrA is known to modulates the protein's conformational dynamics (Arunkumar et al., 2009). In a prevailing model of protein allostery, CzrA may selectively sample functional allosteric ensembles that increase or decrease its binding affinity for the *CzrO* operator (DNA) or the metal ion. The 2KJC ensemble of structures of Zn_2·CzrA and NMR data reveal that the zinc-bound protein adopts an open conformation that significantly reduces the DNA-binding affinity while the closed conformation of CzrA is associated with the CzrA·DNA complex. An initial objective of our investigation was to probe for the existence of open and closed conformational states of CzrA. In the open conformation, the DNA-reading heads of the αR helices are widely separated while the β-wings lift upwards, bringing the wing tips closer. Ser54 lies at the DNA-binding interface of the αR helical region of CzrA. As such, the interprotomer Ser54–Ser54′ Cα-distances provide a useful metric to measure the interprotomer distances between the αR helical DNA-reading heads, and may be used as an identifier for each conformational state. The interprotomer αR distance in the DNA-binding interface differs by ~10 Å among the different allosteric states (Chakravorty et al., 2012). As such, conformations with an interprotomer Ser54 Cα distance in excess of 43 Å were considered to represent an open conformation of CzrA, while a structure with an interprotomer Ser54 Cα distance shorter than 41 Å was considered to be in the closed form of CzrA. A second metric of the conformational state of CzrA exploits the fact that the CzrA dimer adopts a "flat" conformation at the DNA-binding interface in the low-DNA-binding affinity structure, and a "bent" conformation at the interface, which is conducive to high-affinity DNA binding. The two αR–αR′ interhelical angles described by the Ser54–Ser65–Ser54′ and Ser54–Ser65′–Ser54′ Cα "torsion" angles describe the degree to which the αR helices are bent relative to the core of the dimer.

We did not observe any large conformational changes along the protein backbone over 120 ns of MD simulations of the *apo*- and Zn(II)-bound states of CzrA started from their respective crystal structures. The protein structure relaxed in solution and adopted a more open conformation, suggesting that crystal packing effects may have influenced the protein structures in these crystal structures. In our simulations of Apo·CzrA, we found the metal ion recognition sites to be disordered. In contrast, our simulations of Zn_2·CzrA found the metal-binding residues to remain ordered and hold the two subunits together. In these simulations, the protein sampled a maximum interprotomer Ser54 Cα distance of 48 Å, adopting the "flat" or open conformation, that rendered it less suitable for binding to canonical B-DNA. These simulations suggested that the Zn(II)-bound CzrA is structurally similar to the *apo* form of the protein on the timescales of these simulations.

We next simulated the allosteric switching in CzrA in which the protein transitions from a closed DNA-bound conformation towards an open conformation. Based on experimental data and our simulations of Apo·CzrA and Zn_2·CzrA, we hypothesized that CzrA in the closed conformation, will likely transition to an open conformation in the absence of DNA. Furthermore, this structural change will likely be hastened by Zn(II) binding. Based on this hypothesis, we simulated Model 1 of the 2KJB ensemble of NMR structures, solved as the DNA-bound closed conformation of *apo* CzrA, in the *apo*- and Zn(II) ion-bound states. These simulations were in agreement with our hypothesis and found the protein structures becomes more compact and return to the open conformation. In addition, the metal-binding regions become more ordered and less fluxional in both allosteric states. While we did not observe the closed conformation in the simulations of *apo* CzrA for extended periods of time in these simulations, suggesting that DNA binding alone may cause the protein to explore the closed conformations consistently. It is, however, possible that *apo* CzrA may access these closed conformations in simulations over longer timescales.

In contrast to simulations of Apo·CzRA and Zn_2·CzrA, our 120 ns MD simulation of the CzrA·DNA complex, found the protein explore an ensemble of closed conformations that would stabilize DNA binding. We found the protein adopt a bent DNA-binding interface that binds strongly to the palindromic DNA. The αR helices move closer to each other over the course of the simulation, and the residues in the β-wing regions "clamp down" on the DNA. Starting from the GDR structure, the interprotomer Ser54 Cα distance over the last 40 ns of the simulation oscillated between 37 Å and 40 Å, which is close to the ~40 Å distances observed in the 2KJB ensemble of NMR structures (Arunkumar et al., 2009). In line with experimental observations, CzrA binds to DNA in the major groove in a way that places the side chains of Ser54, Ser57, and His58 as interacting strongly with the DNA. The interprotomer αR distances and bend angles in these simulations were lower than those observed in our simulations of the *apo*- and Zn(II) ion-bound states of the protein, indicating, that our MD simulations had successfully captured the low-DNA-binding affinity open conformation of Zn_2·CzrA and the high-DNA-binding affinity closed conformation of the CzrA·DNA complex.

Buoyed by the success of these simulations, we focused our efforts on simulating the closed-to-open transition in DNA-bound CzrA in response to zinc binding. Unlike the other allosteric states, the elusive ternary allosteric state, Zn_2·CzrA·DNA, is a transitory state that has not been experimentally characterized. Starting from a Zn(II)- and DNA-bound structure of CzRA in the closed conformation, we collected 120 ns of MD data of the ternary state of the protein. In a similar vein to our simulations of Zn_2·CzrA starting from the closed conformation of the protein, we found that the presence of Zn(II) ions induced a large conformational change in the protein. The protein sampled the open conformation with the Ser54 Cα distances and the αR–αR′ interhelical angles matching values from our simulation of Zn_2·CzrA. The open conformation caused the DNA molecule to "peel off" from the end of the protein at times, though we did not observed complete disassociation on these timescales. This is in agreement with experiments that found Zn(II) binding to reduce the affinity of CzrA for operator DNA by over four orders of magnitude at 0.4 M NaCl, from $\approx 10^{10}$ to $\approx 10^{6}$ per M. Owing to the prevalence of negative charge on DNA, electrostatic forces are likely to play a dominating role in protein–DNA interactions. Interesting, a charge distribution calculated using semi-empirical methods, found the DNA-binding interface to be less positive in the open conformation as compared to the closed conformation. These results suggest that the quaternary structural changes from the closed-to-open conformation in CzrA works to reduce DNA-binding affinity in terms of both structural incompatibility and electrostatics (Chakravorty et al., 2012). Overall, these simulations document the flexibility of CzrA as it transitions between the allosteric states in response to metal ion binding. To the best of our knowledge, this was the first such computational simulation of allosteric switching in a DNA-bound metal sensor protein. In addition to closing agreeing with NMR data that found the existence of open and closed conformations, our simulations provided new insights into the mechanism of allosteric inhibition in CzrA (Chakravorty, Parker, et al., 2013; Reyes-Caballero et al., 2011).

3.2. Protein mobility

Protein mobility data from NMR experiments suggest that Zn(II) and DNA binding have opposing effects on protein mobility in CzrA. Zinc binding has been found to quench the conformational dynamics in CzrA as part of the mechanism of allosteric regulation. In an effort to validate our simulations and understand the adaptability of CzrA to Zn(II) and DNA binding, we calculated the mobility of every protein residue measured by the root mean square fluctuation (RMSF) of the Cα backbone atoms from our simulations of the allosteric end states of CzrA. In close agreement with NMR data, we found that Zn(II) binding had a drastic impact on protein mobility. While the metal-binding region remained flexible in the *apo-* and DNA-bound states of the protein, the "backbone" stiffens and the α5 helices become well ordered in zinc-bound allosteric forms. In contrast, we found that the charged β-wing regions that interact with DNA are more mobile in more rigid in the CzrA·DNA and Zn_2·CzrA·DNA complexes as compared to Apo·CzrA and Zn_2·CzrA. Our simulations further found that Zn(II) binding "stiffened" the αR region, working to lock the DNA-binding interface into a low-DNA-binding affinity open state with a flat DNA-binding interface.

An essential dynamics analysis on our simulation of Zn_2·CzrA started from a closed conformation of CzrA helped identify the structural changes in the protein that are directly influenced by Zn(II) binding (Amadei, Linssen, & Berendsen, 1993). Using a principal-component analysis method, we diagonalized the covariance matrix for Cα carbon atoms for all protein residues to obtain the first eigenvector of the protein from these simulations. On calculating the mobility of protein residues along the first eigenvector, we found large changes in the β-wing, α2, αR, and α5 regions of CzrA. Conformational changes in these protein regions form the bulk of changes in the protein structure while transitioning between the open- and closed-allosteric forms. A comparative analysis on our simulation of *apo* CzrA in the open conformation did not reveal similar major structural changes associated with its first eigenvector, suggesting that residues in these regions drive the "closed" to open structural change. In support of these findings, a correlation analysis of protein residue motions suggested that both metal-binding sites in CzrA communicate with each other (Chakravorty et al., 2012). Furthermore, zinc binding at the metal-recognition sites is communicated to the DNA-binding interface via networks of residues involved in these motions. Taken together, these results suggest that Zn(II) binding provides the driving force for conformational change in the CzrA homodimer (Chakravorty et al., 2012).

3.3. Conformational ensembles

In an attempt to evaluate the impact of Zn(II) binding on conformational sampling in CzrA, we calculated the distributions of protein backbone RMSD for our simulations of Apo·CzrA and Zn_2·CzrA. In complete agreement with NMR data, we find that Zn_2·CzrA exists in a single conformational state situated in a deep potential well with very steep walls. This ensures that a severe energetic penalty is incurred for moving away from the minima in this allosterically inhibited form. In sharp contrast, the population distribution for Apo·CzrA suggests at least three distinguishable conformational states defined by a shallow potential well with each state separated by a few kcal/mol from the minima. One of these states qualitatively resembles the Zn(II)-bound allosteric state consistent with the idea that *apo* CzrA is capable of making conformational excursions into a Zn(II)-bound like structure. This enables Apo·CzrA to sample more families of conformations than Zn_2·CzrA, which is likely important for binding tightly to Zn(II) or DNA. A detailed clustering analysis to identify unique states sampled over the course of our simulations of *apo*- and Zn(II)-bound CzrA found the existence of three unique states in our simulations of Apo·CzrA, with a population distribution of 51.5%, 16.5%, and 32%, consistent with the findings above (Shao, Tanner, Thompson, & Cheatham, 2007). In contrast, a single state was detected for Zn_2·CzrA that corresponds to state 1 of Apo·CzrA. The calculated RMSD distributions provide a qualitative picture of the conformational space sampled by the protein. Simulations of Apo·CzrA over longer timescales could determine if DNA binds to *apo* CzrA by a mechanism of induced-fit binding or conformational selection (Swint-Kruse & Matthews, 2009).

3.4. Interprotomer interaction energies

Intersubunit interactions play a critical role in the mechanism of allosteric regulation in CzrA. The DNA operator and zinc ions bind across the dimer interface, and cause the *apo*-, Zn(II)-bound, and DNA-bound allosteric forms of CzrA to differ structurally and in their conformational dynamics. To understand the impact of these allosteric stress factors on the protein structure, we calculated the intersubunit interaction energies for the *apo*-, Zn(II)-bound, and DNA-bound allosteric forms of the protein from hundreds of conformations from our MD simulations. These calculations were performed using PM6–DH2 charge model in the MOPAC2009 program (Korth, Pitonak, Rezac, & Hobza, 2010; Stewart, 2008). For each snapshot, the heat of formation of the dimer and each separated monomer was calculated using the MOZYME linear scaling method and the COSMO solvent model with a dielectric constant of 78.4 (Schüürmann & Klamt, 1993; Stewart, 2008). In order to calculate the interaction energies of the Zn(II)-bound structures, we used the model association process, in which the Zn atoms in the separated monomers are stabilized by two bound water molecules. The interaction energy (ΔH_{IE}) between the two monomers was then calculated as the difference between the energies of the dimer ($\Delta H_{\mathrm{Dimer}}$) and each protomer ($\Delta H_{\mathrm{Protomer}\text{-}n}$, where $n=1, 2$):

$$\Delta H_{\mathrm{IE}} = \Delta H_{\mathrm{Dimer}} - \Delta H_{\mathrm{Protomer\text{-}1}} - \Delta H_{\mathrm{Protomer\text{-}2}}. \qquad (10.1)$$

The distributions of interaction energies for the allosteric forms indicate that while zinc binding strengthens the intersubunit interactions in CzrA, DNA binding weakens the interactions between residues at the dimeric interface. These findings are in close agreements with fused linker experiments that suggested DNA binding forced the α5 helices in CzrA apart. In good agreement with our clustering analyses, these calculations suggest that *apo* CzrA visits conformations that are similar to the DNA-bound state of the protein, suggesting that DNA binding takes place via an induced-fit mechanism, while conformational selection plays a lesser role. Taken together these calculations also demonstrate the flexibility at the dimer interface of transcriptional repressors, thereby, explaining their ability to adjust to changing stress factors such as binding to DNA and metal ions.

4. HYDROGEN-BONDING PATHWAY

Structural information from crystal structures, NMR experiments and our simulations suggest that CzrA functions by a mechanism of conformational modulation where Zn(II)-binding quenches the conformational dynamics of the protein. The crystal structure of Zn_2·CzrA, find the conserved metal-binding residues, His97′ oriented towards the αR helical region. The Nε2-bound hydrogen atom of the conserved metal-binding residue forms a hydrogen bond to the backbone carbonyl group of His67′ in the αR helical region (Fig. 5). This interaction forms a HBP that is propagated by a series of backbone alpha helix hydrogen bonds to the DNA-reading heads of the protein, connecting the metal-recognition sites to the DNA-binding interface. A similar metal ion-mediated HBP was determined to be the dominant allosteric pathway in the copper-sensing transcriptional repressor *M. tuberculosis* CsoR (Ma, Cowart, Scott, & Giedroc, 2009; Ma, Cowart, Ward, et al., 2009). The His67–His97′ interaction is also observed in the 1R1U crystal structure of CzrA, though HSQC data collected by the Giedroc group, however, finds that this hydrogen bond is destabilized in *apo* CzrA and absent in DNA-bound CzrA. As such, the zinc-mediated interprotomer hydrogen bond is proposed to nucleate a HBP that is hypothesized to "lock" the protein in the open conformation (Grossoehme & Giedroc, 2009). While dominant pathways in the allosteric mechanisms of proteins are poorly understood, the HBP has been implicated in the mechanism of allosteric regulation in CzrA and other metal ion-sensing proteins.

Our MD, QM/MM MD, and QM/MM calculations of Zn_2·CzrA found the His67–His97′ hydrogen bond to exist at both metal-binding sites

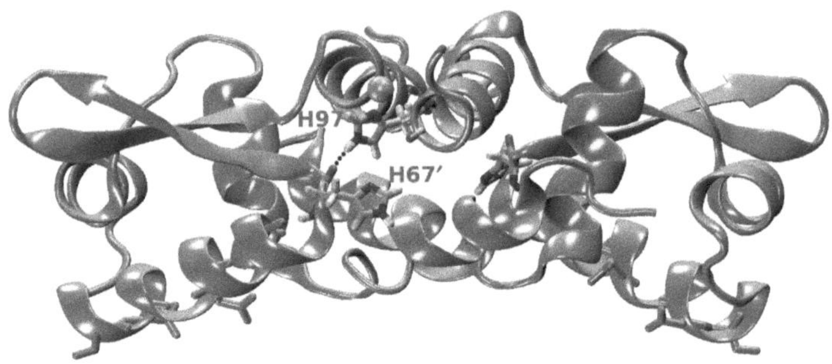

Figure 5 Hydrogen-bonding pathway connecting the metal-binding site to the DNA-binding region in CzrA observed in the 4MR0 structure of the metal-bound protein. The zinc ion is shown as a silver sphere and is bound to His97. The Nε-H of His97 forms a hydrogen bond with the backbone carbonyl group of His67′ of the other protomer. The hydrogen bond extends to the DNA-binding interface via a series of backbone hydrogen bonds in the alpha helix. His97 and His67 are indicated while residues involved in the pathway are colored green (gray in print version). All residues are depicted in stick notation and the protein is shown as a new cartoon.

(Chakravorty et al., 2012). In agreement with experimental data, the HBP was absent in our simulations of DNA-bound CzrA and presumably weaker in the *apo* allosteric state. These findings were further supported by our simulations of Zn_2·CzrA·DNA that though unable to detect the His67–His97′ interaction, found the αR helix of CzrA comes closer to the α5 helix, reducing the hydrogen bond distance in the protein. Finally, our MD simulations of conformational switching propagated from a closed conformation of CzrA, found this HBP to re-emerge in simulations of Zn_2·CzrA that transitioned to an open conformation. In sharp contrast, a MD trajectory of Apo·CzrA from a closed DNA-bound conformation found the protein structure successfully transition to an open conformation without forming the critical H67–H97′ hydrogen bond. In the absence of this hydrogen bond, Apo·CzrA transitioned back to a moderately open conformation, while Zn_2·CzrA maintained the open conformation. These results strongly suggested that this interprotomer HBP primarily functions as a "lock" that holds the CzrA dimer in the low-DNA-binding affinity open conformation, and perhaps functioned as a dominant pathway in the mechanism of allosteric regulation in CzrA.

4.1. HBP in other metal-sensing transcriptional repressors

Similar HBPs have been also been observed in Zn(II)-bound crystal structures of other ArsR/SmtB proteins such as *Synechococcus* SmtB and the Cd/Pb sensor *S. aureus* pI258 CadC (Kandegedara, Thiyagarajan, Kondapalli, Stemmler, & Rosen, 2009; Ye, Kandegedara, Martin, & Rosen, 2005). In the meanwhile, NMR data suggested that a HBP may be present in the zinc-bound form of the Ni/Co sensor *M. tuberculosis* NmtR as well (Lee et al., 2012). These ArsR/SmtB proteins have evolved from a common ancestor but have developed unique selectivity and allosteric responses towards different metal ions (Ma, Jacobsen, & Giedroc, 2009). CzrA, SmtB, NmtR, and CadC share a common Asp-His-His-*X* zinc-binding motif (*X*=His in CzrA and NmtR, *X*=Glu in SmtB and CadC) at their α5 metal-binding sites. In a role similar to that of His67 in CzrA, the metal ion-mediated hydrogen bond is accepted by the backbone carbonyl group of Arg87 in SmtB, Gly74 in NmtR, and Gly84 in CadC. Much like Czra, Zn(II) binding to the α5 metal-recognition sites in SmtB and NmtR triggers an allosteric response that drives the protein to an open conformation that has a low-DNA-binding affinity. In contrast, the α5 sites metal ion-binding sites in CadC are allosterically nonresponsive to Zn(II) binding, and the protein maintains a closed DNA-binding conformation in its zinc-bound state (Arunkumar et al., 2009). Taken together, these factors suggest that HBPs at allosterically functional MBSs in the α5 helix may be a diverged evolutionary characteristic of SmtB, NmtR, and CzrA like ArsR-family proteins that differ from CadC like proteins. In addition to the ArsR/SmtB repressor proteins, a zinc-mediated hydrogen bond has been observed in the MarR family protein, *Streptococcus pneumoniae* AdcR (Guerra, Dann, & Giedroc, 2011). AdcR shares a MarR like fold and has two sets of high-affinity and low-affinity zinc-binding sites. This hydrogen bond between Glu24 of the high-affinity zinc-binding site in AdcR and Asn38 extends to form a HBP that continues to the DNA-binding helices in the protein. Surprisingly, in AdcR, zinc functions as an allosteric *activator* of DNA binding. The presence of such metal ion-mediated interactions at functional allosteric sites across protein families for different metal ions suggest a conserved functional role in metal ion-sensing transcriptional regulator function.

Determining the source of allosteric effects in proteins has been a subject of intense scrutiny. Prior to investigating the role of the metal ion in the HBP allosteric pathways, we first analyzed the dynamic instability at these sites in the *apo* proteins (Chakravorty, Parker, et al., 2013). The HBPs are absent in the crystal structure of *apo*-SmtB (PDB code 1R22), and the NMR structure of *apo*-NmtR (PDB code 2LKP) suggesting that these interactions are metal ion dependent. Similar structural information was, however, unavailable for *apo* forms of CadC and AdcR. MD simulations of the *apo* forms of AdcR, SmtB, NmtR, and CadC provided an ensemble view of the dynamic instability at these hydrogen-bonding sites and facilitate our understanding of the conformational dynamics that are quenched on metal ion binding in the course of allosteric regulation. In the absence of experimental structures of wild-type *apo* forms of CadC and AdcR, models of the *apo* states were built via MD simulations propagated from their zinc-bound structures (PDB codes 1U2W and 3TGN, respectively) after removing zinc (Eicken et al., 2003; Guerra et al., 2011). In close agreement with our simulations of *apo* CzrA, the simulations found the HBP largely broken in the *apo* forms of these proteins, though short-lived conformations with an intact HBP were observed. The calculated free energy profiles helped identify an energetic barrier to forming these interactions in the *apo* form (Chakravorty, Parker, et al., 2013). Entropy calculations and ITC experiments performed on the allosteric forms of CzrA indicate that zinc binding reduces the overall entropy of the protein, including at the His97–His67′

hydrogen-binding site (Chakravorty, Parker, et al., 2013; Grossoehme & Giedroc, 2009). The conformational flexibility at these sites likely allow the transcriptional regulators to visit multiple conformational states that are conducive to DNA or metal ion binding, accounting for their remarkable adaptability. In subsequent sections of this chapter, we discuss our efforts to further characterize this interaction and determine its role in the long-range allosteric mechanism.

4.2. The metal restrained dynamics for NMR structure resolution (MRD-NMR) method for metal ion refinement in NMR structure determination

At this juncture, our efforts to characterize the HBP and to investigate the effect of metal ion binding on the His67–His97′ hydrogen bond interaction were hampered by the absence of high-resolution solution structures of the metal ion-bound protein. While X-ray crystal structures of CzrA provided a high-resolution snapshot of the metal ion-binding sites in these proteins, the protein structures suffered as a result of close packing effects (Chakravorty et al., 2012). In contrast, while solution-state NMR provided an accurate ensemble-guided view of the structure and associated conformational dynamics of the CzrA at an atomistic level, these structures lacked critical information around the metal-binding site as metal ions were not included during the structure refinement process (Arunkumar et al., 2009). In the absence of methods to accurately refine metal ion centers during the NMR structure determination process, this is not uncommon practice. As such there was a need for a broadly applicable general structure refinement protocol that would accurately represents metal ions in an NMR-derived protein structure, ultimately leading to a more precise ensemble of metal-bound protein conformations.

Towards this, we developed the metal ion inclusive restrained dynamics-NMR structure refinement method (MRD-NMR) approach to accurately refine metal ion coordination during NMR structure determination (Chakravorty, Wang, et al., 2013). In this approach, metal ions are modeled using specially developed force field parameters that accurately describe the metal ion coordination and electrostatics in NMR data restrained MD simulations allow the protein structure to sample conformations representative of the entire ensemble on the timescale of tens of nanoseconds. QM/MM MD calculations supplement these simulations and account for changes in metal ion coordination and water exchange while providing extensive sampling. In the MRD-NMR approach for CzrA, zinc ions were modeled in the previously determined NMR structure of the zinc-bound open conformation of CzrA (PDB code 2KJC; Arunkumar et al., 2009). The previously determined NMR structure of Zn(II)–CzrA was calculated using a total of 756 intra- and intersubunit NOE restraints and 112 backbone amide NH residual dipolar coupling (RDC) restraints. As mentioned above, this structure was solved without the inclusion of zinc ions at the metal-recognition sites, and provided a fine test for our MRD-NMR method. In addition to providing an accurate representation of the metal-binding sites, the CzrA structure calculations are using the MRD-NMR method benefit from the inclusion of explicit solvent models, periodic boundary conditions, and an accurate calculation of long-range electrostatics using the particle mesh Ewald approach (Allen & Tildesley, 1987).

4.3. Method

The MRD-NMR approach for metal ion refinement was implemented using the AMBER suite of programs (Case et al., 2010). The approach is readily extended to any software that provides an integrated platform for MD, QM/MM, metal ion force field generation, and NMR structure determination methods. Model 1 of the 2KJC ensemble of NMR structures of Zn_2·CzrA provided a good starting point for our MRD-NMR refinement approach. This structure was determined to comply with NOE restraints while not being overfitted to a smaller number of RDC restraints and was prepared for simulation with zinc ions at each metal-recognition sites using methods described in this chapter. As part of the MRD-NMR protocol, 760 NOE restraints with a force constant of 30 kcal/mol/Å^2 were applied to the solvated protein structure over the course of the energy minimization and equilibration processes. Following the equilibration phase, a NOE-restrained MD simulation was propagated at 300 K for 10 ns using a Langevin thermostat (Allen & Tildesley, 1987). The protein was then gradually cooled to 0 K over 200 ps and further energy minimized to obtain energy minimized NOE-constrained protein structure. No significant NOE violations (>0.3 Å) were observed and the structure passed all structure checks at this point.

The energy minimized NOE-restrained structure was then fitted to 112 NH RDC restraints ($^1D_{HN}$) while maintaining the NOE restraints. The RDC fitting procedure was performed over multiple cycles starting with a force constant of 0.05 kcal/mol/Å^2 applied to every RDC restraint. During this phase, the energy minimized NOE-constrained protein structure was passed through the above refinement process while employing both NOE and RDC restraints. Following the classical MD fitting stage, a QM/MM energy minimization was performed at the SCC-DFTB3/ff99SB level of theory with NOE and RDC restraints in place (Gaus et al., 2011). In consequent RDC fitting cycles, the force constant was incrementally increased by 0.05 kcal/mol/$\mathrm{\AA}^2$ until a correlation of ~1 and an offset of ~0 Hz were achieved between RDC values obtained from the calculated structure and the experimentally determined values. The final RDC fitting cycle was performed with force constants of 0.25 kcal/mol/$\mathrm{\AA}^2$ applied to every RDC restraint. The resulting solution structure of CzrA from QM/MM energy minimized RDC + NOE refinement CzrA maintained attributes that closely resembled the crystal structure. In the final stage of the MRD-NMR structure refinement methodology, 10 independent simulations were propagated from the QM/MM energy minimized NOE + RDC refined structure of CzrA. In each simulation, the structure was equilibrated and propagated for 12 ns of NMR-biased MD at 300 K followed by 1 ns of NMR-biased QM/MM MD refinement. Overall, we successfully determined a structural ensemble of the zinc-bound form of CzrA that correlate well with experimentally observed backbone ${}^1D_{\mathrm{NH}}$ RDC values and NOE constraints. The atomic coordinates of representative structures from this ensemble have been deposited in the PDB (PDB code 2M30).

4.4. Insights from NMR structure refinement

The calculated MRD-NMR structural ensemble provides a detailed and accurate representation of the solution structure of the metal-bound protein that benefits from correctly portraying the first-shell and second-shell coordination effects of the metal ion binding. The MRD-NMR structural ensemble of $Zn_2 \cdot$CzrA found that Zn(II) maintained its coordination with Asp84, His86, His97, and His100 residues in agreement with the geometry observed in the crystal structure and our unrestrained simulations of $Zn_2 \cdot$CzrA (Fig. 6). The NMR-biased QM/MM MD simulations found the individual bond distances to vary within an acceptable range of ± 0.1 Å from the observed distances in the crystal structure. These variations are part of the SCC-DFTB3 Hamiltonian, and may be improved using more rigorous DFT or *ab initio*-based methods.

The MRD-NMR ensemble of structures helped elucidate the mechanism of allosteric regulation in CzrA. The interprotomer zinc-dependent second-shell coordination hydrogen bond between the conserved

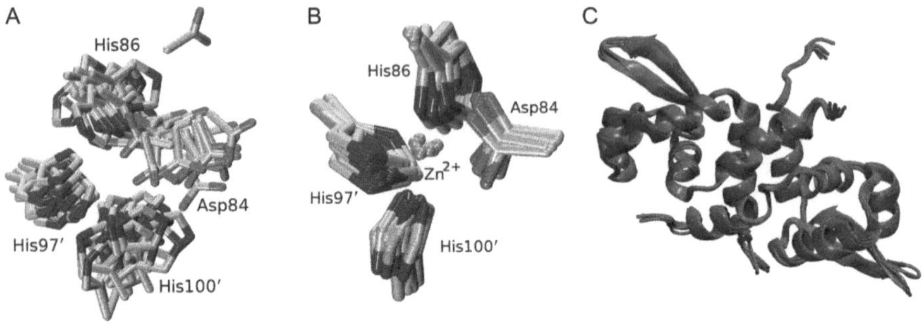

Figure 6 (A) A zinc metal ion-binding site observed in the 2KJC NMR structure of zinc-bound CzrA. The metal ion-binding residues are depicted in stick notation. (B) The zinc metal ion-binding site post-MRD-NMR metal ion-binding site refinement protocol. The metal ion is indicated as a silver sphere. (C) Cartoon structure depiction of CzrA after the MRD-NMR metal ion refinement protocol.

metal-binding ligand His97 and His67′ of the αR helix was observed in every structure belonging to the ensemble though no hydrogen bond restraint was used to restrain His97 and His67′ residues in these calculations. This represented a marked improvement over the 2KJC ensemble of NMR structures that relied on such a bond restraint in order to maintain the His67–His97′ hydrogen bond interaction. In addition to this zinc-dependent hydrogen bond, the ensemble found Asp83, a conserved residue that neighbors the invariant metal-binding residue Asp84, to participate in an interaction observed in the crystal structure of Zn_2·CzrA that was absent in the 2KJC ensemble of NMR structures. The backbone carbonyl group of Asp83 now interacts with the positively charged side chain of the β-wing residue, Lys70, while the backbone amide H of Asp83 interacts with the backbone carbonyl group of Leu68, possibly allowing Asp83 to communicate metal binding to the β-wings.

The MRD-NMR ensemble of structures of Zn_2·CzrA helps validate our MD simulations of CzrA. As observed in our simulations and NMR experiments, CzrA modulates its conformational dynamics to selectively sample functional allosteric ensembles that increase or decrease its binding affinity for the *CzrO* operator (DNA) or metal ions. In favorable agreement with the 2KJC ensemble of NMR structures and our previous data, we find that the MRD-NMR refined ensemble of structures adopt a more open conformation when compared to that of the 1R1V crystal structure of Zn_2·CzrA (Arunkumar et al., 2009; Chakravorty, Wang, et al., 2013). The ensemble samples interprotomer Ser54–Ser54′ Cα-distances and Ser54–Ser65–Ser54′ and Ser54–Ser65′–Ser54′ Cα-angles that are indicative of the weakly DNA binding "open" conformation. The distribution of these distances and angles for the NOE + RDC restrained MRD-NMR ensemble are in agreement with unconstrained MD simulations of Zn_2· CzrA. Interestingly, while we find that conformations visited over the course of our unconstrained simulations of Zn_2· CzrA compiled with the experimentally derived NOE restraints (all deviations <0.5 Å), they failed to reproduce the RDC values. Upon analyzing the distribution of deviations from the experimentally observed RDC values, we found that the structural ensembles from these simulations successfully matched the experimentally observed RDC data in a time-averaged manner as would be experienced in an experimental NMR setting. These data suggest that MD simulations on biologically relevant timescales can capture the structural characteristics of the system under investigation.

5. QUANTIFYING METAL-ION MEDIATED ALLOSTERY

Upon successfully generating an accurate ensemble of structures to represent the Zn_2· CzrA allosteric form, we focused our efforts to determine the energetics driving the formation of the metal-mediated HBP that plays a dominant role in the mechanism of allosteric mechanism in CzrA. We next performed *ab initio* QM calculations to evaluate these interactions in the zinc-bound forms of CzrA, AdcR, NmtR, SmtB, and CadC (Chakravorty, Parker, et al., 2013). In these calculations, basis set superposition error-corrected interaction energies were calculated using the counterpoise method, in an effort to determine the interaction energies were calculated between His67 and the metal bound residues, Zn•D84H86H97′H100′ for the zinc-bound crystal structure geometry of CzrA using the counterpoise method. These calculations were performed on representative model systems at the second order Møller–Plesset perturbation level of theory while employing the augmented Dunning correlation-consistent polarized aug-cc-pVDZ and aug-cc-pVTZ basis sets using the Gaussian09 electronic structure theory calculation program (Balabanov & Peterson, 2005; Dunning, 1989; Frisch et al., 2009). These data were then used to extrapolate to the complete basis set limit (Fu, Li, & Merz, 2011). In order to incorporate sampling and solvation effects, these interaction energies were calculated for additional conformations of Zn_2· CzrA for a number of conformations from crystal structures and the MRD-NMR ensemble of QM-refined NMR protein conformations and QM/MM (M06-2X/LACVP*/OPLS 2005) calculations (Cramer & Truhlar, 2009; Jorgensen & Julian, 2005; Qsite,

2011). We found the His97–His67′ interfragment interaction energy to be on the order of −20 kcal/mol for Zn_2·CzrA. Charge distribution calculations performed on the entire zinc-bound protein structure (M06L-LC/3-21G*/LANL2DZ) found that zinc binding locally polarizes residues around the metal-binding sites, strengthening neighboring interactions (Chakravorty, Parker, et al., 2013; Cramer & Truhlar, 2009). Similar calculations performed on *apo* CzrA for a putative His97–His67′ hydrogen bond by removing the metal ion and appropriately modifying the protonation states, finds the interaction energy to be on the order of ~10 kcal/mol. These data suggest that zinc binding helps strengthen the His97–His67′ interaction by more than 10 kcal/mol compared to conformations of Apo·CzrA with a structurally intact HBP. It is noted that while conformations of Apo·CzrA with an intact HBP are transiently possible, NMR experiments, and our MD simulations find this interaction to be destabilized or absent. The large interaction energy in response to metal ion binding likely overcomes the dynamic instability observed in the *apo* protein. Taken together with entropic effects, this stabilization provides a rationale for the absence of this hydrogen bond in the absence of the metal ion.

In an effort to understand the role of this hydrogen bond in other transcriptional regulator proteins, we calculated the interaction energies for zinc-bound crystal structure geometries of AdcR, SmtB, and CadC. In the absence of an experimentally determined structure of Zn_2·NmtR, we determined a structure of Zn_2·NmtR derived from MD and QM/MM MD simulations starting from a solvated and well-equilibrated structure of Apo·NmtR (Lee et al., 2012). In our models, the bond distances and relative orientations of the zinc coordinating residues at the α5 metal ion-binding sites of NmtR are similar to those observed in CzrA (Chakravorty et al., 2012; Lee et al., 2012). In models derived from these calculations, the Nε2 face of His104 points towards the backbone carbonyl group of Gly74 much like the His97–His67′ interaction observed in CzrA.

Much like in the case of CzrA, we found that the interaction energies were considerably stronger in the metal-bound states at the functional allosteric metal-binding sites of AdcR, NmtR, and SmtB. In contrast, the interaction energies calculated for the nonfunctional site in CadC are comparable to those calculated for the *apo* CzrA model. In CadC, zinc binding to the α5 metal ion-binding sites fails to strengthen the interprotomer interaction compared to *apo* CadC. As such, the stabilization offered by metal ion binding is not sufficiently strong to overcome the protein's intrinsic conformational dynamics and lock the protein into a weak DNA-binding open conformation that is observed in the Zn(II)-bound forms of other ArsR-family sensors. Indeed in some CadCs, the allosterically inhibited α5 Zn(II) sites are not retained, and structural analysis of one of these by NMR is also consistent with a closed conformation (Thomson & Gray, 1998). While SmtB and CadC share identical Zn(II)-binding motifs, the Arg87–Glu120 interaction in contributes ~25 kcal/mol to the uniquely high-interaction energies observed in Zn(II)-SmtB. This strong Arg-Glu salt–bridge interaction is present in *apo* SmtB as well, and possibly plays a role in forming the HBP on binding zinc. In sharp contrast, the imidazole side chain of His67 contributes only ~1 kcal/mol towards this interaction energy and thus cannot be responsible for the modest interaction energy observed in Zn(II)–CadC. CzrA and NmtR share identical zinc-binding motifs, but significantly differ in their zinc-binding affinities and allosteric responses. While Zn(II) is a strong negative allosteric regulator of CzrA, it has a weaker allosteric effect on NmtR as a result of not coordinating the N-terminal residues (Ramos et al., 2005). No significant difference is observed between their model system interaction energies, emphasizing the additional role that conformational dynamics and protein structure play in the allosteric mechanism. Remarkably, the interaction energies for Zn_2·AdcR are of a magnitude that is similar to other functional allosteric MBSs, despite the fact that Zn(II) is an allosteric activator of AdcR.

5.1. Energy decomposition

In an effort to understand the determinants of these interactions, we performed symmetry adapted perturbation theory (SAPT) calculations. SAPT is a powerful analytical tool that decomposes interaction energies into the physical components of electrostatics, exchange repulsion, induction (polarization), and London dispersion forces. These calculations were performed using density-fitted SAPT0 as employed in the PSI4 program for crystal structure and QM/MM MD-derived geometries of CadC, AdcR, CzrA, SmtB, and NmtR at a comparable level of theory (Chakravorty, Parker, et al., 2013; Thanthiriwatte, Hohenstein, Burns, & Sherrill, 2011; Turney et al., 2012; Zhao et al., 2012). These calculations found the interprotomer His97–His67′ like hydrogen bond to be the primary noncovalent interaction at these metal-recognition sites (Fig. 7). While most interactions were predominantly governed by electrostatics, exchange repulsion worked to destabilize this interaction. This was particularly notable for the crystal structure geometries of CadC, where the strong electrostatic interactions are nearly

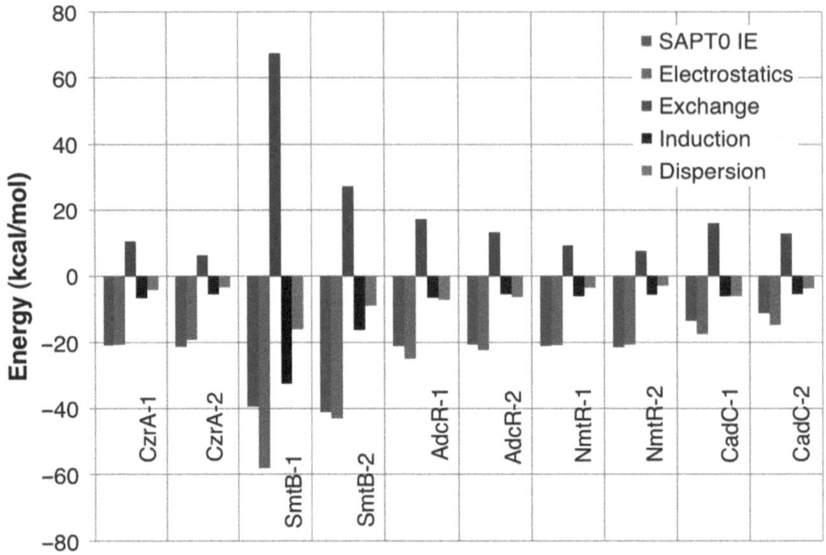

Figure 7 SAPT0 interaction energy values calculated for crystal structure and QM/MM MD conformations of CzrA, AdcR, SmtB, and CadC. The contribution of dispersion, electrostatics, induction, and exchange terms are shown.

canceled by the higher exchange repulsion. Much like the interaction energy calculations, the SmtB fragment contained a diverse set of interactions. These included the crystallographically observed interaction between the positively charged guanidinium and the π-cloud of the histidine imidazole, is nearly equal parts electrostatics and induction. Taken collectively, our calculations suggested that zinc-mediated interactions may be a common feature of the allosteric mechanisms of metal-sensing regulators.

6. MUTANT ANALYSES TO UNDERSTAND THE HBP

6.1. V66A/L68VCzrA

In an effort to experimentally understand the role of the metal ion-mediated interaction in the mechanism of allosteric regulation in CzrA, the Giedroc group has structurally characterized the V66A/L68V double mutant form of CzrA (Campanello et al., 2013). The V66A/L68V double mutant form of CzrA (*mut*CzrA) was designed to create a void around the site of the H67–H97 interaction in CzrA. As a result of this mutation it was expected that Zn(II) binding will no longer be able to quench the conformational dynamics of the mutant form. In agreement with this hypothesis, ITC experiments observed a sharp reduction in the allosteric free energy of coupling in response to zinc binding in V66A/L68V CzrA and $^{1}H,^{15}N$-HSQC spectrum identified changes in the region of the mutation in the Zn(II)-bound double mutant compared to wild-type to Zn_2·CzrA (Campanello et al., 2013; Eicken et al., 2003). In contrast to this, crystal structures of the zinc-bound double mutant and wild-type Zn_2·CzrA were essentially identical with a backbone Cα RMSD of 038 Å (Campanello et al., 2013). In addition, the His67–His97 and the HBP remained intact in the double mutant. Closer inspection around the site of the mutation indicated a small cavity suggestive of poorer packing in the protein core for the double mutant relative to wild-type CzrA, suggesting that poorer packing in *mut*CzrA directly controls the magnitude of the free energy of allosteric coupling. In the absence of definitive structural data, it remained unclear if the mutation directly impacted the His67–His97′ interaction or some other allosteric pathway in the protein.

Based on our simulations of *apo* proteins belonging to the ArsR/SmtB family, we hypothesized that the poor packing observed in Zn_2 V66A/L68V CzrA might result in greater conformational disorder near the critical H67–H97′ hydrogen bond site, thus disrupting the HBP and eliminating the ability of the zinc ion to quench the double mutant protein's conformational dynamics. This was of significance as our calculations and ITC experiments suggested that the mechanism of allosteric regulation involved reducing the entropy of the system as a result of Zn(II) binding (Campanello et al., 2013; Grossoehme & Giedroc, 2009). In addition, the crystal structures of Zn(II)-bound wild-type and V66/L68V CzrAs only provide a static picture of the conformational ensemble accessible to each. Having observed the impact of crystal packing effects on the crystal structure of Zn(II)–CzrA, we hypothesized that the crystal structure of the V66A/L68V double mutant was probably locked in a conformation as a result of crystal packing effects.

Owing to the similarities between the metal-binding regions of wild-type and the double mutant forms of CzrA, we once again used our "bonded" model approach to model Zn(II) ions in the metal-binding regions of CzrA. To identify the changes in the conformational dynamics of these proteins on timescales accessible to full atomistic MD simulations in water, we generated five independent MD trajectories of 140 ns each were propagated from five independently equilibrated structures for Zn_2 V66A/L68V CzrA and compare them to our simulations of wild-type Zn_2·CzrA. Simulations of the V66A/L68 double mutant were propagated from the experimentally determined Zn(II)-bound crystal structure of *mut*CzrA (PDB code 4GGG). Additional mutations in the crystal structure (Asp14Glu, Ala18Glu, Arg73Ala, and Lys92Ala) resulting from poor electron density were resolved, allowing us to directly compare the effect of these mutations on zinc binding to CzrA.

In all, we analyzed 700 ns of MD data to provide reasonable statistics in understanding the nature of allosteric regulation in Zn_2 V66A/L68V CzrA (Chakravorty & Merz, 2014). We find that while wild-type Zn_2·CzrA samples a single conformational well over the course of these simulations and is characterized by an intact His97–His67′ hydrogen bond, the double mutant clearly samples two discreet conformational ensembles, one of which is characterized by an intact His97 $NH\varepsilon^2\cdots O{=}C$ His67′ hydrogen bond, with the other characterized by the presence of a solvent molecule interrupting the donor and acceptor, thus indicative of reduced stability. In addition, we see an average lengthening of the second H-bond in the pathway between Leu68 NH (Val68) and carbonyl oxygen atom of Leu63 in the αR helix. These local changes in structure manifest themselves globally in that the average interprotomer Ser54–Ser54′ distance *decreases* in the double mutant. Thus, the αR helices in the double mutant appear capable of sampling a Zn(II)-bound conformation that more resembles the "closed" DNA-binding competent conformation of CzrA. A quantitative analysis of the correlated and anticorrelated motions of V66A/L68V CzrA cavity mutant reveals areas of much stronger connectivity and significant flexibility, with the possible exception that the allosteric zinc-binding sites on the dimer are no longer correlated strongly to one another in the mutant. These simulations collectively suggest that the ability of zinc to quench the conformational mobility and "stiffen" the molecule is lost to a significant degree in V66A/L68V CzrA.

Our simulations found that the enhanced internal dynamics and poorer packing associated with the allosterically inhibited Zn(II) state of V66A/L68V CzrA relative to wild-type CzrA impact the magnitude of the allosteric free energy of coupling. While the poorer side-chain packing likely reduced the enthalpic contribution to allostery from zinc binding, the enhanced conformational entropy at the site of the critical site of His97–His67 interaction resulted in a larger entropy that could not be driven down by zinc binding alone. Our MD simulations are consistent with experiments and suggest that a change in His97–His67 interaction reduces the stability of the allosterically inhibited zinc-bound form of a protein. This increased internal mobility in turn allows Zn_2·CzrA to access conformational states that approach the DNA-bound state, thus increasing the affinity of Zn_2·CzrA for the DNA operator.

6.2. Unnatural amino acid mutations

We further analyzed the impact of the His97–His67′ hydrogen bond on CzrA function by performing *in silico* mutagenesis experiments that specifically targeted the H67–H97′ interactions (Chakravorty & Merz, 2014). Towards this, we built protein models in which we replaced the backbone carbonyl C=O group of H67 with a hydroxyl group, thus eliminating the H67–H97′ hydrogen bond. In a second experiment, we replaced the backbone —NH group of L68 with a —CH—OH group to create L68/L68-OH CzrA. We hypothesized the —OH group would interact strongly with the adjacent backbone —C=O group of His67, thus eliminating the H67–H97′ interaction. While obtaining crystallographic and NMR protein structures for unnatural amino acid mutants are challenging work, these mutants provide the most direct line of evidence in support of our work. Simulations of the zinc-bound forms of the two synthetic amino acid mutant forms of CzrA were started from the structure of appropriately modified wild-type Zn_2·CzrA and were propagated for 100 ns. Simulations of L68/L68-OH Zn_2·CzrA were rapidly unable to detect the found the H97–H67 interaction. Moreover, simulations of both mutant forms found the zinc-bound forms to be more mobile than the corresponding zinc-bound wild-type protein and experience a larger conformational spread. Overall, our simulations found strong support for the hypothesis that the H97–H67′ interaction plays a critical role in rigidifying the protein in a low-DNA-binding affinity open conformation in response to zinc binding, suggesting that metal ion-mediated HBP is indeed a dominant long-range pathway in the mechanism of allosteric regulation by zinc binding in CzrA.

7. CONCLUSIONS

In this chapter, we presented details of our computational studies of the allosteric forms of the paradigm ArsR/SmtB family zinc sensor CzrA that helped explain the mechanism of allosteric regulation in the protein. Our simulations provided new insights into the mechanism of allosteric regulation in CzrA and are in close agreement with the extensive experimental findings for this system. We find that large quaternary structural conformational transitions of CzrA play an essential role in the ability of CzrA to function as an allosterically regulated transcriptional repressor. In complete agreement with experimental data, our simulations reveal that *apo-* and Zn(II)-bound CzrA each adopt a stable open conformation while high-affinity DNA-binding stabilizes the closed conformation. Our simulations of allosteric switching in CzrA successfully capture this structural transition between closed-to-open conformations upon Zn(II) binding to CzrA bound to the DNA, prior to dissociation. We analyzed the structure and conformational dynamics of CzrA, and by doing so obtained a better understanding of the relationship between Zn(II) binding at a distal location and the associated reduction in DNA-binding affinity.

We identified groups of residues that are involved in hierarchical networks of correlated and anticorrelated motions that clearly connect the metal-binding and DNA-binding sites in functionally important ways. These networks are present in all allosteric forms of the protein, but are greatly attenuated in the Zn(II)-bound state. They are indicative of communications channels that help to account for the structural sensitivity of CzrA to both metal and operator DNA binding. Our simulations support the existence of a functionally important HBP that connects the first coordination shell of the Zn(II) complex to the DNA-binding site through the αR helices (Arunkumar et al., 2009). This pathway was stable only in Zn_2·CzrA, but was destabilized or absent in all other allosteric forms, including the intermediate Zn_2·CzrA·DNA complex. We find that the formation of a second-shell coordination interaction is an energetically important feature of allosteric activation or inhibition of DNA binding by metal sensor proteins. Our calculations reveal that allosterically functional metal-binding sites have stronger zinc-mediated interaction energies (~20 kcal/mol or greater) that allow metal ion binding to lock the transcriptional repressor protein into an open conformation. Our data suggest the possibility that this pathway "rigidifies" the structure CzrA, and ultimately prevents it from closing to a high-DNA-binding affinity conformation. The essential dynamics reveal that Zn(II) binding to a closed conformation of the protein drives the β-wing regions and the core of the protein to adopt a more open quaternary structure. We show that the structural changes in CzrA on Zn(II) binding are accompanied by changes in the electrostatic potential energy surface, particularly in the DNA-binding interface, in a manner that should destabilize the protein–DNA complexes. This study provides a novel energetic and dynamic perspective into the evolutionary adaptation of zinc-mediated hydrogen bonds at functional allosteric sites in bacterial metalloregulatory proteins.

In addition to providing insights into protein allostery, the studies described in this chapter represent technical and developmental advances in the field of computational chemistry. The MD simulations described here were likely the first to observe allosteric switching in a protein–DNA complexes on a biologically relevant timescale (Chakravorty et al., 2012). As such, we have demonstrated that it is indeed possible to study protein allostery using well-validated computational methods (Chakravorty, Parker, et al., 2013; Chakravorty et al., 2012; Chakravorty, Wang, et al., 2013; Lee et al., 2012). Our efforts to understand the mechanism of allosteric regulation in CzrA, also led to the development of the MRD-NMR method that may be utilized to refine metal ion centers in NMR protein structures (Chakravorty, Wang, et al., 2013). The method couples the NMR structure determination protocol with metal ion force field-based MD simulations and QM/MM MD calculations, allowing for an accurate description of the metal ion-binding site, while providing for adequate sampling. Our DFT calculations investigating the effect of zinc ion binding on charge distribution in CzrA, to the best of our knowledge represented the largest calculation in terms of system size to be studied using long-range corrected DFT calculations (Chakravorty, Wang, et al., 2013). We further found that our classical bonded model approach accurately modeled the metal ion and its effect on protein structure in the NMR structure calculation process. Finally, the studies described in the text were the result of a unique collaboration in "real time" between an experimental and theoretical group and potentially lay the foundation for future studies in a similar collaborative environment.

Drug efforts targeting transcriptional regulators have been hampered by the lack of a traditional drug binding active site. The ArsR/SmtB family of proteins is ubiquitous in pathogenic prokaryotes, and is critical to their survival. Our studies suggest that the conserved energetics, structural characteristics, and conformational dynamics of ArsR/SmtB proteins may be exploited to selectively impair the MH machinery in pathogenic organisms. The unique conformational dynamics of the various allosteric forms of CzrA provide multiple targets for drug molecules that may stabilize or destabilize an allosteric form of these proteins. For example, in CzrA the large amount of conformational fluidity when bound to DNA may be exploited to bind drug molecules in the metal-binding region of the protein, preventing the protein from achieving the low-DNA-binding affinity conformational form. Conversely, the DNA-binding interface of the protein may be affected to prevent DNA binding, thus preventing the protein from repressing the expression of the CzrB cation diffusion. Such drugs may be designed to exploit the conserved interactions in these ArsR/SmtB proteins. It is possible that by functioning as likely agonists or antagonists, these xenobiotic-mediated homeostatis effectors will stabilize or destabilize an allosteric form of these proteins, moving the homeostatic window. This in turn will negatively influence repressor activity and will throw the MH machinery off-balance and impairing the organism in the process (Hood & Skaar, 2012). Such drugs would provide a new class of antibiotics that could specifically target the transcriptional machinery in these organisms.

In summary, our simulations provide new perspectives into allosteric switching by a metal sensor protein that are not possible from an experimental study alone. Our simulations qualitatively show that Zn(II) binding restricts the protein's ability to sample conformational space. In all, large-scale global conformational changes superimposed on local dynamics dominate the biological function of CzrA, and highlight the general role of metal ion mediate hydrogen bonds in transcriptional regulators.

ACKNOWLEDGMENTS

We thank the High-Performance Computing Center at the University of Florida and the Louisiana Optical Network Initiative for providing the computational resources used in this study. We thank Yue Yang, Benjamin Roberts, Greg Campanello, Pengfei Li, Bing Wang, Trent Parker, David P. Giedroc, Xue Li, Chul Won Lee, Peter Ortoleva, Yinglong Miao, Abhishekh Singha Roy, John Faver, Charles David Sherrill, and Sarah Gordon for helpful discussions. We gratefully acknowledge financial support for this project from the United States National Institutes of Health (K. M. M.; GM044974 and GM066859) and the Louisiana Alliance for Simulation Guided Materials Analysis (National Science Foundation and the Louisiana Board of Regents (D. K. C.; NSF-EPSCOR LEQS-EPS-1003897).

REFERENCES

Allen, M. P., & Tildesley, D. J. (1987). *Computer simulations of liquids*. Oxford: Clarendon Press.

Amadei, A., Linssen, A. B. M., & Berendsen, H. J. C. (1993). Essential dynamics of proteins. *Proteins, 17*, 412–425.

Amin, E. A., Wood, R. J., Johnson, R. L., & Young-Dixon, B. J. (2008). COMP 161-Evaluation of density functionals, semiempirical methods, SCC-DFTB and molecular mechanics force fields for prolyl-leucyl-glycinamide (PLG) analogs designed as dopamine D2 receptor modulators. In *Abstracts of papers of the American Chemical Society 235th national meeting, New Orleans, LA*.

Arunkumar, A. I., Campanello, G. C., & Giedroc, D. P. (2009). Solution structure of a paradigm ArsR family zinc sensor in the DNA-bound state. *Proceedings of the National Academy of Sciences of the United States of America, 106*, 18177–18182.

Arunkumar, A. I., Pennella, M. A., Kong, X. M., & Giedroc, D. P. (2007). Resonance assignments of the metal sensor CzrA in the apo-, Zn-2- and DNA-bound (42 kDa) states. *Biomolecular NMR Assignments, 1*, 99–101.

Balabanov, N. B., & Peterson, K. A. (2005). Systematically convergent basis sets for transition metals. I. All-electron correlation consistent basis sets for the 3d elements Sc-Zn. *Journal of Chemical Physics*, *123*, 064107.

Busenlehner, L. S., Pennella, M. A., & Giedroc, D. P. (2003). The SmtB/ArsR family of metalloregulatory transcriptional repressors: Structural insights into prokaryotic metal resistance. *FEMS Microbiology Reviews*, *27*, 131–143.

Campanello, G. C., Ma, Z., Grossoehme, N. E., Guerra, A. J., Ward, B. P., Dimarchi, R. D., et al. (2013). Allosteric inhibition of a zinc-sensing transcriptional repressor: Insights into the arsenic repressor (ArsR) family. *Journal of Molecular Biology*, *425*, 1143–1157.

Case, D. A., Darden, T. A., Cheatham, T. E. I., Simmerling, C. L., Wang, J., Duke, R. E., et al. (2010). *AMBER 11*. San Francisco, CA: University of California.

Chakravorty, D. K., & Hammes-Schiffer, S. (2010). Impact of mutation on proton transfer reactions in ketosteroid isomerase: Insights from molecular dynamics simulations. *Journal of the American Chemical Society*, *132*, 7549–7555.

Chakravorty, D. K., Kumarasiri, M., Soudackov, A. V., & Hammes-Schiffer, S. (2008). Implementation of umbrella integration within the framework of the empirical valence bond approach. *Journal of Chemical Theory and Computation*, *4*, 1974–1980.

Chakravorty, D. K., & Merz, K. M. (2014). Probing protein allostery in a zinc sensor protein. Unpublished work.

Chakravorty, D. K., Parker, T. M., Guerra, A. J., Sherrill, C. D., Giedroc, D. P., & Merz, K. M., Jr. (2013). Energetics of zinc-mediated interactions in the allosteric pathways of metal sensor proteins. *Journal of the American Chemical Society*, *135*, 30–33.

Chakravorty, D. K., Soudackov, A. V., & Hammes-Schiffer, S. (2009). Hybrid quantum/classical molecular dynamics simulations of the proton transfer reactions catalyzed by ketosteroid isomerase: Analysis of hydrogen bonding, conformational motions, and electrostatics. *Biochemistry*, *48*, 10608–10619.

Chakravorty, D. K., Wang, B., Lee, C. W., Giedroc, D. P., & Merz, K. M., Jr. (2012). Simulations of allosteric motions in the zinc sensor CzrA. *Journal of the American Chemical Society*, *134*, 3367–3376.

Chakravorty, D. K., Wang, B., Lee, C. W., Guerra, A. J., Giedroc, D. P., & Merz, K. M. (2013). Solution NMR refinement of a metal ion bound protein using metal ion inclusive restrained molecular dynamics methods. *Journal of Biomolecular NMR*, *56*, 125–137.

Chakravorty, D. K., Wang, B., Ucisik, M. N., & Merz, K. M., Jr. (2011). Insight into the cation–pi interaction at the metal binding site of the copper metallochaperone CusF. *Journal of the American Chemical Society*, *133*, 19330–19333.

Cornell, W. D., Cieplak, P., Bayly, C. I., & Kollman, P. A. (1993). Application of Resp charges to calculate conformational energies, hydrogen-bond energies, and free-energies of solvation. *Journal of the American Chemical Society*, *115*, 9620–9631.

Cramer, C. J., & Truhlar, D. G. (2009). Density functional theory for transition metals and transition metal chemistry. *Physical Chemistry Chemical Physics*, *11*, 10757–10816.

Darden, T., York, D., & Pedersen, L. (1993). Particle mesh Ewald—An N·Log(N) method for Ewald sums in large systems. *Journal of Chemical Physics*, *98*, 10089–10092.

Dunning, T. H. (1989). Gaussian-basis sets for use in correlated molecular calculations. 1. The atoms boron through neon and hydrogen. *Journal of Chemical Physics*, *90*, 1007–1023.

Eicken, C., Pennella, M. A., Chen, X. H., Koshlap, K. M., VanZile, M. L., Sacchettini, J. C., et al. (2003). A metal–ligand-mediated intersubunit allosteric switch in related SmtB/ArsR zinc sensor proteins. *Journal of Molecular Biology*, *333*, 683–695.

Frisch, M. J., Trucks, G. W., Schlegel, H. B., Scuseria, G. E., Robb, M. A., Cheeseman, J. R., et al. (2009). *Gaussian 09, A.01*. Wallingford, CT: Gaussian, Inc.

Fu, Z., Li, X., & Merz, K. M., Jr. (2011). Accurate assessment of the strain energy in a protein-bound drug using QM/MM X-ray refinement and converged quantum chemistry. *Journal of Computational Chemistry*, *32*, 2587–2597.
Gaus, M., Cui, Q. A., & Elstner, M. (2011). DFTB3: Extension of the self-consistent-charge density-functional tight-binding method (SCC-DFTB). *Journal of Chemical Theory and Computation*, *7*, 931–948.
Giedroc, D. P., & Arunkumar, A. I. (2007). Metal sensor proteins: Nature's metalloregulated allosteric switches. *Dalton Transactions*, *29*, 3107–3120.
Gordon, J. C., Myers, J. B., Folta, T., Shoja, V., Heath, L. S., & Onufriev, A. (2005). H++: A server for estimating pK(a)s and adding missing hydrogens to macromolecules. *Nucleic Acids Research*, *33*, W368–W371.
Grossoehme, N. E., & Giedroc, D. P. (2009). Energetics of allosteric negative coupling in the zinc sensor S. aureus CzrA. *Journal of the American Chemical Society*, *131*, 17860–17870.
Guerra, A. J., Dann, C. E., & Giedroc, D. P. (2011). Crystal structure of the zinc-dependent MarR family transcriptional regulator AdcR in the Zn(II)-bound state. *Journal of the American Chemical Society*, *133*, 19614–19617.
Guerra, A. J., & Giedroc, D. P. (2012). Metal site occupancy and allosteric switching in bacterial metal sensor proteins. *Archives of Biochemistry and Biophysics*, *519*, 210–222.
Hood, M. I., & Skaar, E. P. (2012). Nutritional immunity: Transition metals at the pathogen–host interface. *Nature Reviews. Microbiology*, *10*, 525–537.
Hoops, S. C., Anderson, K. W., & Merz, K. M. (1991). Force-field design for metalloproteins. *Journal of the American Chemical Society*, *113*, 8262–8270.
Hornak, V., Abel, R., Okur, A., Strockbine, B., Roitberg, A., & Simmerling, C. (2006). Comparison of multiple Amber force fields and development of improved protein backbone parameters. *Proteins*, *65*, 712–725.
Jorgensen, W. L., Chandrasekhar, J., Madura, J. D., Impey, R. W., & Klein, M. L. (1983). Comparison of simple potential functions for simulating liquid water. *Journal of Chemical Physics*, *79*, 926–935.
Jorgensen, W. L., & Tirodo-Rives, J. (2005). Potential energy functions for atomic-level simulations of water and organic and biomolecular systems. *Proceedings of the National Academy of Sciences of the United States of America*, *102*, 6665–6670.
Joung, I. S., & Cheatham, T. E., 3rd. (2008). Determination of alkali and halide monovalent ion parameters for use in explicitly solvated biomolecular simulations. *The Journal of Physical Chemistry B*, *112*, 9020–9041.
Kandegedara, A., Thiyagarajan, S., Kondapalli, K. C., Stemmler, T. L., & Rosen, B. P. (2009). Role of bound Zn(II) in the CadC Cd(II)/Pb(II)/Zn(II)-responsive repressor. *Journal of Biological Chemistry*, *284*, 14958–14965.
Korth, M., Pitonak, M., Rezac, J., & Hobza, P. (2010). A transferable H-bonding correction for semiempirical quantum-chemical methods. *Journal of Chemical Theory and Computation*, *6*, 344–352.
Lee, C. W., Chakravorty, D. K., Chang, F. M., Reyes-Caballero, H., Ye, Y., Merz, K. M., Jr., et al. (2012). Solution structure of Mycobacterium tuberculosis NmtR in the apo state: Insights into Ni(II)-mediated allostery. *Biochemistry*, *51*, 2619–2629.
Li, P. F., & Merz, K. M. (2014). Taking into account the ion-induced dipole interaction in the nonbonded model of ions. *Journal of Chemical Theory and Computation*, *10*, 289–297.
Li, P. F., Roberts, B. P., Chakravorty, D. K., & Merz, K. M. (2013). Rational design of particle mesh Ewald compatible Lennard-Jones parameters for +2 metal cations in explicit solvent. *Journal of Chemical Theory and Computation*, *9*, 2733–2748.
Ma, Z., Cowart, D. M., Scott, R. A., & Giedroc, D. P. (2009). Molecular insights into the metal selectivity of the copper(i)-sensing repressor CsoR from Bacillus subtilis. *Biochemistry*, *48*, 3325–3334.

Ma, Z., Cowart, D. M., Ward, B. P., Arnold, R. J., DiMarchi, R. D., Zhang, L. M., et al. (2009). Unnatural amino acid substitution as a probe of the allosteric coupling pathway in a mycobacterial Cu(I) sensor. *Journal of the American Chemical Society, 131*, 18044–18045.

Ma, Z., Jacobsen, F. E., & Giedroc, D. P. (2009). Coordination chemistry of bacterial metal transport and sensing. *Chemical Reviews, 109*, 4644–4681.

Pennella, M. A., Arunkumar, A. I., & Giedroc, D. P. (2006). Individual metal ligands play distinct functional roles in the zinc sensor Staphylococcus aureus CzrA. *Journal of Molecular Biology, 356*, 1124–1136.

Pennella, M. A., Shokes, J. E., Cosper, N. J., Scott, R. A., & Giedroc, D. P. (2003). Structural elements of metal selectivity in metal sensor proteins. *Proceedings of the National Academy of Sciences of the United States of America, 100*, 3713–3718.

Peters, M. B., Yang, Y., Wang, B., Fusti-Molnar, L., Weaver, M. N., & Merz, K. M. (2010). Structural survey of zinc-containing proteins and development of the zinc amber force field (ZAFF). *Journal of Chemical Theory and Computation, 6*, 2935–2947.

Ponomarev, S. Y., Click, T. H., & Kaminski, G. A. (2011). Electrostatic polarization is crucial in reproducing Cu(I) interaction energies and hydration. *The Journal of Physical Chemistry B, 115*, 10079–10085.

Ponomarev, S. Y., & Kaminski, G. A. (2011). Polarizable Simulations with Second order Interaction Model (POSSIM) force field: Developing parameters for alanine peptides and protein backbone. *Journal of Chemical Theory and Computation*, 7, 1415–1427.

Qsite, 5.7 ed., Schrodinger Inc. NY 2011.

Ramos, J. L., Martinez-Bueno, M., Molina-Henares, A. J., Teran, W., Watanabe, K., Zhang, X., et al. (2005). The TetR family of transcriptional repressors. *Microbiology and Molecular Biology Reviews, 69*, 326–356.

Reyes-Caballero, H., Campanello, G. C., & Giedroc, D. P. (2011). Metalloregulatory proteins: Metal selectivity and allosteric switching. *Biophysical Chemistry, 156*, 103–114.

Schüürmann, G., & Klamt, A. (1993). COSMO: A new approach to dielectric screening in solvents with explicit expressions for the screening energy and its gradient. *Journal of the Chemical Society, Perkin Transactions 2, 2*, 799–805.

Seabra, G. D., Walker, R. C., Elstner, M., Case, D. A., & Roitberg, A. E. (2007). Implementation of the SCC-DFTB method for hybrid QM/MM simulations within the amber molecular dynamics package. *Journal of Physical Chemistry A, 111*, 5655–5664.

Sedlak, R., Janowski, T., Pitonak, M., Rezac, J., Pulay, P., & Hobza, P. (2013). The accuracy of quantum chemical methods for large noncovalent complexes. *Journal of Chemical Theory and Computation, 9*, 3364–3374.

Senn, H. M., & Thiel, W. (2009). QM/MM methods for biomolecular systems. *Angewandte Chemie, 48*, 1198–1229 (International Ed. in English).

Settergren, N. M., Buhlmann, P., & Amin, E. A. (2008). Assessment of density functionals, semiempirical methods, and SCC-DFTB for protonated creatinine geometries. *Journal of Molecular Structure: THEOCHEM, 861*, 68–73.

Shao, J. Y., Tanner, S. W., Thompson, N., & Cheatham, T. E. (2007). Clustering molecular dynamics trajectories: 1. Characterizing the performance of different clustering algorithms. *Journal of Chemical Theory and Computation, 3*, 2312–2334.

Stewart, J. J. P. (2008). *MOPAC 2009*. Colorado Springs, CO: Stewart Computational Chemistry.

Swint-Kruse, L., & Matthews, K. S. (2009). Allostery in the LacI/GalR family: Variations on a theme. *Current Opinion in Microbiology, 12*, 129–137.

Tao, J. M., & Perdew, J. P. (2003). New meta-generalized gradient approximation for exchange and correlation. *Abstracts of Papers of the American Chemical Society, 225*, U469–U470.

Thanthiriwatte, K. S., Hohenstein, E. G., Burns, L. A., & Sherrill, C. D. (2011). Assessment of the performance of DFT and DFT-D methods for describing distance dependence of hydrogen-bonded interactions. *Journal of Chemical Theory and Computation*, 7, 88–96.

Thomson, A. J., & Gray, H. B. (1998). Bio-inorganic chemistry. *Current Opinion in Chemical Biology*, *2*, 155–158.

Turney, J. M., Simmonett, A. C., Parrish, R. M., Hohenstein, E. G., Evangelista, F. A., Fermann, J. T., et al. (2012). PSI4: An open-source ab initio electronic structure program. *WIREs Computational Molecular Science*, *2*, 556–565.

Tus, A., Rakipovic, A., Peretin, G., Tomic, S., & Sikic, M. (2012). BioMe: Biologically relevant metals. *Nucleic Acids Research*, *40*, W352–W357.

Ucisik, M. N., Chakravorty, D. K., & Merz, K. M., Jr. (2013). Structure and dynamics of the N-terminal domain of the Cu(I) binding protein CusB. *Biochemistry*, *52*, 6911–6923.

Wadt, W. R., & Hay, P. J. (1985). Ab initio effective core potentials for molecular calculations—Potentials for main group elements Na to Bi. *Journal of Chemical Physics*, *82*, 284–298.

Waldron, K. J., & Robinson, N. J. (2009). How do bacterial cells ensure that metalloproteins get the correct metal? *Nature Reviews. Microbiology*, 7, 25–35.

Waldron, K. J., Rutherford, J. C., Ford, D., & Robinson, N. J. (2009). Metalloproteins and metal sensing. *Nature*, *460*, 823–830.

Weaver, M. N., Yang, Y., & Merz, K. M. (2009). Assessment of the CCSD and CCSD(T) coupled-cluster methods in calculating heats of formation for Zn complexes. *Journal of Physical Chemistry A*, *113*, 10081–10088.

Wu, J. C., Piquemal, J. P., Chaudret, R., Reinhardt, P., & Ren, P. (2010). Polarizable molecular dynamics simulation of Zn(II) in water using the AMOEBA force field. *Journal of Chemical Theory and Computation*, *6*, 2059–2070.

Yang, Y., Chakravorty, D. K., & Merz, K. M., Jr. (2010). Finding a needle in the haystack: Computational modeling of Mg^{2+} binding in the active site of protein farnesyltransferase. *Biochemistry*, *49*, 9658–9666.

Yang, Y., Weaver, M. N., & Merz, K. M. (2009). Assessment of the "6-31+G** +LANL2DZ" mixed basis set coupled with density functional theory methods and the effective core potential: Prediction of heats of formation and ionization potentials for first-row-transition–metal complexes. *Journal of Physical Chemistry A*, *113*, 9843–9851.

Ye, J., Kandegedara, A., Martin, P., & Rosen, B. P. (2005). Crystal structure of the Staphylococcus aureus pI258 CadC Cd(II)/Pb(II)/Zn(II)-responsive repressor. *Journal of Bacteriology*, *187*, 4214–4221.

Zhao, C., Parrish, R. M., Smith, M. D., Pellechia, P. J., Sherrill, C. D., & Shimizu, K. D. (2012). Do deuteriums form stronger CH–pi interactions? *Journal of the American Chemical Society*, *134*, 14306–14309.

Zhao, Y., & Truhlar, D. G. (2008). Density functionals with broad applicability in chemistry. *Accounts of Chemical Research*, *41*, 157–167.

CHAPTER SEVEN

Insights in the Mechanism of Action and Inhibition of *N*-Acylethanolamine Acid Amidase by Means of Computational Methods

Alessio Lodola[1], Silvia Rivara, Marco Mor
Dipartimento di Farmacia, Università degli Studi di Parma, Parma, Italy
[1]Corresponding author: e-mail address: alessio.lodola@unipr.it

Contents

Abstract

Computer-aided approaches are widely used in modern medicinal chemistry to improve the efficiency of the discovery phase. *N*-Acylethanolamine acid amidase (NAAA) is a cysteine amidase belonging to the N-terminal nucleophile (Ntn) hydrolases that primarily degrades anti-inflammatory and analgesic lipid amide palmitoylethanolamide. In this chapter, we review our contribution to (i) the determination of the reaction mechanism of amide hydrolysis catalyzed by cysteine Ntn-hydrolases and (ii) the discovery and optimization of active-site-directed inhibitors of NAAA characterized by a β-lactone warhead. The combination of different computational tools, ranging from homology modeling, docking, and mechanistic simulations based on hybrid quantum mechanics/molecular mechanics potentials, contributed to the elucidation of the mechanism of action and inhibition of NAAA enzyme and to the design of more potent inhibitors.

Advances in Protein Chemistry and Structural Biology, Volume 96
ISSN 1876-1623
http://dx.doi.org/10.1016/bs.apcsb.2014.06.007

1. INTRODUCTION

N-Acylethanolamine acid amidase (NAAA) is a cysteine hydrolase involved in the hydrolysis of fatty acid ethanolamides (FAEs) with a marked preference for palmitoylethanolamide (PEA; Fig. 1; Tsuboi et al., 2005; Ueda, Yamanaka, & Yamamoto, 2010). In the last decade, we have witnessed a growing interest in NAAA driven by studies showing that its main substrate PEA has significant anti-inflammatory and analgesic properties (Mazzari, Canella, Petrelli, Marcolongo, & Leon, 1996). Most of these effects depend on the ability of PEA to bind and activate the peroxisome proliferator-activated receptor-α (Lo Verme et al., 2005, 2006), even if other mechanism may have a role (Petrosino, Iuvone, & Di Marzo, 2010), such as the activation of TRPV1, GPR55, and GPR129.

Recent investigations have shown that endogenous levels of PEA undergo a marked reduction during inflammation in various rodent cells and tissues. Further studies have related the decrement of the cellular levels of PEA associated to a proinflammatory condition to the downregulation of a specific phospholipase, known as NAPE-PLD, involved in the biosynthesis of PEA and of other biologically active FAEs such as oleoylethanolamide and anandamide (AEA; Zhu et al., 2011). This bunch of observations has thus suggested that pharmacological approaches aimed at restoring the tissue levels of PEA, for instance through the direct inhibition of the enzyme responsible for the deactivating hydrolysis of PEA, might represent a novel approach to control inflammation and pain. The catabolism of PEA is controlled by NAAA and by another key intracellular amidase called fatty acid amide hydrolase (FAAH; McKinney & Cravatt, 2005) which primarily hydrolyses AEA and for which potent inhibitors (Deng, 2010; Kathuria et al., 2003; Mor et al., 2004) have been advanced to clinical trials, while the search for potent and selective NAAA inhibitors is still at its beginning (Bandiera, Ponzano, & Piomelli, 2014).

In the present review, we summarize the results of our research aimed at identifying and characterizing the mechanism of action of potent NAAA

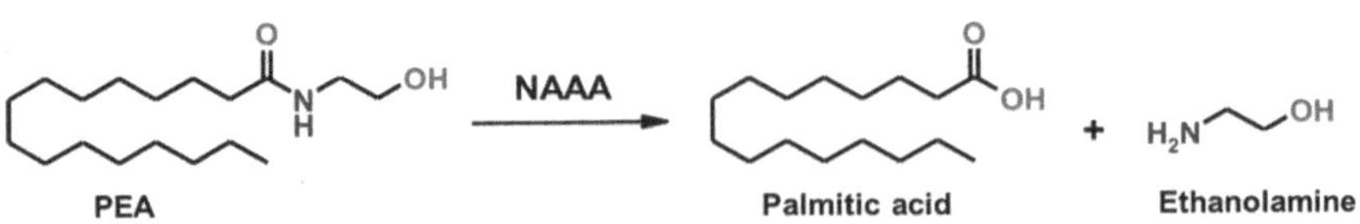

Figure 1 Hydrolysis of PEA catalyzed by NAAA.

inhibitors featured by a β-lactone scaffold by means of computational and experimental methods. Due to the importance of enzyme reactivity in the design and optimization of active-site-directed inhibitors, we will also focus our attention on the catalytic mechanism of the cysteine N-terminal nucleophile (Ntn)-hydrolases, a family of enzymes to which NAAA and other important pharmacological targets belong.

2. GENERAL INFORMATION AND CATALYTIC MECHANISM OF CYSTEINE NTN-HYDROLASES

Despite its functional similarity to other amidases including FAAH, NAAA shares no homology with this enzyme or other enzymes of the amidase signature family. While FAAH is a membrane serine amidase which localizes in the endoplasmic reticulum and possesses a moderate preference for cleaving AEA, NAAA is a soluble cysteine amidase localized in the lysosome that preferentially hydrolyses PEA over other FAEs (Tsuboi, Takezaki, & Ueda, 2007).

NAAA belongs to the choloylglycine hydrolase family of hydrolases, which are characterized by their ability to cleave nonpeptide amide bonds. NAAA is an Ntn-hydrolase and, as such, is proteolytically processed to a shorter active form upon incubation at acidic pH (Zhao, Tsuboi, Okamoto, Nagahata, & Ueda, 2007). Processing of rodent NAAA renders cysteine 131 (Cys131) the N-terminal amino acid and the putative catalytic residue. Site-directed mutagenesis experiments have confirmed the importance of Cys131 for catalysis by NAAA (Solorzano et al., 2009; Wang et al., 2008). Kinetic experiments have also shown that the hydrolytic activity of NAAA displays a bell-shaped pH dependency with an optimal activity at pH 5, consistent with its localization in the lysosomes.

While the three-dimensional (3D) structure of NAAA is still unknown, the crystallographic coordinates of a choloylglycine hydrolase prototype, namely conjugate bile acid hydrolase (CBAH; Rossocha, Schultz-Heienbrok, von Moeller, Coleman, & Saenger, 2005), have been recently resolved and used to build a homology model of NAAA. This model, which was validated by mutagenesis experiments (Rossocha et al., 2005), suggests that CBAH and NAAA share a relatively well-conserved active site (Fig. 2) where, in addition to the *N*-terminal cysteine Cys2 of CBAH (Cys131 in NAAA), other residues appear to be essential for the amidase activity of the enzymes. These are Arg18 (147) and Asp21 (150), which interacted with the sulfhydryl and α-amino groups of Cys2 (131) and have been proposed to

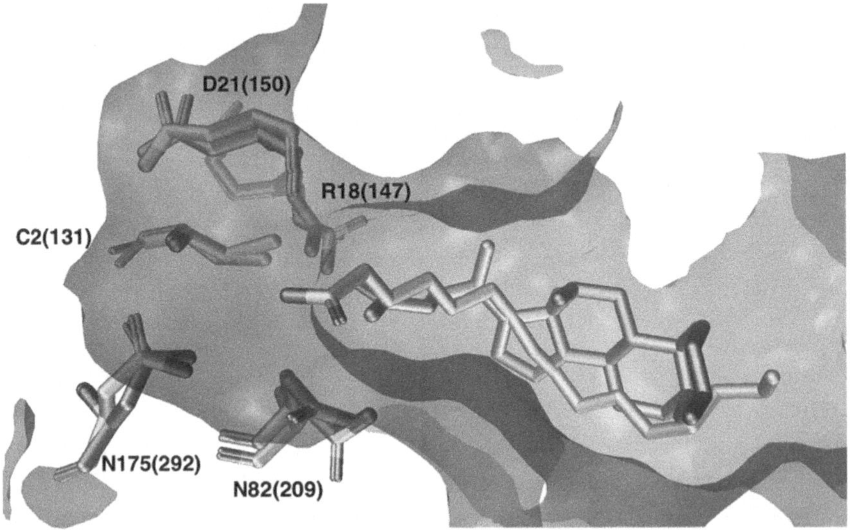

Figure 2 Superposition of the active-site residues of Ntn members NAAA (cyan (light gray in the print version) carbons) and CBAH (orange (dark gray in the print version) carbons, PDB: 2BJF). Palmitic acid (docked in NAAA) and deoxycholic acid (observed in the X-ray structure of CBAH) are also displayed. Numbers close to the amino acid one letter code refer to CBAH, those in parenthesis to NAAA.

modulate their pK_a values; Asn82 (209) and Asn175 (292), which form the putative oxyanion hole concur to the recognition of the substrate and stabilize transition state (TS) and intermediates along the pathways of acylation and deacylation of the catalytic cysteine.

A general catalytic mechanism for Ntn-hydrolases has been proposed by Oinonen and Rouvinen on the basis of structural data (Oinonen & Rouvinen, 2000). According to this hypothesis, the N-terminal catalytic residue of these enzymes lives in the neutral form. The reaction starts with a proton transfer between the nucleophile (—OH or —SH) of the catalytic residue and the α-amino group of the same amino acid (Fig. 3). Once deprotonated, the nucleophile attacks the carbonyl carbon of the substrate, leading to the formation of an anionic tetrahedral intermediate (TI). The reaction becomes complete when the α-amino group of the catalytic residue donates a proton to the nitrogen of the scissile amide bond. In this way, an acylenzyme adduct is formed and the amino leaving group of the substrate is released (Fig. 3). However, despite being very suggestive, this mechanism is in contrast with experimental evidence on cysteine Ntn-hydrolases. For instance, since the optimal activity of cysteine Ntn-hydrolases occurs at acidic pH, it is unlikely that the α-amino group of the N-terminal Cys would act as a general base during the catalysis.

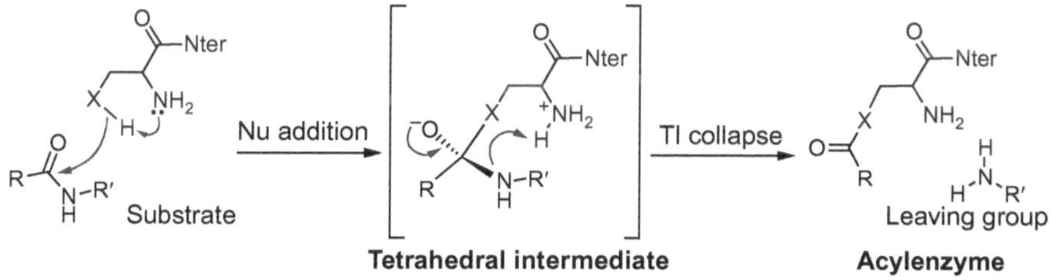

Figure 3 Mechanism proposed by Oinonen and Rouvinen for Ntn-hydrolase.

Computer simulations were used to characterize the ionization state of the N-terminal cysteine, using CBAH as the reference model (Lodola et al., 2012). The pK_a values of Cys-NH_3^+ and Cys-SH groups were predicted to be ~7.5 and ~4 (Li, Robertson, & Jensen, 2005). Calculations also showed that while the ionization state of the α-amino group (Cys-NH_3^+) was scarcely influenced by enzyme environment when compared to solution, the thiol group (Cys-SH) resulted significantly more acidic than in solution (calculated pK_a in CBAH = 4, experimental pK_a in solution = 8; Li et al., 2005). A further analysis revealed that the shift in the pK_a value was mainly due to electrostatic interactions between the terminal cysteine sulfur atom and the conserved residues Arg18 (147), Asp21 (150), Asn82 (209), and Asn175 (292), suggesting that enzyme structure has evolved to stabilize the thiolate anion. Since CBAH and NAAA display an optimal hydrolytic activity close to pH 5, we proposed that their N-terminal cysteine participates to catalysis in zwitterionic form, i.e., Cys-S^-/Cys-NH_3^+ ion pair. This hypothesis is supported by the bell-shaped correlation between hydrolytic activity and pH reported for CBAH in Fig. 4. In acidic conditions (i.e., pH < 5), the catalytic activity decreases as the fraction of the thiolate nucleophile decreases. On the other hand, in basic conditions, the fraction of Cys-NH_3^+ decreases leading to a progressive loss of catalytic activity.

To elucidate the catalytic mechanism of CBAH, the first step of amide substrate hydrolysis was investigated by means of a hybrid quantum mechanics/molecular mechanics technique (Field, Bash, & Karplus, 1990) in combination with the path collective variable method (Branduardi, Gervasio, & Parrinello, 2007). Based on the free energy profile obtained with this approach, the catalytic mechanism of cysteine Ntn-hydrolase and specifically the process of cysteine acylation can be divided into five regions (Fig. 5): (i) the Michaelis complex (i.e., the reactants; region 1); (ii) a distorted Michaelis complex featuring a pyramidalized amide nitrogen (region 2); (iii) the TS, dominated by protonation of amide nitrogen (region 3);

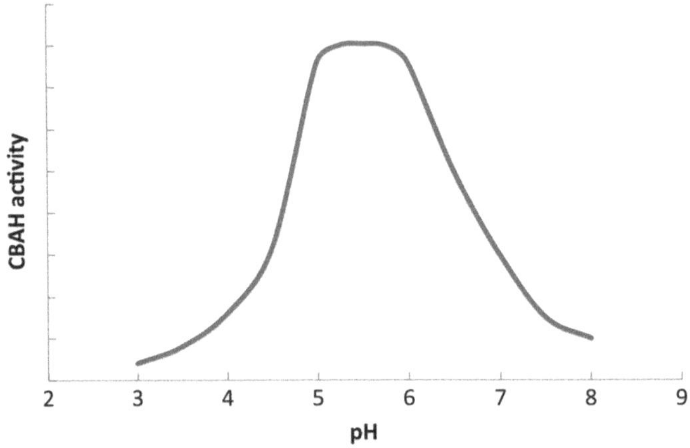

Figure 4 pH dependence of CBAH activity according to the data reported by Gopal-Srivastava and Hylemon (1988).

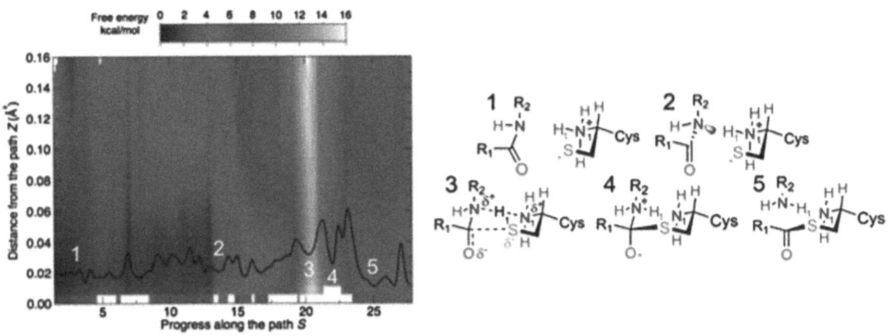

Figure 5 Free energy surface for amide hydrolysis catalyzed by CBAH in the *S* (progress along the path) and *Z* (distance from the path) space from umbrella sampling simulations. The minimum free energy path is displayed with a continuous black line. Configurations 1–5 are crucial structures identified along the reaction path, *S*.

(iv) the TI which did not correspond to a well-defined free energy minimum (region 4); and (v) the acylenzyme adduct (i.e., the products; region 5). The free energy surface (FES) of Fig. 5 also shows that, while the profile is flat in regions **1** and **2**, a significant increment in the free energy is observed approaching region **3** (the TS area), which is characterized by a saddle-like surface. Region **4** accounts for the nucleophilic attack and spontaneously evolves to **5**, the acylenzyme adduct, through the expulsion of the protonated leaving group.

The activation free energy of 13 kcal/mol for this catalytic process (i.e., measured as difference between states **1** and **3**) was in excellent agreement with experimental data (experimental activation free energy of 15 kcal/mol; Kumar et al., 2006), indicating the likelihood of the described mechanism. A detailed inspection of the geometries of the key points identified on the FES showed that the protonation of the amide leaving group and the nucleophile attack were tightly concerted. As consequence, the overall catalytic process required the formation of a zwitterionic TI (i.e., configuration **4**) rather than an anionic one as a crossroad for acylenzyme formation (Fig. 5). Overall computational and experimental investigations suggest that (i) the catalytic N-terminal cysteine of CBAH and NAAA can participate to catalysis via its zwitterionic form and (ii) the lowest free energy reaction pathway does not show the formation of an anionic TI, in analogy with other cysteine hydrolases (Harrison, Burton, & Hillier, 1997). Calculations also revealed the presence of a zwitterionic TI characterized by a cyclic "chair-like" structure, which might represent the catalytic signature of this class of enzyme.

3. INHIBITION OF NAAA BY β-LACTONES

The elucidation of the primary structure of NAAA and its classification as a cysteine hydrolase belonging to the Ntn family of enzymes oriented our research of NAAA inhibitors toward compounds containing chemical groups susceptible to nucleophilic attack by the thiol group of the catalytic cysteine. The screening of a series of compounds containing cysteine-reactive warheads led to the identification of 2-oxo-3-oxetanylcarbamic acid benzyl ester (SD-41; Fig. 6), also a weak NAAA inhibitor (Solorzano et al., 2009). A chemical exploration around this compound led to the identification of (*S*)-*N*-(2-oxo-3-oxetanyl)-3-phenylpropanamide [(*S*)-OOPP; Fig. 6], a rather potent NAAA inhibitor with an IC_{50} of 420 nM (Solorzano et al., 2009).

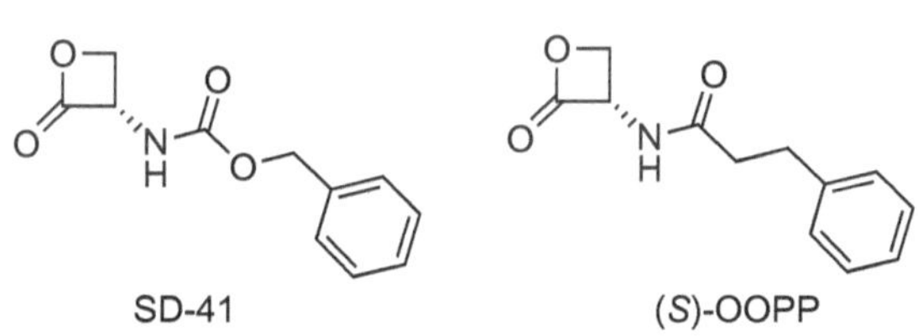

Figure 6 Structures of SD-41 and (*S*)-OOPP identified as NAAA inhibitors.

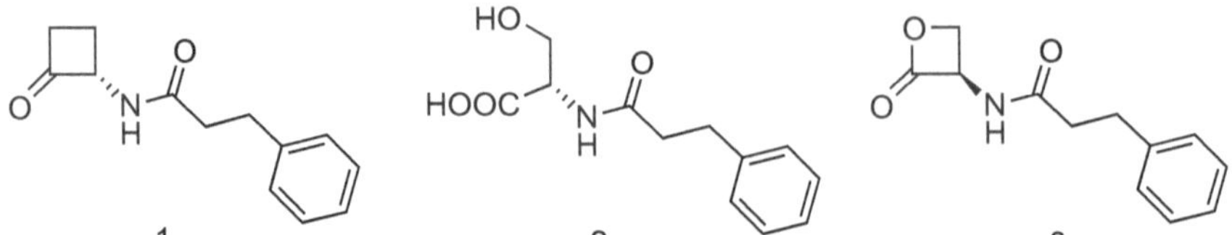

1 2 3

Figure 7 Structures of inactive (**1**, **2**) or poorly active (**3**) derivatives of (*S*)-OOPP.

A structure–activity relationship (SAR) investigation demonstrated the importance of both reactivity and recognition in the inhibition of NAAA by (*S*)-OOPP derivatives (Solorzano et al., 2010). Replacement of the β-lactone with a cyclobutanone or with an open-chain analogue led to the inactive compounds **1** and **2** (Fig. 7), indicating the essential role of the β-lactone ring for obtaining active compounds. On the other hand, the markedly diminished potency displayed by the *R* enantiomer of (*S*)-OOPP, **3** highlighted a selective recognition between enzyme and inhibitor.

On the basis of these observations, it was initially proposed that (*S*)-OOPP interacted with NAAA active site by covalent binding to the sulfhydryl group of the *N*-terminal Cys131 through its 2-oxo-3-oxetane ring. Two alternative routes were proposed according to previous studies on β-lactone reactivity (Fig. 8; Lall, Karvellas, & Vederas, 1999): route *a*, in which the nucleophilic attack occurs at the 2-carbonyl giving a thioester as the result of NAAA acylation, and route *b* where the nucleophilic attack occurs at the 4-methylene group, resulting in NAAA alkylation.

The 3D model of NAAA was used to build intermediates for routes *a* and *b* (Fig. 6) and thus to propose a likely mechanism of inhibition for this class of compounds. A putative reaction intermediates resulting from (a) nucleophilic attack of the sulfhydryl group of Cys131 on the lactone carbonyl and (b) alkylation of the cysteine residue by the 4-methylene group of the lactone ring were built using (*S*)-OOPP as reference compound. Visual inspection of these covalent intermediates suggested acylation rather than alkylation as an effective mechanism of NAAA inhibition. Indeed, following the acylation mechanism, the phenylpropionyl chain of (*S*)-OOPP fits into the lipophilic pocket lined by the aromatic ring of Tyr151 (Fig. 9), which has an important role in the recognition of NAAA substrates. In the alkylation reaction, the propionyl chain of (*S*)-OOPP reaches out of the catalytic pocket and does not show any interaction with the surrounding polar groups or the substrate-binding lipophilic cavity (model not shown).

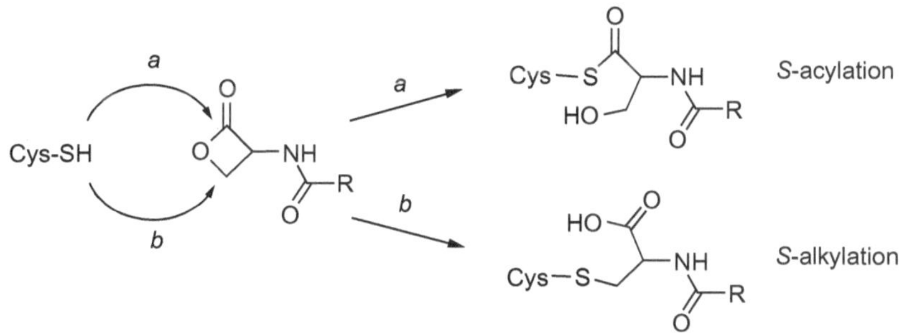

Figure 8 Proposed mechanisms for covalent inactivation of NAAA by β-lactone derivatives.

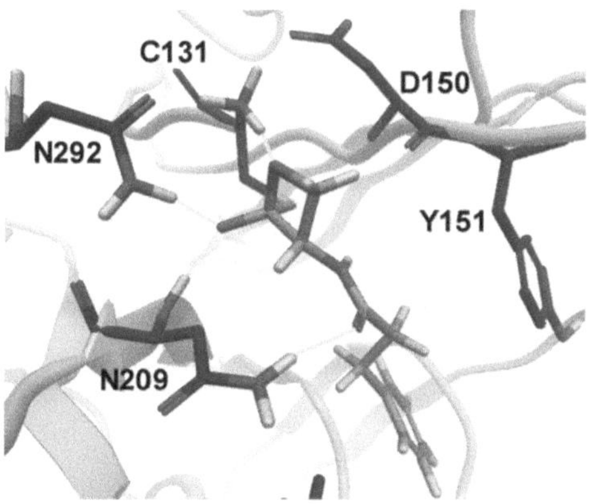

Figure 9 Putative tetrahedral intermediate resulting from the nucleophile attack of the sulfhydryl group of NAAA Cys131 on the lactone carbonyl of (*S*)-OOPP.

The finding that NAAA inhibition by (*S*)-OOPP is partially reversible (Solorzano et al., 2009) confirms the initial speculation, proposing acylation of Cys131 as the main mechanism of NAAA inhibition. Notably, LC-MS experiments performed by the Makriyannis group (West et al., 2012) as well as by the Piomelli group (Armirotti et al., 2012) indicated that these β-lactone-based compounds inhibited NAAA through covalent modification of its catalytic cysteine. Piomelli and coworkers identified acylation and not alkylation as the mechanism of action for the reference β-lactone NAAA inhibitor URB913 (ARN077) by applying a high-resolution mass spectrometry approach.

To support the design of (*S*)-OOPP derivatives with improved NAAA inhibitory potency, we performed docking studies aimed at identifying a noncovalent complex that could highlight the role played by the

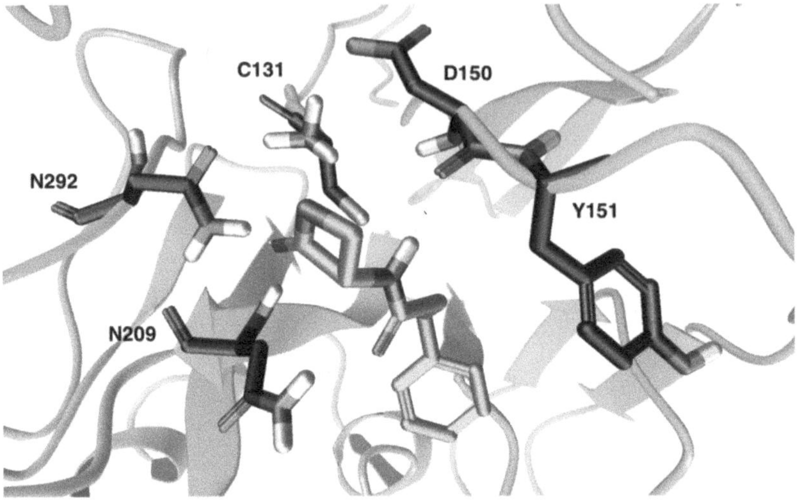

Figure 10 Putative noncovalent complex involving NAAA and (*S*)-OOPP as obtained by docking simulations.

phenylpropionamide moiety in the recognition process. Automated docking of (*S*)-OOPP into the NAAA model gave only a solution in which the lactone ring was sufficiently close to Cys131 to allow a reactive event while the phenylpropionamide chain was actively contributing to the recognition phase. In this binding mode (Fig. 10), the lactone carbonyl points its oxygen toward the putative oxyanion hole, formed by the backbone NH of Asn209 and the side chain NH of Asn292, and juxtaposes its carbon to the Cys131 sulfhydryl group of NAAA. Moreover, the amide fragment of (*S*)-OOPP makes two hydrogen bonds with the backbone of Asp150, as well as with the side chain of Asn209, which has been demonstrated to be essential for enzyme functioning (Li et al., 2005; Rossocha et al., 2005). In addition, the phenylpropionyl chain of (*S*)-OOPP fits into the lipophilic pocket lined by the aromatic ring of Tyr151, which has been recognized by mutagenesis studies to be involved in the catalytic process (Solorzano et al., 2009). More importantly, this theoretical model also implied that inhibitory potency could be improved by modulating the lipophilicity of the phenylpropionic chain of (*S*)-OOPP.

Staring from this interaction model, a series of (*S*)-OOPP derivatives characterized by a more rigid acylating group at the exocyclic nitrogen atom at the α-amino-β-lactone portion were prepared and tested on rat NAAA. This exploration led to the identification of several NAAA inhibitors (compounds **4–13** in Table 1), some of which were featured by good inhibitory potency (Solorzano et al., 2010). Among the tested compounds, the

Table 1 IC_{50} of *N*-(2-oxo-3-oxetanyl)arylamides (**4–13**) on rat NAAA (Solorzano et al., 2010)

Compounds	Structure	IC_{50} (nM)
4	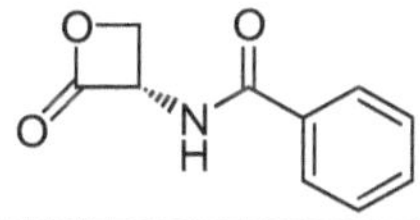	1700
5	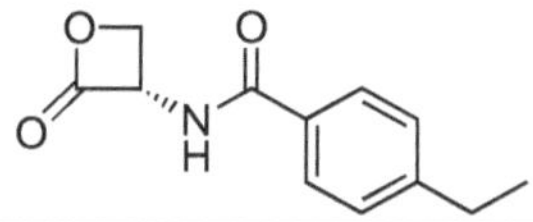	102
6	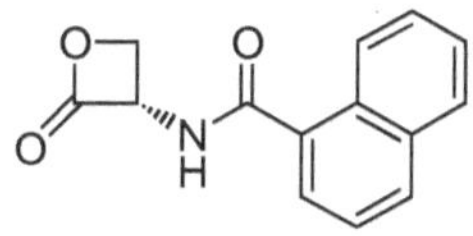	50,000
7	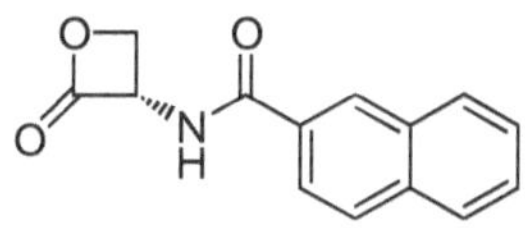	160
8	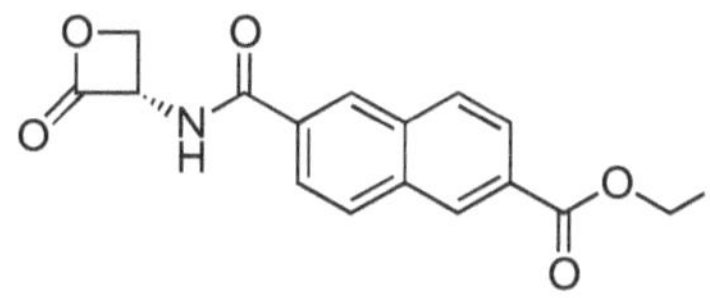	300
9	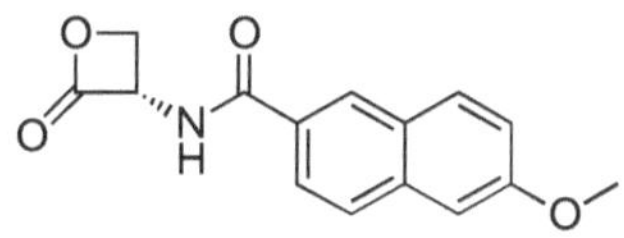	400
10	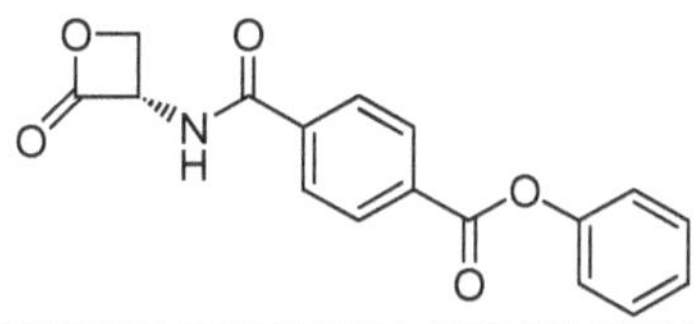	300
11	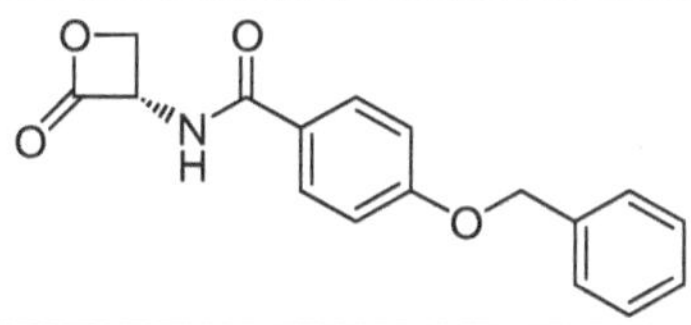	90
12	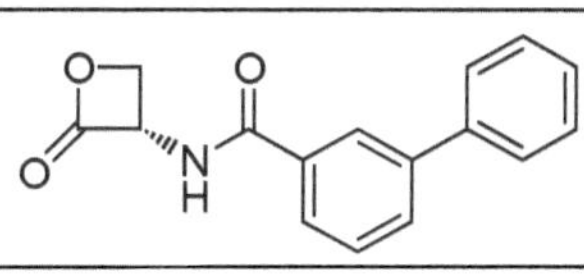	4400
13	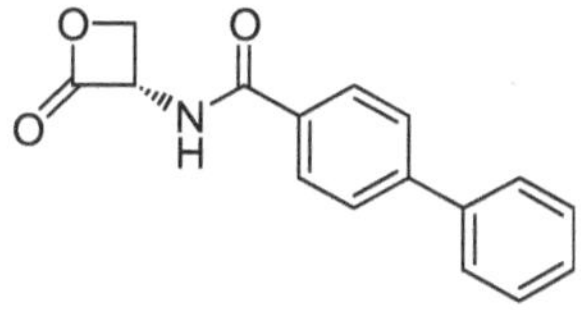	115 ± 13

benzyloxyphenyl derivative **11** and the *para*-biphenyl derivative **13** emerged as rather potent inhibitors having IC_{50} in the nanomolar range. To provide a quantitative description of the SARs for a homogeneous subclass of compounds, the set of (*S*)-*N*-(2-oxo-3-oxetanyl)arylamides reported in Table 1 was analyzed using the three-dimensional quantitative structure–activity relationship (3D-QSAR) tool, CoMSIA (Klebe & Abraham, 1999; Klebe, Abraham, & Mietzner, 1994), after superposition of their common lactone rings in minimum-energy conformations.[1] Statistical analysis was performed employing the steric field only, since the electrostatic field did not offer sufficient variation to get significant results. Using partial least-squares analysis to correlate pIC_{50} values and the steric field, we obtained the model with the best predictive power using two latent variables (LVs). The two-LV model, ($R^2=0.932$, $s=0.259$, $N=10$, $F=48.34$, $Q^2_{LOO}=0.536$, SDEP$=0.568$), is graphically represented in Fig. 11. Analysis of the colored volumes points out that allowed regions beyond the first aromatic ring (green in Fig. 11) shape a flat and straight pocket, so that bent structures fall in disallowed regions, responsible of potency loss (yellow in Fig. 11). The different inhibitory potencies observed for the two biphenyl derivatives **12** and **13** give evidence that specific molecular recognition, and not only lipophilic partitioning, takes place at this portion of the enzyme active site.

4. RECENT ADVANCES IN THE FIELD OF ACTIVE-SITE-DIRECTED NAAA INHIBITORS

The α-amino-β-lactone scaffold, although essential for the inhibitory activity of *N*-(2-oxo-3-oxetanyl)amides, is also responsible for their low chemical stability. Compounds incorporating an α-amino-β-lactone portion, including the reference compound (*S*)-OOPP, promptly react with bionucleophiles such as albumin and are readily hydrolyzed in aqueous media (Duranti et al., 2012). As a consequence, their use in pharmacological studies remains restricted to topical administration and requires caution in

[1] Inhibitor models were built and optimized using the MMFF94s force field to an energy gradient of 0.01 kcal/mol/Å. The minimum-energy conformations having a value of approximately 180° for the torsion angle between C2–C3 in the 2-oxo-3-oxetane ring and N–C in the amide fragment were selected, and their 2-oxo-3-oxetane rings superimposed. The CoMSIA steric field was calculated within a lattice of 1 Å of grid resolution, whose extension was at least 4 Å beyond every molecule boundary in each direction. Regression analyses were performed applying the partial least-squares (PLS) algorithm in Sybyl. The optimal number of latent variables (LVs) was selected by means of the cross-validation technique, on the basis of the Q_2 value calculated with the SAMPLS algorithm.

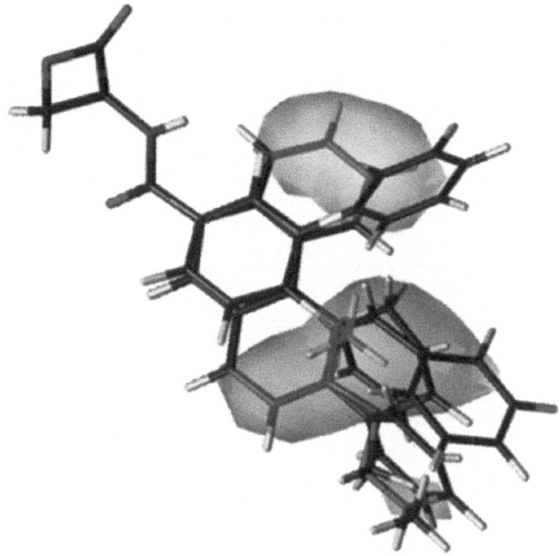

Figure 11 Superposition of *N*-(2-oxo-3-oxetanyl)amides employed for 3D-QSAR analysis and coefficients of the CoMSIA steric fields; green (gray in the print version): coefficients >0.008; yellow (light gray in the print version): coefficients <−0.008.

their handling and storage. Interestingly, recent data demonstrated that the introduction of an α-carbamic acid ester side chain and a β-methyl group on the β-lactone ring resulted in potent NAAA inhibitors with enhanced chemical stability (Duranti et al., 2012). This exploration has thus led to (2*S*,3*R*)-2-methyl-4-oxo-3-oxetanylcarbamic acid 5-phenylpentyl ester (URB913/ARN077; Fig. 12), which (i) is a potent *in vitro* rat NAAA inhibitor (IC_{50} = 127 nM), (ii) has an improved chemical stability profile, (iii) has a reduced propensity to react with bovine serum albumin, and (iv) is rapidly cleaved by plasma hydrolases. This compound is a useful new tool for investigating the role of NAAA in inflammation and analgesia after topical administration (Sasso et al., 2013). Only recently, the first class of orally available NAAA inhibitors derived from the α-amino-β-lactone series has been reported. These compounds were obtained through the isosteric replacement of the β-lactone group with a β-lactam one (Fiasella et al., 2014).

Among the synthesized compounds, *N*-[(*S*)-2-oxoazetidin-3-yl] nonanamide (Fig. 12) showed good NAAA inhibitory potency (IC_{50} = 340 nM), excellent chemical stability in buffer, and good plasma stability. Importantly, this compound showed good oral bioavailability in rats

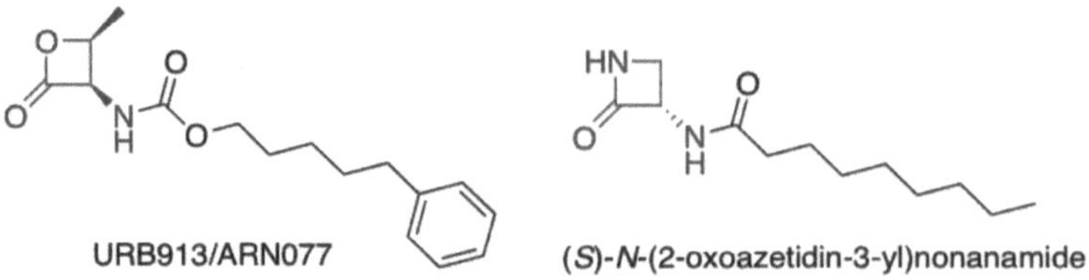

Figure 12 Structures of second-generation NAAA inhibitors.

and thus may help to further characterize the functional roles of NAAA and assess the therapeutic potential of systemically available NAAA inhibitors.

ACKNOWLEDGMENT

We thank our coauthors and collaborators in the relevant work reviewed here.

REFERENCES

Armirotti, A., Romeo, E., Ponzano, S., Mengatto, L., Dionisi, M., Karacsonyi, C., et al. (2012). β-Lactones inhibit *N*-acylethanolamine acid amidase by *S*-acylation of the catalytic *N*-terminal cysteine. *ACS Medicinal Chemistry Letters*, *3*, 422–426.

Bandiera, T., Ponzano, S., & Piomelli, D. (2014). Advances in the discovery of *N*-acylethanolamine acid amidase inhibitors. *Pharmacological Research*. http://dx.doi.org/10.1016/j.phrs.2014.04.011.

Branduardi, D., Gervasio, F. L., & Parrinello, M. (2007). From A to B in free energy space. *The Journal of Chemical Physics*, *126*, 054103.

Deng, H. (2010). Recent advances in the discovery and evaluation of fatty acid amide hydrolase inhibitors. *Expert Opinion on Drug Discovery*, *5*, 961–993.

Duranti, A., Tontini, A., Antonietti, F., Vacondio, F., Fioni, A., Silva, C., et al. (2012). *N*-(2-Oxo-3-oxetanyl)-carbamic acid esters as *N*-acylethanolamine acid amidase inhibitors: Synthesis and structure-activity and structure-property relationships. *Journal of Medicinal Chemistry*, *55*, 4824–4836.

Fiasella, A., Nuzzi, A., Summa, M., Armirotti, A., Tarozzo, G., Tarzia, G., et al. (2014). 3-Aminoazetidin-2-one derivatives as *N*-acylethanolamine acid amidase (NAAA) inhibitors suitable for systemic administration. *ChemMedChem*, *9*(7), 1602–1614. http://dx.doi.org/10.1002/cmdc.201300546.

Field, M. J., Bash, P. A., & Karplus, M. (1990). A combined quantum mechanical and molecular mechanical potential for molecular dynamics simulations. *Journal of Computational Chemistry*, *11*, 700–733.

Gopal-Srivastava, R., & Hylemon, P. B. (1988). Purification and characterization of bile salt hydrolase from Clostridium perfringens. *The Journal of Lipid Research*, *29*, 1079–1085.

Harrison, M. J., Burton, N. A., & Hillier, I. H. (1997). Catalytic mechanism of the enzyme papain: Predictions with a hybrid quantum mechanical/molecular mechanical potential. *Journal of the American Chemical Society*, *119*, 12285–12291.

Kathuria, S., Gaetani, S., Fegley, D., Valiño, F., Duranti, A., Tontini, A., et al. (2003). Modulation of anxiety through blockade of anandamide hydrolysis. *Nature Medicine*, *9*, 76–81.

Klebe, G., & Abraham, U. (1999). Comparative molecular similarity index analysis (CoMSIA) to study hydrogen-bonding properties and to score combinatorial libraries. *Journal of Computer-Aided Molecular Design*, *13*, 1–10.

Klebe, G., Abraham, U., & Mietzner, T. (1994). Molecular similarity indices in a comparative analysis (CoMSIA) of drug molecules to correlate and predict their biological activity. *Journal of Medicinal Chemistry*, *37*, 4130–4146.

Kumar, R. S., Brannigan, J. A., Prabhune, A. A., Pundle, A. V., Dodson, G. G., Dodson, E. J., et al. (2006). Structural and functional analysis of a conjugated bile salt hydrolase from *Bifidobacterium longum* reveals an evolutionary relationship with penicillin V acylase. *Journal of Biological Chemistry*, *281*, 32516–32525.

Lall, M. S., Karvellas, C., & Vederas, J. C. (1999). β-lactones as a new class of cysteine proteinase inhibitors: Inhibition of hepatitis A virus 3C proteinase by *N*-Cbz-serine β-lactone. *Organic Letters*, *1*, 803–806.

Li, H., Robertson, A. D., & Jensen, J. H. (2005). Very fast empirical prediction and rationalization of protein pK_a values. *Proteins*, *61*, 704–721.

Lodola, A., Branduardi, D., De Vivo, M., Capoferri, L., Mor, M., Piomelli, D., et al. (2012). A catalytic mechanism for cysteine N-terminal nucleophile hydrolases, as revealed by free energy simulations. *PLoS One*, *7*, e32397.

Lo Verme, J., Fu, J., Astarita, G., La Rana, G., Russo, R., Calignano, A., et al. (2005). The nuclear receptor peroxisome proliferator-activated receptor-α mediates the anti-inflammatory actions of palmitoylethanolamide. *Molecular Pharmacology*, *67*, 15–19.

Lo Verme, J., Russo, R., La Rana, G., Fu, J., Farthing, J., Mattace-Raso, G., et al. (2006). Rapid broad-spectrum analgesia through activation of peroxisome proliferator-activated receptor-α. *Journal of Pharmacology and Experimental Therapeutics*, *319*, 1051–1061.

Mazzari, S., Canella, R., Petrelli, L., Marcolongo, G., & Leon, A. (1996). *N*-(2-Hydroxyethyl)hexadecanamide is orally active in reducing edema formation and inflammatory hyperalgesia by down-modulating mast cell activation. *European Journal of Pharmacology*, *300*, 227–236.

McKinney, M. K., & Cravatt, B. F. (2005). Structure and function of fatty acid amide hydrolase. *Annual Review of Biochemistry*, *74*, 411–432.

Mor, M., Rivara, S., Lodola, A., Plazzi, P. V., Tarzia, G., Duranti, A., et al. (2004). Cyclohexylcarbamic acid 3′- or 4′-substituted biphenyl-3-yl esters as fatty acid amide hydrolase inhibitors: Synthesis, quantitative structure–activity relationships, and molecular modeling studies. *Journal of Medicinal Chemistry*, *47*, 4998–5008.

Oinonen, C., & Rouvinen, J. (2000). Structural comparison of Ntn-hydrolases. *Protein Science*, *9*, 2329–2337.

Petrosino, S., Iuvone, T., & Di Marzo, V. (2010). *N*-palmitoyl-ethanolamine: Biochemistry and new therapeutic opportunities. *Biochimie*, *92*, 724–727.

Rossocha, M., Schultz-Heienbrok, R., von Moeller, H., Coleman, J. P., & Saenger, W. (2005). Conjugated bile acid hydrolase is a tetrameric N-terminal thiol hydrolase with specific recognition of its cholyl but not of its tauryl product. *Biochemistry*, *44*, 5739–5748.

Sasso, O., Moreno-Sanz, G., Martucci, C., Realini, N., Dionisi, M., Mengatto, L., et al. (2013). Antinociceptive effects of the *N*-acylethanolamine acid amidase inhibitor ARN077 in rodent pain models. *Pain*, *154*, 350–360.

Solorzano, C., Antonietti, F., Duranti, A., Tontini, A., Rivara, S., Lodola, A., et al. (2010). Synthesis and structure-activity relationships of *N*-(2-oxo-3-oxetanyl)amides as *N*-acylethanolamine-hydrolyzing acid amidase inhibitors. *Journal of Medicinal Chemistry*, *53*, 5770–5781.

Solorzano, C., Zhu, C., Battista, N., Astarita, G., Lodola, A., Rivara, S., et al. (2009). Selective *N*-acylethanolamine-hydrolyzing acid amidase inhibition reveals a key role for endogenous palmitoylethanolamide in inflammation. *Proceedings of the National Academy of Sciences of the United States of America*, *106*, 20966–20971.

Tsuboi, K., Sun, Y. X., Okamoto, Y., Araki, N., Tonai, T., & Ueda, N. (2005). Molecular characterization of *N*-acylethanolamine-hydrolyzing acid amidase, a novel member of the choloylglycine hydrolase family with structural and functional similarity to acid ceramidase. *Journal of Biological Chemistry*, *280*, 11082–11092.

Tsuboi, K., Takezaki, N., & Ueda, N. (2007). The *N*-acylethanolamine-hydrolyzing acid amidase (NAAA). *Chemistry & Biodiversity*, *4*, 1914–1925.

Ueda, N., Yamanaka, K., & Yamamoto, S. (2010). Purification and characterization of an acid amidase selective for *N*-palmitoylethanolamine, a putative endogenous anti-inflammatory substance. *Journal of Biological Chemistry*, *276*, 35552–35557.

Wang, J., Zhao, L.-Y., Uyama, T., Tsuboi, K., Tonai, T., & Ueda, N. (2008). Amino acid residues crucial in pH regulation and proteolytic activation of *N*-acylethanolamine-hydrolyzing acid amidase. *Biochimica et Biophysica Acta*, *1781*, 710–717.

West, J. M., Zvonok, N., Whitten, K. M., Vadivel, S. K., Bowman, A. L., & Makriyannis, A. (2012). Biochemical and mass spectrometric characterization of human *N*-acylethanolamine-hydrolyzing acid amidase inhibition. *PLoS One*, *7*, e43877.

Zhao, L.-Y., Tsuboi, K., Okamoto, Y., Nagahata, S., & Ueda, N. (2007). Proteolytic activation and glycosylation of *N*-acylethanolamine-hydrolyzing acid amidase, a lysosomal enzyme involved in the endocannabinoid metabolism. *Biochimica et Biophysica Acta*, *1771*, 1397–1405.

Zhu, C., Solorzano, C., Sahar, S., Realini, N., Fung, E., Sassone-Corsi, P., et al. (2011). Proinflammatory stimuli control *N*-acylphosphatidylethanolamine-specific phospholipase D expression in macrophages. *Molecular Pharmacology*, *79*, 786–792.

CHAPTER EIGHT

CHARMM-GUI PDB Manipulator for Advanced Modeling and Simulations of Proteins Containing Nonstandard Residues

Sunhwan Jo*, Xi Cheng†, Shahidul M. Islam*, Lei Huang*, Huan Rui*, Allen Zhu*, Hui Sun Lee†, Yifei Qi†, Wei Han‡, Kenno Vanommeslaeghe§, Alexander D. MacKerell Jr.§, Benoît Roux*,[1], Wonpil Im†,[1]

*Department of Biochemistry and Molecular Biology, Gordon Center for Integrative Science, University of Chicago, Chicago, Illinois, USA

†Department of Molecular Biosciences and Center for Bioinformatics, The University of Kansas, Lawrence, Kansas, USA

‡Beckman Institute and Center for Biophysics and Computational Biology, University of Illinois at Urbana–Champaign, Urbana, Illinois, USA

§Department of Pharmaceutical Sciences, School of Pharmacy, University of Maryland, Baltimore, Maryland, USA

[1]Corresponding authors: e-mail address: roux@uchicago.edu; wonpil@ku.edu

Contents

Advances in Protein Chemistry and Structural Biology, Volume 96
ISSN 1876-1623
http://dx.doi.org/10.1016/bs.apcsb.2014.06.002

Abstract

CHARMM-GUI, http://www.charmm-gui.org, is a web-based graphical user interface to prepare molecular simulation systems and input files to facilitate the usage of common and advanced simulation techniques. Since it is originally developed in 2006, CHARMM-GUI has been widely adopted for various purposes and now contains a number of different modules designed to setup a broad range of simulations including free energy calculation and large-scale coarse-grained representation. Here, we describe functionalities that have recently been integrated into CHARMM-GUI *PDB Manipulator*, such as ligand force field generation, incorporation of methanethiosulfonate spin labels and chemical modifiers, and substitution of amino acids with unnatural amino acids. These new features are expected to be useful in advanced biomolecular modeling and simulation of proteins.

1. INTRODUCTION

Computer modeling and simulation of biologically important molecules have come a long way since its inception. For the last two decades, researchers have made considerable efforts to develop modeling and simulation methodologies to elucidate the mysteries in biological phenomena (Bernèche & Roux, 2001; Dror, Dirks, Grossman, Xu, & Shaw, 2012; MacKerell, 2004; Rhee, Sorin, Jayachandran, Lindahl, & Pande, 2004; Roux & MacKinnon, 1999). The advances in biomolecular modeling and simulation allow researchers not only to explain why certain phenomena occur, but also to predict what would happen in different situations and design new experiments to test the developed hypothesis (Kuhlman et al., 2003; Ostmeyer, Chakrapani, Pan, Perozo, & Roux, 2013; Shukla, Meng, Roux, & Pande, 2014). Combined with tremendous increase in computing power, biomolecular modeling and simulation are becoming routine tools used in many research laboratories.

The rapid development of computational methodologies also made a large number of utilities available for both experts and beginners. As an illustration, more than 700 utilities are currently listed in a directory of computer-aided drug design tools maintained by Swiss Institute of Bioinformatics (http://www.click2drug.org). Among these tools, the emergence of web-based modeling tools is notable. Compared to downloadable software, web-based tools do not require program installation or upgrades, and can be operated on a web browser, making the tool platform-independent and easy to use. Among various web-based modeling tools, the WHAT IF server particularly stands out. It was designed in 1997 and has continued to operate until now. The WHAT IF server provides various services to users ranging from structural analyses, such as solvent accessibility and coordinate manipulation, to homology modeling and ligand docking (Hekkelman et al., 2010).

After the success of the WHAT IF server, various web-based modeling and simulation services have been developed and made publically available. CHARMM-GUI is a unique web-based set of tools developed by Im and coworkers (Jo, Kim, Iyer, & Im, 2008). It was originally developed in 2006 and its primary goal was to prepare molecules for molecular modeling and to provide a set of CHARMM inputs for molecular dynamics (MD) simulations, so that users can download the inputs and perform simulations on their local machines or remote computational resources. CHARMM-GUI is powered by CHARMM (Brooks et al., 2009), a highly flexible academic research program used worldwide for simulations of many-particle systems, particularly macromolecules of biological interest. What makes CHARMM-GUI unique is that it sets out to simplify and generalize the protocol for building simulation systems. Molecular simulation systems are often comprised of several components, and in-depth knowledge on modeling software is necessary to understand and build a sophisticated simulation system. In addition, generalizing the system building protocols is challenging because customization is often necessary. For example, a protein–membrane complex system contains water, ions, protein, and lipid components, and it would require considerable efforts even for an expert to build a realistic and physiologically relevant simulation system. CHARMM-GUI *Membrane Builder* successfully encapsulates such a sophisticated process into a well-defined protocol that can reproducibly generate a realistic protein–membrane complex or a membrane-only system within minutes (Jo, Kim, & Im, 2007; Jo, Lim, Klauda, & Im, 2009). Currently, there are several web-based tools that provide similar functionality (Ghahremanpour, Arab, Aghazadeh, Zhang, & van der Spoel, 2014; Miller et al., 2008; Schmidt & Kandt, 2012; Skjevik, Madej, Walker, & Teigen, 2012; Staritzbichler, Anselmi, Forrest, & Faraldo-Gómez, 2011), but CHARMM-GUI *Membrane Builder* remains the interface containing the most comprehensive list of lipid molecules available for generating realistic biological membrane simulation systems.

Since its original development, CHARMM-GUI has been widely used in the biomolecular modeling and simulation community, and it has grown into a platform of web-based tools for simulations: *Ligand Binder* for free energy perturbation molecular dynamics (FEP/MD) simulations for protein–ligand binding affinity calculations (Jo, Jiang, Lee, Roux, & Im, 2013), *Micelle Builder* for protein–micelle complex simulation system generation (Cheng, Jo, Lee, Klauda, & Im, 2013), *GCMC/BD* ion simulator for Brownian dynamics of ions across ion channels (Lee, Jo, Rui, et al., 2012), *Glycan Reader* for preparation of simulation systems containing carbohydrates or proteoglycans (Jo, Song, Desaire, MacKerell, & Im, 2011), and recently, *PACE CG Builder* for coarse-grained simulation system preparation (Qi et al., 2014).

Here, we describe the newest functionalities that have been integrated into CHARMM-GUI *PDB Manipulator*. First, ligand force field (FF) generation is discussed in detail. Generating a consistent and accurate FF for a new compound is often one of the major hurdles in computational modeling of protein–ligand interactions. Thus, the ability to read or automatically generate ligand FF parameters would be undoubtedly useful to study protein–ligand interactions. Second, incorporation of methanethiosulfonate (MTS) spin labels and chemical modifiers is discussed. Distance measurements using spin labels and introducing MTS reagents to modify chemical environments around specific residues are widely used in protein structure and function studies, so that making spin labeled and/or chemically modified models would be useful for related computational studies. Third, substituting regular amino acids with unnatural amino acids (UAAs) is discussed. As researchers are more interested in more complex biological problems, these functions would also be useful. Each of these new functionalities is elaborated in a step-wise manner and further discussed with common issues (or mistakes) in using these functionalities.

2. LIGAND FF GENERATION

A molecular mechanics FF is one of the most important components of modern biomolecular modeling and simulation (MacKerell, 2004). These FFs are typically derived and validated using physicochemical properties of (small) target molecules, which are obtained either experimentally or computationally. Parameterizing molecular mechanics FFs is often challenging because it requires careful curation of experimental and computational data as well as optimization of a large number of parameters.

There are several popular contemporary FFs that are readily available, such as AMBER (Cornell et al., 1995; Weiner et al., 1984), CHARMM (Best et al., 2012; MacKerell et al., 2002), GROMOS (Christen et al., 2005), and OPLS (Jorgensen, Maxwell, & Tirado-Rives, 1996). They often cover key biological macromolecules, such as proteins, nucleic acids, carbohydrates, and phospholipids. However, these FFs often do not treat important small molecules. Accurate representation of small molecules in modeling and simulation of protein–ligand interactions is crucial for the studies of rational computer-aided drug discovery as well as proteins with coenzymes, and so on. General FFs, such as MM2 (Allinger, 1977), MM3 (Allinger, Yuh, & Lii, 1989; Lii & Allinger, 1989a,1989b), MM4 (Allinger, Chen, Lii, & Durkin, 2003), MMFF94 (Halgren, 1996), and CFF93 (Rappé, Casewit, Colwell, Goddard, & Skiff, 1992), were developed specifically for small organic compounds, but performed poorly on biomolecules and intermolecular interactions in bulk phase, making them ill-suited for studying protein–ligand interactions. Combining one of the former biomolecular FFs with one of the latter general FFs is not an appropriate solution due to imbalances in the nonbonded interactions, as previously discussed (MacKerell, 2004).

The number of possible drug-like molecules is estimated to be roughly 10^{60} (Bohacek, McMartin, & Guida, 1996). Hence, it is not currently feasible to develop a comprehensive molecular mechanics FF set that accurately covers such a large chemical space. However, to account for important small molecules in a biomolecular context, typically those of medicinal chemistry interest, generalized FFs for small molecules have been developed such as the CHARMM general force field (CGenFF) (Vanommeslaeghe et al., 2010) and the general AMBER force field (GAFF; Wang, Wolf, Caldwell, Kollman, & Case, 2004). In addition, algorithms that can suggest topologies and FF parameters have also been developed based on such a generalized FF. PRODRG (Schüttelkopf & van Aalten, 2004) is a web-based interface that provides topology and parameters compatible with the GROMOS FF. The CGenFF program (Vanommeslaeghe & MacKerell, 2012; Vanommeslaeghe, Raman, & MacKerell, 2012; available to nonprofit users through the cgenff.paramchem.org interface) and SwissParam (Zoete, Cuendet, Grosdidier, & Michielin, 2011) provide the same function for the CHARMM FF. Antechamber (Wang, Wang, Kollman, & Case, 2006) and MATCH (Yesselman, Price, Knight, & Brooks, 2012) provide a command-line utility for automated topology and parameter generation for the AMBER and CHARMM FFs, respectively. There are also new tools such as the VMD plugin ffTK (Mayne, Saam, Schulten, Tajkhorshid, & Gumbart, 2013) and the general automated atomic model parameterization

(GAAMP) web server (Huang & Roux, 2013). The GAAMP method is described in Sections 2.3 and 4.1.

CHARMM-GUI adopts the standard CHARMM36 FF. Although the CHARMM36 FF currently covers a large array of biomolecules (Best et al., 2012; Denning, Priyakumar, Nilsson, & MacKerell, 2011; Guvench et al., 2008; Klauda et al., 2010) and various small organic molecules (Vanommeslaeghe et al., 2010), it is still far short to cover a huge number of small molecules of interest. To remedy this situation, CHARMM-GUI provides several options to generate or read a small molecule FF. In this section, we describe how to use CHARMM-GUI to prepare simulations when FF parameters are missing for small molecules of interest.

2.1. Reading small molecules that do not exist in the CHARMM FF

As shown in Fig. 1, CHARMM-GUI provides three options for small molecule FF parameters: (1) CGenFF, (2) Antechamber, and (3) custom CHARMM FF. When a custom CHARMM FF is available to users, it can be uploaded into CHARMM-GUI. Note that the uploaded CHARMM FF must have separate topology and parameter files. In addition, if new atoms are defined, the atom numbers in the MASS section should not overlap with those of the standard CHARMM36 FF.

If custom CHARMM FF parameters are not available for a small molecule of interest, either the CGenFF or Antechamber option can be used to automatically generate customized FF parameters for the molecule. To use the automatic FF generation option, the molecular representation of the small molecule in Mol2 file format is required. Users can upload a Mol2 format file, or, if a Mol2 format file is not available, a SDF format file can be

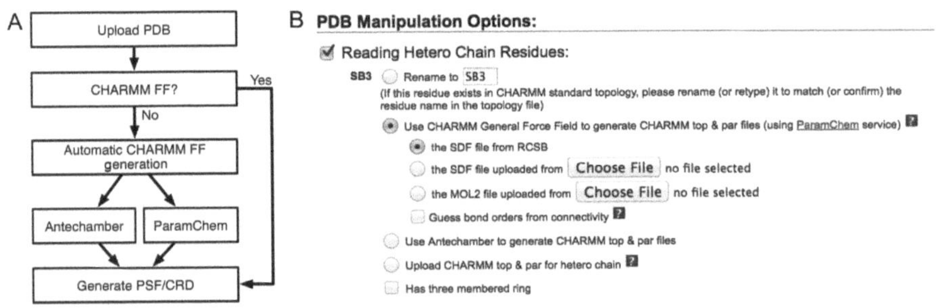

Figure 1 (A) A flowchart for reading small molecules that do not exist in the CHARMM FF. (B) A screenshot of options provided by CHARMM-GUI for reading small molecules that are not available in the CHARMM FF.

used instead (which is then converted to a Mol2 format file using OpenBabel software; O'Boyle et al., 2011). If users do not have those files, a SDF file could be obtained from the RCSB ligand database. However, careful attention must be paid because the ligand residue name in the original PDB file is used to match the small molecule in the RCSB database, and they are not necessarily the same molecule (see Section 2).

The CGenFF program is based on the CGenFF, which is part of the standard CHARMM36 FF parameter set, and Antechamber provides GAFF-based FF parameters for small molecules that can be used in CHARMM. Although Antechamber-generated GAFF FF can reproduce hydration energies reasonably well, it is generally desirable to remain internally consistent and avoid mixing the parameters of different FFs. Thus, using the CGenFF option is recommended if the small molecule is studied together with other biomolecules using the CHARMM36 FF.

2.2. Issues in using automated ligand FF generation procedures

When the automatic FF generation option is chosen to combine the small molecule FF with the existing FF parameters, there are three main issues that users need to be cautious about. First, the order of atoms listed in the PDB and the Mol2 (or SDF) files must match. CHARMM-GUI reorders the atoms in the PDB file based on the list of atoms in the Mol2 (or SDF) file, after the small molecule FF is generated by either the CGenFF program or Antechamber. If the atom names in the topology and the PDB files do not match, atomic positions can be mixed when CHARMM reads the coordinate. Note that, if no minimizations are performed, the atoms might look normal even with such a problem, but the problem becomes obvious after a minimization, often leading to failure of the minimization.

Second, if a user chooses to use a SDF format file, CHARMM-GUI downloads a SDF file from the RCSB ligand database (Rose et al., 2011). In this case, the residue name in the originally uploaded PDB and the RCSB database's entry ID must match. For example, there is an entry for R-wafarin in RCSB (http://www.rcsb.org/pdb/ligand/ligandsummary.do?hetId=RWF), which has an entry ID of "RWF." If a PDB file containing residue "RWF" and the option for downloading a SDF file from RCSB is selected, CHARMM-GUI downloads the R-wafarin structure from RCSB. Currently, however, there is no way for users to specify which SDF file to download from RCSB, and the residue name used in the original PDB file is used to fetch a SDF file from RCSB. If the residue name and the

RCSB ligand entry ID do not match, users need to either change the residue name in the PDB file or manually download a SDF file from other sources (RCSB, PubChem, etc.), modify it, and upload the corrected SDF file.

Third, the CHARMM-GUI automatic parameter generation option takes a Mol2 (or SDF) file for molecular representation without any further modification. Therefore, it is very important for users to explicitly add hydrogen atoms according to the desired protonation state. In addition, the bond orders provided by users are also important because they are used to determine proper atom types by the CGenFF program and Antechamber.

2.3. Further optimization strategy

The automatic parameter generation is not perfect. For example, the approach in the CGenFF program is based on the similarity between the atom types that define each required parameter and those in existing parameters. The dissimilarity is quantified in terms of a penalty score that is returned to the user in the output toppar stream file. Thus, a lower penalty score indicates better quality parameters, though users should be aware that the transferability of empirical FF parameters is limited and care should be taken to test the assigned parameters. Antechamber uses the AM1-BCC protocol to assign atomic charges, but there is still room for improvement in atomic charges and internal energy (bond/angle/torsion) parameters. A recent benchmark by Brooks III and coworkers (Knight, Yesselman, & Brooks, 2013) shows that the performance of parameters generated by Antechamber and the CGenFF program is similar on average in terms of hydration free energies estimated using implicit solvent models, but also demonstrates areas where improvements are possible.

The initial FF parameters obtained from the automatic algorithms can be improved in a step-wise manner as detailed in Vanommeslaeghe et al. (2010). Briefly, the optimal geometry of the target molecule is first generated in a quantum mechanics (QM) calculation. Second, given the geometry, water molecules are placed around hydrogen bond donors and acceptors, and the interaction energy is calculated on a QM level. Third, the optimal molecular mechanics charges are determined to reproduce the QM interaction energy following the appropriate scaling. Lastly, bonded parameters such as bonds, angles, and dihedrals are optimized using a QM geometry and vibrational analysis and/or QM-level potential energy scans. Note that the optimization is typically performed to only those parameters that are guessed (i.e., those with penalty scores) as the original CGenFF parameters are already optimized.

Optimization of FF parameters requires considerable efforts even for a small molecule. There are several tools available that help users to perform the optimization. Two promising tools are ffTK (Mayne et al., 2013) and GAAMP (Huang & Roux, 2013). ffTK is a VMD plugin that provides a GUI for each and every aspect of the aforementioned optimization process, so users do not have to write necessary inputs manually. The GAAMP gateway, http://gaamp.lcrc.anl.gov, is a web-based interface that starts with initial FF parameters generated by the CGenFF program or Antechamber, and automatically performs the FF optimization without any user intervention. Figure 2 shows an example of a dihedral angle optimization of a small molecule (SB3; CHEMBL ID 126568) performed by the GAAMP server. The optimization for this size of molecule took more than a day, which is relatively long considering the fact that the CGenFF program returns the topology and parameters for the small molecule almost instantly. However, GAAMP can improve the atomic charges and torsion angle parameters considerably without any user intervention (Fig. 2B).

2.4. Practical applications to binding free energy calculations

The efficacy of the automatic FF generation option integrated into CHARMM-GUI is illustrated by a recent study of large-scale computational docking and simulation. In this study, Im and coworkers (Lee, Jo, Lim, & Im, 2012) have developed an integrated methodology for accurate prediction of the binding mode and affinity of drug-like molecules

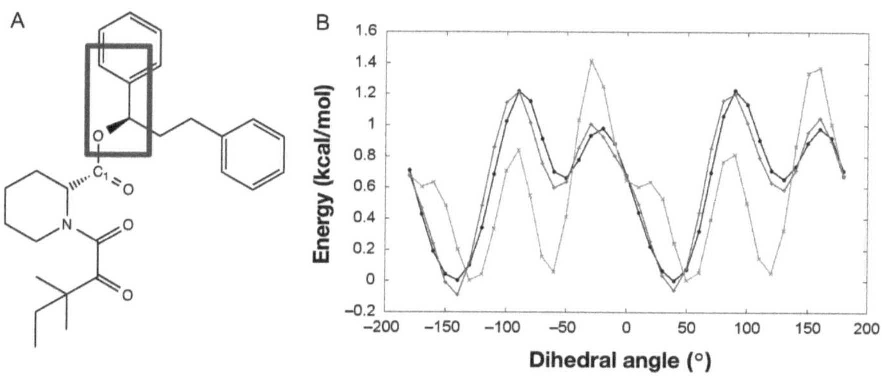

Figure 2 (A) Structure of the FKBP ligand SB3. (B) Example optimization result from the GAAMP server for the dihedral angle highlighted in (A): QM (black), optimized (red; dark gray in the print version), and initial (green; gray in the print version).

(Fig. 3B). They have applied a computational docking approach to generate a large number of decoy structures and selected a few docking pose candidates using clustering approaches. Because these molecules are not available in the standard CHARMM36 FF, the authors first used the CHARMM-GUI automatic parameterization option to obtain ligand FFs. The selected docking poses were then subjected to short-MD simulations to filter out irrelevant poses, followed by the FEP/MD simulations prepared by CHARMM-GUI *Ligand Binder* (Jo et al., 2013) for further ranking of the poses (Fig. 3).

The target small molecules are antagonists of MDM2 and MDMX. Figure 3A shows the chemical structures of the small molecules used in their study, and the FF parameters were generated using the CGenFF option without any further modification. The calculated binding free energies for MDM2 complexes were overestimated compared to experimental measurements (Fig. 3C) mainly due to the difficulties in sampling highly flexible apo-MDM2 conformations within the simulation timescale. Nonetheless, the FEP/MD binding free energy calculations are more promising in discriminating binders from nonbinders than commonly used docking scores (Fig. 3B and C). In addition, the FEP/MD calculations provide detailed information on the different energetic contributions to ligand binding, leading to a better understanding of the sensitivity and specificity of protein–ligand interactions. Therefore, CHARMM-GUI *PDB Manipulator* is expected to be useful as a platform that can rapidly prepare necessary FF parameters of small molecules of interest with the help of other tools. Setting up such sophisticated simulations can allow researchers to tackle more complex biological problems of protein–ligand interactions.

3. MTS REAGENTS

MTS reagents are often used for protein structure and function studies. Their use includes labeling and blocking groups, cross-linking groups, affinity-labeling groups, and reporter groups for chemical modification of peptides and proteins. MTS reagents are introduced to a specific site in a protein through site-directed mutagenesis (Hubbell, Mchaourab, Altenbach, & Lietzow, 1996). These reagents react very rapidly and specifically with cysteine residues, converting cysteine sulfhydryls to cysteine disulfide bonds. MTS reagents of cysteine residues may produce a measurable change in different protein functional states, which can be measured by various biophysical techniques. For example, MTSSL (1-oxyl-2,2,5,

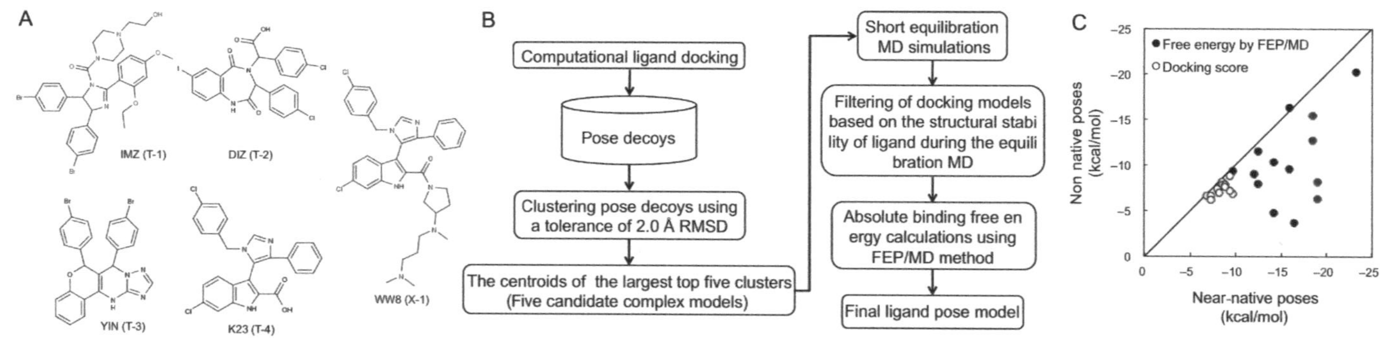

Figure 3 (A) Ligand structures. (B) Schematic of the docking and FEP/MD protocol used by Im and coworkers (Lee, Jo, Lim, et al., 2012). (C) The correlation between binding affinity of near-native poses and the nonnative poses. The FEP/MD method can discriminate near-native and nonnative poses better than a docking score. *The figures are reproduced with permission from the* Journal of Chemical Information and Modeling.

5-tetramethylpyrroline-3-methyl MTS; CYR1 in Fig. 4) is an MTS reagent that is widely used as a spin-label probe in electron spin resonance (ESR) spectroscopy. MTSSL has an unpaired electron, which offers a very strong signal in the ESR spectrum that provides valuable information about the structure, dynamics, and function of a protein system. In particular, site-specific mutagenesis with MTS reagents has proved to be a very useful technique in characterizing the structure–function relationship of membrane proteins, such as ion channels and transporter proteins, as well as enzymes and receptors (Chen, LiuChen, & Rudnick, 1997; Choi et al., 2000; Forrest et al., 2008; Hvorup et al., 2007; Jeschke, 2012; Kazmier et al.,

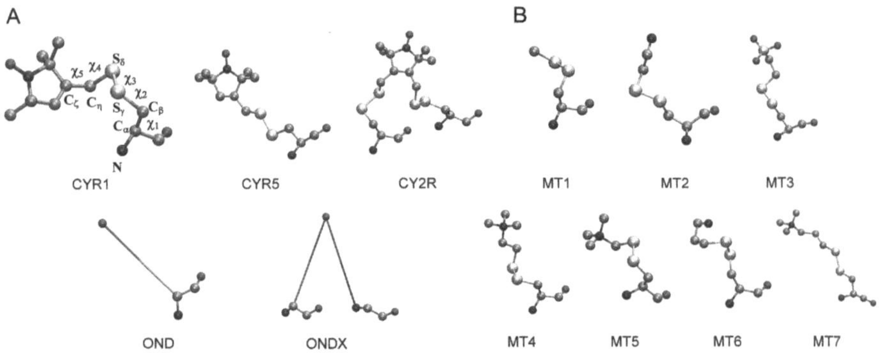

Figure 4 MTS side chains available in CHARMM-GUI. For clarity, hydrogen atoms are not shown and the oxygen, nitrogen, carbon, and sulfur atoms are represented in red (dark gray in the print version), blue (black in the print version), cyan (gray in the print version), and yellow (light gray in the print version) colors, respectively. (A) Monofunctional spin-label side chains, CYR1 and CYR5, resulting from linking MTSSL (1-oxyl-2,2,5,5-tetramethylpyrroline-3-methyl methanethiosulfonate) and PROXYL-MTS ((1-oxyl-2,2,5,5-tetramethylpyrrolidin-3-yl) methyl methanethiosulfonate) to cysteine through a disulfide bond, a bifunctional spin-label side chain, CY2R, resulting from selectively introducing MTSSL (1-oxyl-2,2,5,5-tetramethylpyrroline-3-methyl methanethiosulfonate) to two nearby cysteines through disulfide bonds, and the dummy spin labels, OND and ONDX, that mimic the dynamics of the CYR1 and CY2R spin-label side chains, respectively. In the case of ONDX, $X=3$ or 4 depending on the position of the second residue to which it is attached. The five dihedral angles connecting Cα atom of the protein backbone to nitroxide ring are also shown in CYR1. (B) The chemically modified MTS side chains, MT1, MT2, MT3, MT4, MT5, MT6, and MT7, resulting from linking methyl methanethiolsulfonate (MMTS), 2-aminoethyl methanethiosulfonate (MTSEA), sodium-(2-sulfonatoethyl) methanethiosulfonate (MTSES), 2-trimethylammonium-ethyl methanethiosulfonate (MTSET), (trimethylammonium)methyl methanethiosulfonate (MTSMT), 3-aminopropyl methanethiosulfonate (MTSPA), and 3-(trimethylammonium)propyl methanethiosulfonate (MTSPT) reagents, respectively, to cysteine through a disulfide bond.

2014; Perozo, Cortes, & Cuello, 1998; Raghuraman, Islam, Mukherjee, Roux, & Perozo, 2014; Roberts et al., 2008; Roberts, Lewis, Ballou, Olson, & Shafer, 1986; Tombola, Pathak, & Isacoff, 2006). Since many biophysical experiments are routinely performed with these MTS reagents, it is often necessary to introduce them into proteins for the purpose of MD simulation. Keeping this in mind, the FF parameters for a number of MTS reagents have been incorporated into CHARMM-GUI, which is expected to help users to readily prepare initial systems and simulation input files for MD simulation with selected MTS reagents.

3.1. MTS reagents available in CHARMM-GUI

There are two sets of MTS reagents available in CHARMM-GUI: the nitroxide spin labels (Fig. 4A) and the traditional chemical modifiers (Fig. 4B). There are two natural monofunctional spin-label side chains (CYR1, Berliner, Grunwald, Hankovszky, & Hideg, 1982, and CYR5, McHaourab, Kalai, Hideg, & Hubbell, 1999), a natural bifunctional (two-residue) spin-label side chain (CY2R, Fleissner et al., 2011), and a monofunctional and two bifunctional dummy spin-label side chains (OND, Islam, Stein, McHaourab, & Roux, 2013, and ONDX, Islam & Roux, 2014, with $X=3$ or 4). All these spin-label side chains were parameterized by Roux and coworkers (Islam & Roux, 2014; Islam et al., 2013; Sezer, Freed, & Roux, 2008). In addition, a total of seven MTS chemical modifiers (MT1, MT2, MT3, MT4, MT5, MT6, and MT7) parameterized by the GAAMP method (Huang & Roux, 2013) are currently available.

The step-by-step procedure to introduce MTS reagents into a protein in solution follows:

1. Go to CHARMM-GUI, http://www.charmm-gui.org
2. Select "*Input Generator*"
3. Select "*Quick MD Simulator*"
4. Download or upload your PDB file, and then "Next Step"
5. Select "Next Step" if you do not need to change anything
6. In *PDB Manipulator*, select "Add MTS reagents: nitroxide spin labels" and/or "Add MTS reagents: chemical modifier" (Fig. 5A).
   6.1. With "Add MTS reagents: nitroxide spin labels," choose the spin labels (CYR1, CYR5, CY2R, OND, OND3, and OND4) and "SEGID," "RESID," and "Amino acid" from the drop-down menu. This will modify the selected residue(s) with the natural MTS spin-label side chain(s) or add the dummy spin-label(s) to the

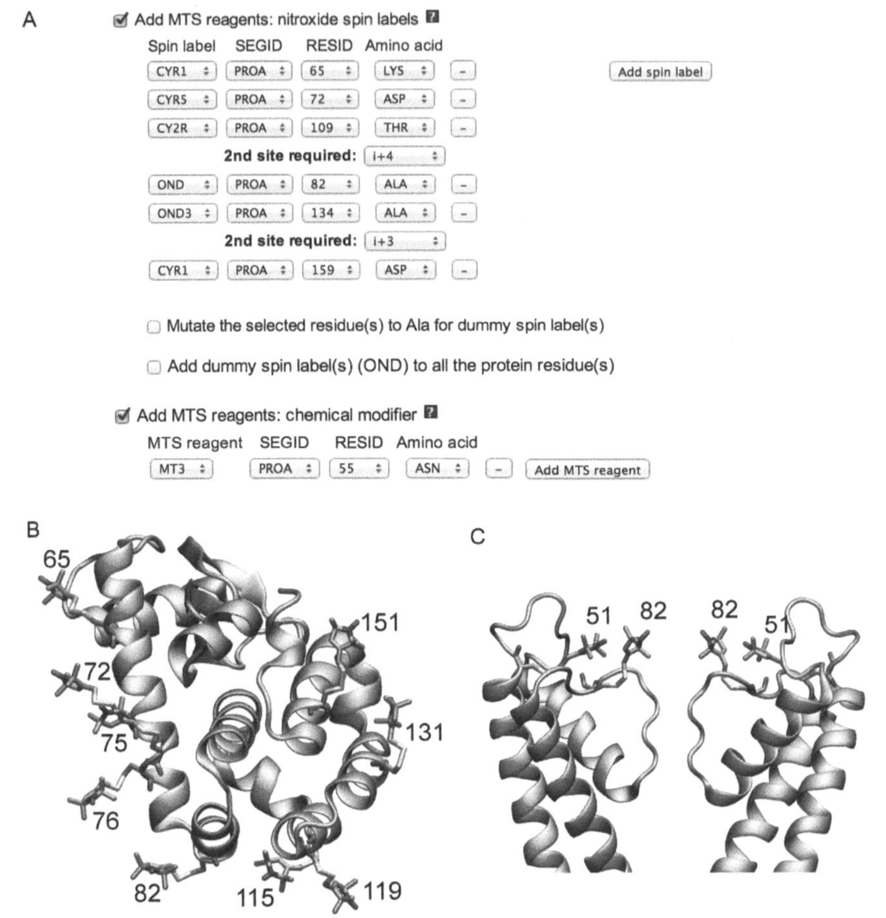

Figure 5 (A) A snapshot to introduce MTS spin labels and chemical modifiers into some specific residues in CHARMM-GUI. (B) Cartoon representations of T4 lysozyme with nine spin-label side chains at positions 65, 72, 75, 76, 82, 115, 119, 131, and 151. (C) Cartoon representations of the diagonal subunits of KcsA K^+ channel with two spin-label side chains at positions 51 and 82 on each subunit. Only diagonal subunits of tetrameric KcsA K^+ channel are shown for clarity.

selected residue(s). For CY2R, select $i+4$, $i+3$, $i+2$, or custom to attach this spin-label to a second residue. Two more options are also available:

Option 1: Selecting "Mutate the selected residue(s) to Ala for dummy spin label(s)" will mutate all the selected wild-type residues to alanine before attachment of the dummy spin-label(s).

Option 2: Selecting "Add dummy spin label(s) (OND) to all the protein residue(s)" will add dummy spin labels to all the residues. This option is particularly useful to conduct an MD simulation of the dummy spin labels (MDDS; Islam & Roux, 2014; Islam et al., 2013) to calculate spin-pair distance distributions between two dummy spin labels.

6.2. With "Add MTS reagents: chemical modifier," select the MTS reagents (MT1, MT2, MT3, MT4, MT5, MT6, and MT7) and "SEGID," "RESID," and "Amino acid" from the drop-down menu. This will modify the selected residue(s) with the MTS reagent(s).

7. Follow the instructions (essentially the same procedure of "*Quick MD Simulator*") to build a system in solution.

Note that the help windows are available in both step **6.1** and step **6.2** of the above illustration. The MTS reagents are also available in *Membrane Builder* (Jo et al., 2007, 2009) and other CHARMM-GUI modules. In particular, the dummy spin labels are available in *PACE CG Builder* (Qi et al., 2014).

3.2. Initial conformation generation of the MTS reagents in CHARMM-GUI

MTS reagents are highly flexible because they have multiple dihedral angles that can accommodate a number of conformational states. For example, as shown in Fig. 4, CYR1 spin-label side chain has five dihedral angles denoted by $\chi_1, \chi_2, \chi_3, \chi_4$, and χ_5 along the flexible bonds, C_α-C_β-S_γ-S_δ-C_η-C_ζ. From a careful study of the spin labels inserted in 37 positions in T4 lysozyme, Roux and coworkers recently proposed the most important rotamers for this spin-label side chain (Islam et al., 2013). The dihedral angles χ_1 and χ_2 can adopt threefold conformations, +60° (or gauche+), 180° (or trans), and −60° (or gauche), which are denoted by *p*, *t*, and *m*, respectively, and the dihedral angle χ_3 can adopt only two stable conformations, *p* (+90°) and *m* (−90°). Therefore, eight stable rotamers of the spin-label side chain have been identified along the dihedral angles χ_1, χ_2, and χ_3: *tpp*, *mtp*, *mmp*, *mmm*, *mtm*, *ttm*, *tpm*, and *ptm*. These observations are consistent with the available information from X-ray crystallography. χ_4 and χ_5 have been found extremely flexible, which is also consistent with the fact that no reliable X-ray crystallographic information is available for these dihedral angles.

Finding suitable MTS spin-label side-chain conformations in a specific protein site may require a substantially long-MD simulation to sample each conformational degree of freedom along its dihedral angles. Therefore, we have implemented an efficient approach in CHARMM-GUI to build a reasonable initial conformation for the MTS reagents (except CY2R, OND, and ONDX) in a protein structure. This approach is discussed below:

1. The conformations of the MTS spin-label side chains and the MTS chemical modifiers are characterized by three dihedral angles, χ_1, χ_2, and χ_3, along the flexible side chain C_α-C_β-S_γ-S_δ, as shown in Fig. 4. χ_4 and χ_5 dihedral angles are ignored due to the fact that they are extremely flexible and they will most likely converge to the right rotameric state with a relatively short-MD simulation.
2. Each of the eight rotameric states (*tpp*, *mtp*, *mmp*, *mmm*, *mtm*, *ttm*, *tpm*, and *ptm*) is sequentially added to a selected site in a protein, followed by 200-step minimization with steepest descent (SD) and adopted basis Newton–Raphson (ABNR) algorithms, respectively. During the minimization, all the protein atoms are kept fixed and only the selected MTS reagent atoms are allowed to move with flat bottom harmonic restraints on χ_1, χ_2, and χ_3 of the MTS side chain with a force constant of 500 kcal/(mol·rad^2) and a width of 2.5°. The minimum energy conformation of the MTS side chain (among the eight rotameric states) is then taken for further processing.
3. Step 2 is repeated for all the selected spin-label sites. Finally, all the MTS side chains with their minimum energy conformations (determined in step 2) and the surrounding atoms (up to 5 Å from the MTS reagent) are minimized for 200 steps with the SD and ABNR energy minimization, respectively.

This approach has shown to provide very reliable starting conformations of the MTS side chains in proteins. The performance of the above approach is examined using two protein systems, T4 lysozyme (Fig. 5B) and KcsA K^+ channel (Fig. 5C).

3.3. Comparison with available rotameric data from X-ray crystallography

Comparison with the available information from X-ray crystallographic structures offers a good route to examine the validity of the initial rotamers generated by CHARMM-GUI. Several crystallographic structures of MTSSL spin labels inserted into positions 65, 72, 75, 76, 82, 115, 119, 131, and 151 in T4 lysozyme (Fig. 5B) are available from the literature

(Fleissner, 2007; Fleissner et al., 2011; Fleissner, Cascio, & Hubbell, 2009; Guo, Cascio, Hideg, & Hubbell, 2008; Guo, Cascio, Hideg, Kalai, & Hubbell, 2007; Langen, Oh, Cascio, & Hubbell, 2000; Sezer, Freed, & Roux, 2009; Toledo Warshaviak, Cascio, Khramtsov, & Hubbell, 2010). However, the X-ray data are mostly limited to dihedrals χ_1 and χ_2, since the electron density along the flexible side chain C_α-C_β-S_γ-S_δ-C_η-C_ζ is missing beyond the S_δ atom. Using CHARMM-GUI, CYR1 was introduced individually at the same positions in T4 lysozyme. The comparison of spin-label structures from the PDB and CHARMM-GUI is given in Table 1.

The χ_1 and χ_2 rotamers from CHARMM-GUI are broadly consistent with the information from the X-ray crystal structures and previous MD simulations for five positions: a *tp* rotamer is observed at 72 and 115, an *mt* rotamer is observed at 75, and an *mm* rotamer is observed at 82 and 131. The clearest disagreement is shown in spin labels at positions 65, 76, 119, and 151 with *tp*, *tm*, *mm*, and *mm* rotamers, respectively, in the X-ray structure, while CHARMM-GUI provides *mt*, *tt*, *tp*, and *mt* rotamers,

Table 1 Comparison of rotamers of the CYR1 spin-label side chain along χ_1 and χ_2 obtained from X-ray structures and CHARMM-GUI

Residue	PDB ID	Expl. rotamer	CHARMM-GUI
65	3K2R[a–c]	*tp*	*mt*
72	MD[d]	*mm*, *tp*	*tp*
75	Fleissner[a]	*mt*	*mt*
76	3K2R[c]	*tm*	*tt*
82	1ZYT[a,e]	*mm*	*mm*
115	2IGC, 2OU8[a,f]	*mm*, *tp*	*tp*
119	3L2X[f]	*mm*	*tp*
131	2CUU, 3G3V[e,g] MD[d]	*mm*, *tp* *mm*, *tp*, *mt*, *tt*	*mm*
151	3G3X[e]	*mm*	*mt*

[a]Fleissner (2007).
[b]Langen et al. (2000).
[c]Toledo Warshaviak et al. (2010).
[d]Sezer et al. (2009).
[e]Fleissner et al. (2009, 2011).
[f]Guo et al. (2007).
[g]Guo et al. (2008).

respectively. However, X-ray crystallographic structure of the spin-label at position 65 may not be right because the spin-label is involved in crystal contacts (Langen et al., 2000). Also, CHARMM-GUI provides the rotameric states of *t* and *m* along the χ_1 dihedral angle of the spin labels at positions 76 and 151, respectively, which agree with their X-ray crystallographic structure. Overall, CHARMM-GUI can provide reliable initial conformations of the MTS side chains, which can be used in further MD simulations.

KcsA K^+ channel has four equivalent subunits forming a tetramer. We found that simple minimization with no rotameric search approach yielded multiple (different) conformations of the spin-label side chains at positions 51 and 82 in four subunits of KcsA. However, the rotameric search approach implemented in CHARMM-GUI was able to provide very similar initial conformations of the spin labels at all four subunits (Fig. 5B). All four spin-label side chains in position 51 were in the *mtp* rotamer along the dihedrals χ_1, χ_2, and χ_3. Spin labels introduced in position 82 were also in the same rotamer conformation (*mtm*) in all four subunits. This result shows that CHARMM-GUI is able to provide very good starting conformation of the MTS side chains even for a complex system such as KcsA K^+ channel.

3.4. Practical applications to protein structure refinement using ESR data

Recently, two state-of-the-art MD simulation methods, the conventional MDDS (Islam & Roux, 2014; Islam et al., 2013) and the restrained-ensemble (RE) simulation method (Islam et al., 2013; Roux & Islam, 2013), have been developed to analyze ESR/DEER (double electron electron resonance) data. The MDDS simulation provides distance distributions between spin labels that can be directly compared to the distance distributions obtained from ESR/DEER spectroscopy and can be used to assess the reliability or correctness of an already existing model structure of the system. On the other hand, the RE method drives a model structure towards the desired, refined structure using a global ensemble-based energy restraint that forces the spin–spin distance distribution histograms calculated from a multiple-copy MD simulation to match those obtained from ESR/DEER. The simplified nitroxide dummy spin labels, shown in Fig. 4A, are used mostly for both MDDS and RE simulations. Since the dummy spin labels do not interact with each other, multiple dummy atoms can be introduced to a single protein structure. CHARMM-GUI can be used to prepare the required files for both MDDS and RE simulations. For example, 51 distance distributions are available from 37 spin labeled positions in T4 lysozyme.

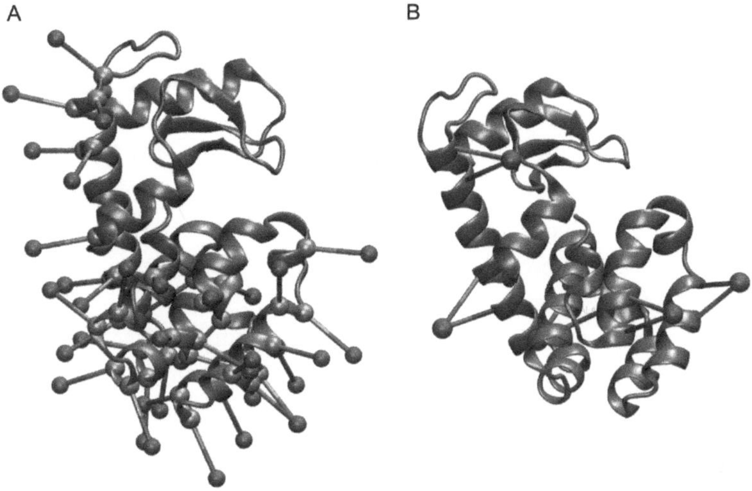

Figure 6 Cartoon representation of T4 lysozyme with dummy spin labels. (A) T4 lysozyme with 37 OND dummy spin labels at positions 59–62, 64, 65, 72, 75, 76, 79, 82, 83, 85, 86, 89, 90, 93, 94, 108, 109, 112, 115, 116, 119, 122, 123, 127, 128, 131, 132, 134, 135, 140, 151, 154, 155, and 159. (B) T4 lysozyme with five OND4 dummy spin labels at positions 60–64, 72–76, 108–112, 131–135, and 151–155.

CHARMM-GUI was used to attach the dummy spin labels at these sites in T4 lysozyme (Fig. 6A). The generated files are then used for MDDS simulation from which the spin-label distance distributions can be calculated to compare with those obtained from the ESR/DEER spectroscopy. Bifunctional dummy spin labels (OND4) were also inserted into five positions 60–64, 72–76, 108–112, 131–135, and 151–155 in T4 lysozyme (Fig. 6B). The generated structures can be used for the purpose of MDDS or RE simulations. Thus, CHARMM-GUI *PDB Manipulator* is expected to be useful in preparing necessary input files that can be used to understand protein structure and function in conjunction with ESR/DEER spectroscopy.

4. UNNATURAL AMINO ACIDS

Introducing UAAs into a protein is a powerful method to investigate the role of specific residues in protein structure and function, as well as to design proteins with novel functions or enhanced catalytic activity (Chin et al., 2003; Hackenberger & Schwarzer, 2008; Liu & Schultz, 2010; Lu, Yeung, Sieracki, & Marshall, 2009; Noren, Anthonycahill, Griffith, & Schultz, 1989). As demonstrated by Ahern and coworkers (Pless, Galpin,

Niciforovic, & Ahern, 2011), UAA mutagenesis combined with electrophysiology and voltage-clamp fluorometry could be employed to study the effects of negative electrostatic potentials from Glu293 to Asp316 on the channel gating and voltage sensing of Shaker K^+ channels. More applications can be found in a review by Pless and Ahern (2013). In principle, simulation studies of any of these chemically modified systems could be carried out to complement the experimental information. However, accurate molecular mechanics models are needed for an ever-growing number of possible UAAs.

As mentioned in Section 2, several FFs, including AMBER (Cornell et al., 1995; Weiner et al., 1984), CHARMM (Best et al., 2012), GROMOS (Christen et al., 2005), and OPLS (Jorgensen et al., 1996), are widely used in protein simulations, but there are generally no parameter sets specifically optimized for UAAs. This is becoming a major obstacle to model and study proteins containing UAAs in MD simulations. Although one can use Antechamber or the CGenFF program to assign parameters for an arbitrary compound, as discussed in Section 3, the obtained FF parameters often need further optimization as partial atomic charges and torsion parameters are not readily transferable across various molecules. The assigned parameters from analogs may be reasonably good in some cases, but they may not be so reliable in others. Another issue is that the topology and parameter files have to be manually edited to attach the unnatural side chain and protein backbone together. To overcome these difficulties, the GAAMP method was used to develop the FF parameters for UAAs (with negligible human efforts), and CHARMM-GUI *PDB Manipulator* now provides a facile way to model proteins containing UAAs.

4.1. UAAs available in CHARMM-GUI

The details about the GAAMP method can be found in Huang and Roux's work (Huang & Roux, 2013). In the GAAMP method, GAFF or CGenFF is taken as the initial FF. Partial atomic charges and dihedral parameters are optimized to reproduce QM target data. With minor adjustments, such an optimization algorithm has been extended to parameterize arbitrary amino acids, including UAAs, in a manner that is consistent with the backbone from the parent FF. Briefly, the partial atomic charges of the side-chain compound (the side-chain capped with one hydrogen atom) are first determined using the GAAMP charge fitting procedure under the constraint that the charge carried by the hydrogen atom is fixed at zero. As a second step,

CHARMM format topology, parameter, and coordinate files for the full molecule, comprising the side-chain molecule, and the backbone of an alanine dipeptide, are generated automatically. As a third step, the soft dihedrals around rotatable bonds within the side chain are identified and the parameters are optimized. During the side-chain dihedral fitting, the backbone atoms are fixed with the ϕ and φ backbone dihedrals in an α-helical conformation ($-60°$ and $-45°$ for ϕ and φ). The parameters of the resulting model have the CHARMM36 FF for the backbone and the GAAMP-optimized side chain. There are two UAA parameter sets available: one uses GAFF and the other uses CGenFF as initial parameters, respectively. A total of 17 UAAs commonly used in experiments (communicated with Dr. Christopher Ahern) have been parameterized and they are shown in Fig. 7A.

The step-by-step procedure of incorporating UAAs into a protein in solution follows:

1. Go to CHARMM-GUI, http://www.charmm-gui.org
2. Select "*Input Generator*"
3. Select "*Quick MD Simulator*"
4. Download or upload your PDB file, and then "Next Step"
5. Select "Next Step" if you do not need to change anything
6. As illustrated in Fig. 7B, select "Unnatural amino acid substitution," and then a new menu for UAA modeling shows up. Choose the parameters based on "CGenFF" or "GAFF," and then choose the UAA type and the index of amino acid to be substituted by the selected UAA.
7. Follow the instructions (essentially the same procedure of "Quick MD Simulator") to build a system in solution.

Note that in step **6**, one can view the UAA structures in the help window. In addition, the UAA substitution is also available in *Membrane Builder* (Jo et al., 2007, 2009) and other CHARMM-GUI modules.

4.2. Performance evaluation

4.2.1 Testing the parameterization strategy with natural amino acids

To test the performances of the amino acid parameters obtained by GAAMP, we reparameterize all amino acids except glycine and proline to be consistent with the CHARMM36 FF. Three proteins with diverse topologies, shown in Fig. 8, are used to compare the resulting FF (denoted as GAAMP) with the CHARMM36 FF: PDB:1CTF (mixed α-helices and β-sheets), PDB:1MJC (all β-sheets), and PDB:1R69 (all α-helices). Three independent 100-ns MD simulations in aqueous solution were conducted starting from the crystal structure of each protein. These three proteins

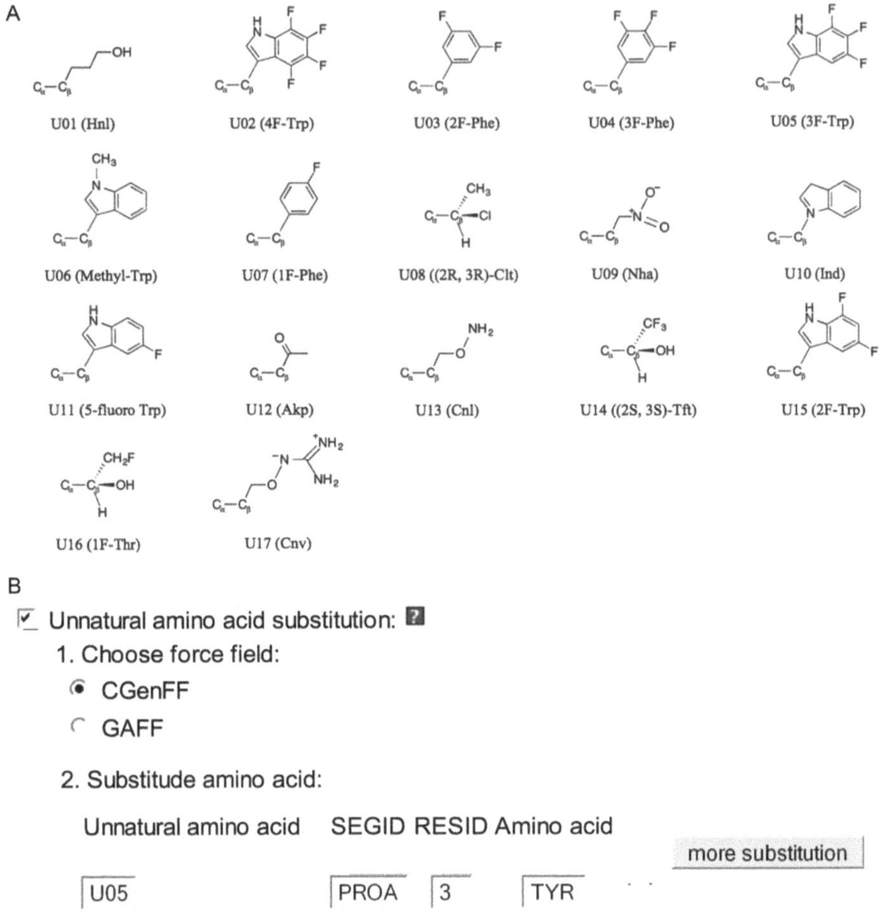

Figure 7 (A) Chemical structures of 17 UAAs available in CHARMM-GUI. (B) A snapshot of CHARMM-GUI to introduce a UAA into a specific residue in CHARMM-GUI.

are stable in 100-ns simulations with both the CHARMM36 and GAAMP FFs with similar conformational fluctuations. The simulation results suggest that the parameters of amino acids generated by GAAMP are compatible with the CHARMM36 FF.

4.2.2 Test cases of UAA-containing systems

Two separate approaches are taken to illustrate the efficacy of UAA functionality in CHARMM-GUI *PDB Manipulator*. The first approach involves simulating UAA dipeptides in solution with the two FF parameter options (i.e., GAFF and CGenFF) currently available in CHARMM-GUI. For each

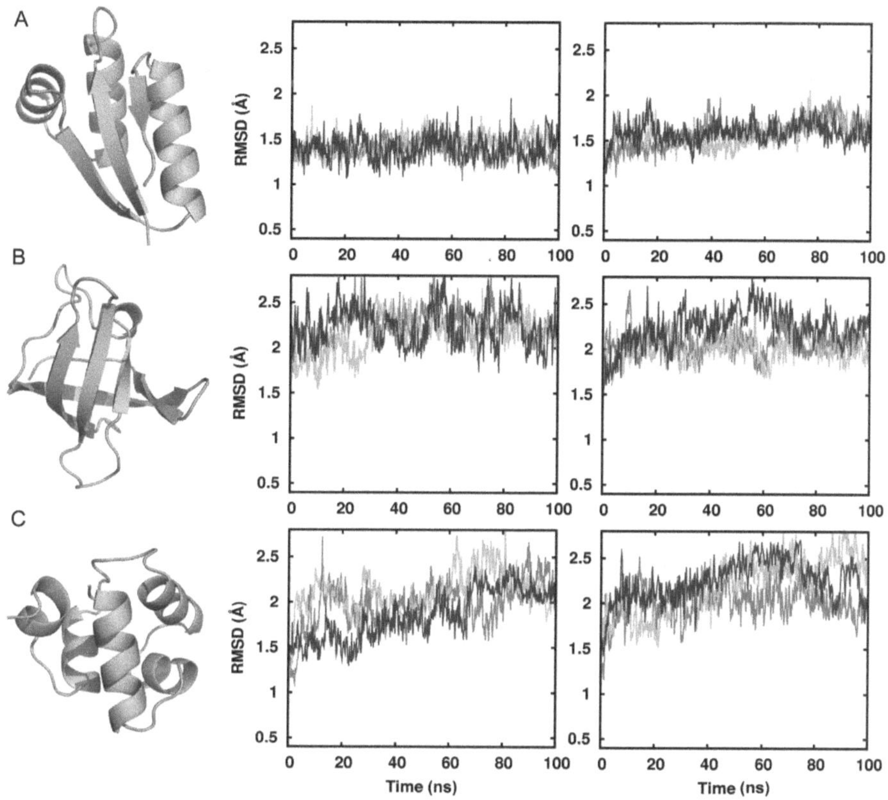

Figure 8 Three proteins with diverse structures (PDBs (A) 1CTF, (B) 1MJC, and (C) 1R69) are shown as ribbon representation generated by PyMol (Schrödinger, 2010). The traces of Cα-RMSD of 100-ns MD simulations for three replicas (with different colors (with different gray shades in the print version)) are compared between using the CHARMM36 FF (middle) and using the GAAMP FF (right) for each protein.

of 17 UAAs, a dipeptide is formed with the same UAA species. This could be performed with CHARMM-GUI by substituting both alanine residues in an alanine dipeptide. Afterwards, the dipeptide is put into a water box and simulated for 35 ns using the simulation options provided in CHARMM-GUI *Quick MD Simulator*. Peptide properties such as torsion angle distributions and hydrogen bonding patterns are calculated and compared between the two FF parameters. A sample result of the torsion angle distributions calculated from the U09 dipeptide simulation is shown in Fig. 9. The two parameter sets produce similar structural behaviors of the dipeptides and the differences are subtle in the case of U09. In spite

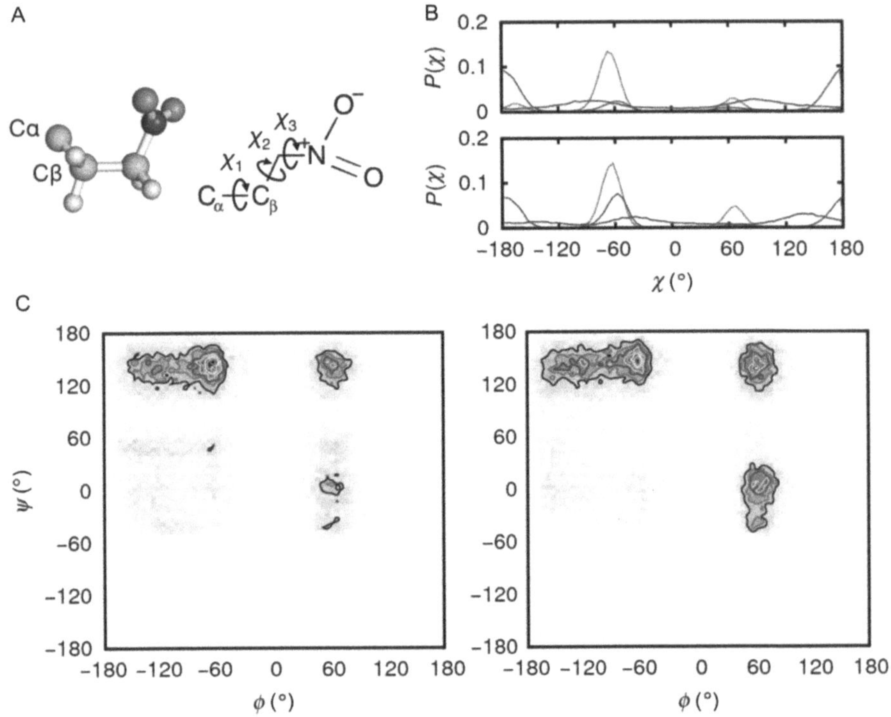

Figure 9 (A) Structure (left) and chemical formula (right) of the unnatural amino acid nitrohomoalanine (residue ID U09 in CHARMM-GUI). The three side-chain torsion angles, χ_1, χ_2, and χ_3, are shown on top of the chemical formula. (B) The torsion angle distributions for χ_1 (red; light gray in the print version), χ_2 (green; dark gray in the print version), and χ_3 (blue; black in the print version) calculated using parameters from GAFF (top) and CGenFF (bottom). (C) The backbone ϕ–ψ angle distributions calculated using parameters from GAFF (left) and CGenFF (right).

of the similarity in the simulation outcome, the CGenFF is recommended as the protein–UAA system is treated with the CHARMM36 FF.

In the second approach, protein systems with UAA substitutions are generated and short-MD simulations are performed to examine if it is indeed possible to generate a system with UAA substitutions in an automatic fashion using CHARMM-GUI and if such systems are relatively stable on a nanosecond timescale. The tested protein–UAA systems include UAA substitutions at different ion coordination sites in the Na^+, K^+-ATPase. More specifically, the Glu residues at the binding site are replaced by nitrohomoalanine (U09, Fig. 9A) one at a time. The protein is then embedded in a POPC bilayer with KCl and NaCl, 0.15 M each, in the aqueous solution. These systems are prepared using *Membrane Builder* (Jo et al., 2007, 2009). After the protein–membrane complex systems are generated, each of them is subjected to a 675-ps equilibration and 5-ns production with options provided by *Membrane Builder*. The trajectory analyses show that the systems are stable during the course of the simulation and the binding site heavy atom RMSD is below 1.2 Å most of the time. This indicates that the current CHARMM-GUI UAA functionality could be used in conjunction with other modules in CHARMM-GUI, such as *Membrane Builder*, and it is expected to be a useful tool in the study of complex biological systems with UAA substitutions.

5. PERSPECTIVES

CHARMM-GUI has been developed to provide a web-based user interface to build various molecular systems and generate input files for CHARMM and NAMD simulations. CHARMM-GUI has recently added a *Drude Prepper* facility that allows the conversion of CHARMM36 PSF and coordinate files into the electronically polarizable Drude FF (Lopes et al., 2013; Savelyev & MacKerell, 2014), which includes parameters for lipids, proteins, DNA, and carbohydrates. *PDB Manipulator* in CHARMM-GUI aims to help users handle the PDB files easily such as mutating/protonating/phosphorylating specific residues, reading ligands, or small molecules, adding MTS reagents and/or UAAs. In this work, we have described some of these functionalities in detail with a hope that users can use them widely for various biomolecular modeling and simulation. As *PDB Manipulator* is the first step of most input generation modules in CHARMM-GUI, these functionalities can be used for system building in solution and membrane (bilayer or micelle) environments, as well as in coarse-grained *PACE CG Builder*. With most side-chain modifications covered (together with future addition of more spectroscopic probes such as FRET and LRET) as well as the ability to handle small molecules, it is our hope that CHARMM-GUI *PDB Manipulator* becomes a useful tool for structure refinement of biomolecules with spectroscopic probes, structure, and dynamics studies with side-chain modifications, and biomolecules' interactions with small molecules in realistic biological solution and membrane environments.

ACKNOWLEDGMENTS

We wish to acknowledge the help from Christopher Ahern with the UAAs, and the help from Eduardo Perozo and Hassane Mchaourab with the spin labels. This work was carried out in the context of the Membrane Protein Structural Dynamics Consortium funded by grant U54-GM087519 (B. R. and W. I.) from the National Institute of Health (NIH), NIH grants GM072558 (B. R. and A. D. M.), GM070855 (A. D. M.), GM051501 (A. D. M.), and XSEDE grant MCB070009 (W. I.).

REFERENCES

Allinger, N. L. (1977). Conformational analysis. 130. MM2. A hydrocarbon force field utilizing V1 and V2 torsional terms. *Journal of the American Chemical Society*, *99*(25), 8127–8134. http://dx.doi.org/10.1021/ja00467a001.

Allinger, N. L., Chen, K.-H., Lii, J.-H., & Durkin, K. A. (2003). Alcohols, ethers, carbohydrates, and related compounds. I. The MM4 force field for simple compounds. *Journal of Computational Chemistry*, *24*(12), 1447–1472. http://dx.doi.org/10.1002/jcc.10268.

Allinger, N. L., Yuh, Y. H., & Lii, J. H. (1989). Molecular mechanics. The MM3 force field for hydrocarbons. 1. *Journal of the American Chemical Society*, *111*(23), 8551–8566. http://dx.doi.org/10.1021/ja3112719.

Berliner, L. J., Grunwald, J., Hankovszky, H. O., & Hideg, K. (1982). A novel reversible thiol-specific spin label: Papain active site labeling and inhibition. *Analytical Biochemistry*, *119*(2), 450–455.

Bernèche, S., & Roux, B. (2001). Energetics of ion conduction through the K^+ channel. *Nature*, *414*(6859), 73–77. http://dx.doi.org/10.1038/35102067.

Best, R. B., Zhu, X., Shim, J., Lopes, P. E. M., Mittal, J., Feig, M., et al. (2012). Optimization of the additive CHARMM all-atom protein force field targeting improved sampling of the backbone φ, ψ and side-chain χ(1) and χ(2) dihedral angles. *Journal of Chemical Theory and Computation*, *8*(9), 3257–3273. http://dx.doi.org/10.1021/ct300400x.

Bohacek, R. S., McMartin, C., & Guida, W. C. (1996). The art and practice of structure-based drug design: A molecular modeling perspective. *Medicinal Research Reviews*, *16*(1), 3–50. http://dx.doi.org/10.1002/(SICI)1098-1128(199601)16:1<3::AID-MED1>3.0.CO;2-6.

Brooks, B. R., Brooks, C. L., III, MacKerell, A. D., Jr., Nilsson, L., Petrella, R. J., Roux, B., et al. (2009). CHARMM: The biomolecular simulation program. *Journal of Computational Chemistry*, *30*(10), 1545–1614. http://dx.doi.org/10.1002/jcc.21287.

Chen, J. G., LiuChen, S., & Rudnick, G. (1997). External cysteine residues in the serotonin transporter. *Biochemistry*, *36*(6), 1479–1486. http://dx.doi.org/10.1021/Bi962256g.

Cheng, X., Jo, S., Lee, H. S., Klauda, J. B., & Im, W. (2013). CHARMM-GUI micelle builder for pure/mixed micelle and protein/micelle complex systems. *Journal of Chemical Information and Modeling*, *53*(8), 2171–2180. http://dx.doi.org/10.1021/ci4002684.

Chin, J. W., Cropp, T. A., Anderson, J. C., Mukherji, M., Zhang, Z. W., & Schultz, P. G. (2003). An expanded eukaryotic genetic code. *Science*, *301*(5635), 964–967.

Choi, Y. B., Tenneti, L., Le, D. A., Ortiz, J., Bai, G., Chen, H. S. V., et al. (2000). Molecular basis of NMDA receptor-coupled ion channel modulation by S-nitrosylation. *Nature Neuroscience*, *3*(1), 15–21.

Christen, M., Huenenberger, P. H., Bakowies, D., Baron, R., Bürgi, R., Geerke, D. P., et al. (2005). The GROMOS software for biomolecular simulation: GROMOS05. *Journal of Computational Chemistry*, *26*(16), 1719–1751. http://dx.doi.org/10.1002/jcc.20303.

Cornell, W. D., Cieplak, P., Bayly, C. I., Gould, I. R., Merz, K. M., Jr., Ferguson, D. M., et al. (1995). A second generation force field for the simulation of proteins, nucleic acids, and organic molecules. *Journal of the American Chemical Society*, *117*(19), 5179–5197.

Denning, E. J., Priyakumar, U. D., Nilsson, L., & MacKerell, A. D., Jr. (2011). Impact of 2'-hydroxyl sampling on the conformational properties of RNA: Update of the CHARMM all-atom additive force field for RNA. *Journal of Computational Chemistry*, *32*(9), 1929–1943. http://dx.doi.org/10.1002/jcc.21777.

Dror, R. O., Dirks, R. M., Grossman, J. P., Xu, H., & Shaw, D. E. (2012). Biomolecular simulation: A computational microscope for molecular biology. *Annual Review of Biophysics*, *41*, 429–452. http://dx.doi.org/10.1146/annurev-biophys-042910-155245.

Fleissner, M. R. (2007). X-ray structures of nitroxide side chains in proteins: a basis for interpreting distance measurements and dynamic studies by electron paramagnetic resonance. Ph.D. Thesis, University of California, Los Angeles, Los Angeles.

Fleissner, M. R., Bridges, M. D., Brooks, E. K., Cascio, D., Kalai, T., Hideg, K., et al. (2011). Structure and dynamics of a conformationally constrained nitroxide side chain and applications in EPR spectroscopy. *Proceedings of the National Academy of Sciences of the United States of America*, *108*(39), 16241–16246. http://dx.doi.org/10.1073/pnas.1111420108.

Fleissner, M. R., Cascio, D., & Hubbell, W. L. (2009). Structural origin of weakly ordered nitroxide motion in spin-labeled proteins. *Protein Science*, *18*(5), 893–908. http://dx.doi.org/10.1002/Pro.96.

Forrest, L. R., Zhang, Y. W., Jacobs, M. T., Gesmonde, J., Xie, L., Honig, B. H., et al. (2008). Mechanism for alternating access in neurotransmitter transporters. *Proceedings of the National Academy of Sciences of the United States of America*, *105*(30), 10338–10343. http://dx.doi.org/10.1073/Pnas.0804659105.

Ghahremanpour, M. M., Arab, S. S., Aghazadeh, S. B., Zhang, J., & van der Spoel, D. (2014). MemBuilder: A web-based graphical interface to build heterogeneously mixed membrane bilayers for the GROMACS biomolecular simulation program. *Bioinformatics (Oxford, England)*, *30*(3), 439–441. http://dx.doi.org/10.1093/bioinformatics/btt680.

Guo, Z. F., Cascio, D., Hideg, K., & Hubbell, W. L. (2008). Structural determinants of nitroxide motion in spin-labeled proteins: Solvent-exposed sites in helix B of T4 lysozyme. *Protein Science*, *17*(2), 228–239. http://dx.doi.org/10.1110/Ps.073174008.

Guo, Z. F., Cascio, D., Hideg, K., Kalai, T., & Hubbell, W. L. (2007). Structural determinants of nitroxide motion in spin-labeled proteins: Tertiary contact and solvent-inaccessible sites in helix G of T4 lysozyme. *Protein Science*, *16*(6), 1069–1086. http://dx.doi.org/10.1110/Ps.062739107.

Guvench, O., Greene, S. N., Kamath, G., Brady, J. W., Venable, R. M., Pastor, R. W., et al. (2008). Additive empirical force field for hexopyranose monosaccharides. *Journal of Computational Chemistry*, *29*(15), 2543–2564. http://dx.doi.org/10.1002/jcc.21004.

Hackenberger, C. P. R., & Schwarzer, D. (2008). Chemoselective ligation and modification strategies for peptides and proteins. *Angewandte Chemie-International Edition*, *47*(52), 10030–10074.

Halgren, T. A. (1996). Merck molecular force field. I. Basis, form, scope, parameterization, and performance of MMFF94. *Journal of Computational Chemistry*, *17*(5–6), 490–519. http://dx.doi.org/10.1002/(SICI)1096-987X(199604)17:5/6<490::AID-JCC1>3.0.CO;2-P.

Hekkelman, M. L., te Beek, T. A. H., Pettifer, S. R., Thorne, D., Attwood, T. K., & Vriend, G. (2010). WIWS: A protein structure bioinformatics web service collection. *Nucleic Acids Research*, *38*(Suppl. 2), W719–W723. http://dx.doi.org/10.1093/nar/gkq453.

Huang, L., & Roux, B. (2013). Automated force field parameterization for nonpolarizable and polarizable atomic models based on ab initio target data. *Journal of Chemical Theory and Computation*, *9*(8), 3543–3556. http://dx.doi.org/10.1021/Ct4003477.

Hubbell, W. L., Mchaourab, H. S., Altenbach, C., & Lietzow, M. A. (1996). Watching proteins move using site-directed spin labeling. *Structure*, *4*(7), 779–783. http://dx.doi.org/10.1016/S0969-2126(96)00085-8.

Hvorup, R. N., Goetz, B. A., Niederer, M., Hollenstein, K., Perozo, E., & Locher, K. P. (2007). Asymmetry in the structure of the ABC transporter-binding protein complex BtuCD-BtuF. *Science*, *317*(5843), 1387–1390. http://dx.doi.org/10.1126/science.1145950.

Islam, S. M., & Roux, B. (2014). *Molecular dynamics of dummy spin-labels (MDDS) for structure and function study of proteins*. (in preparation).

Islam, S. M., Stein, R. A., McHaourab, H. S., & Roux, B. (2013). Structural refinement from restrained-ensemble simulations based on EPR/DEER data: Application to T4 lysozyme. *The Journal of Physical Chemistry B, 117*(17), 4740–4754. http://dx.doi.org/10.1021/jp311723a.

Jeschke, G. (2012). DEER distance measurements on proteins. *Annual Review of Physical Chemistry, 63*(63), 419–446. http://dx.doi.org/10.1146/Annurev-Physchem-032511-143716.

Jo, S., Jiang, W., Lee, H. S., Roux, B., & Im, W. (2013). CHARMM-GUI ligand binder for absolute binding free energy calculations and its application. *Journal of Chemical Information and Modeling, 53*(1), 267–277. http://dx.doi.org/10.1021/ci300505n.

Jo, S., Kim, T., & Im, W. (2007). Automated builder and database of protein/membrane complexes for molecular dynamics simulations. *PLoS ONE, 2*(9), e880. http://dx.doi.org/10.1371/journal.pone.0000880.

Jo, S., Kim, T., Iyer, V. G., & Im, W. (2008). CHARMM-GUI: A web-based graphical user interface for CHARMM. *Journal of Computational Chemistry, 29*(11), 1859–1865. http://dx.doi.org/10.1002/jcc.20945.

Jo, S., Lim, J. B., Klauda, J. B., & Im, W. (2009). CHARMM-GUI membrane builder for mixed bilayers and its application to yeast membranes. *Biophysical Journal, 97*(1), 50–58.

Jo, S., Song, K. C., Desaire, H., MacKerell, A. D., Jr., & Im, W. (2011). Glycan reader: Automated sugar identification and simulation preparation for carbohydrates and glycoproteins. *Journal of Computational Chemistry, 32*(14), 3135–3141. http://dx.doi.org/10.1002/jcc.21886.

Jorgensen, W. L., Maxwell, D. S., & Tirado-Rives, J. (1996). Development and testing of the OPLS all-atom force field on conformational energetics and properties of organic liquids. *Journal of the American Chemical Society, 118*(45), 11225–11236. http://dx.doi.org/10.1021/ja9621760.

Kazmier, K., Sharma, S., Quick, M., Islam, S. M., Roux, B., Weinstein, H., et al. (2014). Conformational dynamics of ligand-dependent alternating access in LeuT. *Nature Structural & Molecular Biology, 21*(5), 472–479. http://dx.doi.org/10.1038/nsmb.2816.

Klauda, J. B., Venable, R. M., Freites, J. A., O'Connor, J. W., Tobias, D. J., Mondragon-Ramirez, C., et al. (2010). Update of the CHARMM all-atom additive force field for lipids: Validation on six lipid types. *Journal of Physical Chemistry B, 114*(23), 7830–7843. http://dx.doi.org/10.1021/jp101759q.

Knight, J. L., Yesselman, J. D., & Brooks, C. L., III (2013). Assessing the quality of absolute hydration free energies among CHARMM-compatible ligand parameterization schemes. *Journal of Computational Chemistry, 34*(11), 893–903. http://dx.doi.org/10.1002/jcc.23199.

Kuhlman, B., Dantas, G., Ireton, G. C., Varani, G., Stoddard, B. L., & Baker, D. (2003). Design of a novel globular protein fold with atomic-level accuracy. *Science (New York, NY), 302*(5649), 1364–1368. http://dx.doi.org/10.1126/science.1089427.

Langen, R., Oh, K. J., Cascio, D., & Hubbell, W. L. (2000). Crystal structures of spin labeled T4 lysozyme mutants: Implications for the interpretation of EPR spectra in terms of structure. *Biochemistry, 39*(29), 8396–8405. http://dx.doi.org/10.1021/Bi000604f.

Lee, H. S., Jo, S., Lim, H.-S., & Im, W. (2012). Application of binding free energy calculations to prediction of binding modes and affinities of MDM2 and MDMX inhibitors. *Journal of Chemical Information and Modeling, 52*(7), 1821–1832. http://dx.doi.org/10.1021/ci3000997.

Lee, K. I., Jo, S., Rui, H., Egwolf, B., Roux, B., Pastor, R. W., et al. (2012). Web interface for brownian dynamics simulation of ion transport and its applications to beta-barrel pores. *Journal of Computational Chemistry, 33*(3), 331–339. http://dx.doi.org/10.1002/jcc.21952.

Lii, J. H., & Allinger, N. L. (1989a). Molecular mechanics. The MM3 force field for hydrocarbons. 2. Vibrational frequencies and thermodynamics. *Journal of the American Chemical Society*, *111*(23), 8566–8575. http://dx.doi.org/10.1021/ja00205a002.

Lii, J. H., & Allinger, N. L. (1989b). Molecular Mechanics. The MM3 force field for hydrocarbons. 3. The van der Waals' potentials and crystal data for aliphatic and aromatic hydrocarbons. *Journal of the American Chemical Society*, *111*(23), 8576–8582. http://dx.doi.org/10.1021/ja00205a003.

Liu, C. C., & Schultz, P. G. (2010). Adding new chemistries to the genetic code. *Annual Review of Biochemistry*, *79*, 413–444.

Lopes, P. E. M., Huang, J., Shim, J., Luo, Y., Li, H., Roux, B., et al. (2013). Force field for peptides and proteins based on the classical Drude oscillator. *Journal of Chemical Theory and Computation*, *9*(12), 5430–5449. http://dx.doi.org/10.1021/ct400781b.

Lu, Y., Yeung, N., Sieracki, N., & Marshall, N. M. (2009). Design of functional metalloproteins. *Nature*, *460*(7257), 855–862.

MacKerell, A. D., Jr. (2004). Empirical force fields for biological macromolecules: Overview and issues. *Journal of Computational Chemistry*, *25*(13), 1584–1604. http://dx.doi.org/10.1002/jcc.20082.

MacKerell, A. D., Brooks, B., Brooks, C. L., Nilsson, L., Roux, B., Won, Y., et al. (2002). CHARMM: The energy function and its parameterization. *Encyclopedia of computational chemistry*. Hoboken, NJ: John Wiley & Sons, Ltd.

Mayne, C. G., Saam, J., Schulten, K., Tajkhorshid, E., & Gumbart, J. C. (2013). Rapid parameterization of small molecules using the force field toolkit. *Journal of Computational Chemistry*, *34*(32), 2757–2770. http://dx.doi.org/10.1002/jcc.23422.

McHaourab, H. S., Kalai, T., Hideg, K., & Hubbell, W. L. (1999). Motion of spin-labeled side chains in T4 lysozyme: Effect of side chain structure. *Biochemistry*, *38*(10), 2947–2955. http://dx.doi.org/10.1021/bi9826310.

Miller, B. T., Singh, R. P., Klauda, J. B., Hodoscek, M., Brooks, B. R., & Woodcock, H. L. (2008). CHARMMing: A new, flexible web portal for CHARMM. *Journal of Chemical Information and Modeling*, *48*(9), 1920–1929. http://dx.doi.org/10.1021/ci800133b.

Noren, C. J., Anthonycahill, S. J., Griffith, M. C., & Schultz, P. G. (1989). A general-method for site-specific incorporation of unnatural amino-acids into proteins. *Science*, *244*(4901), 182–188.

O'Boyle, N. M., Banck, M., James, C. A., Morley, C., Vandermeersch, T., & Hutchison, G. R. (2011). Open babel: An open chemical toolbox. *Journal of Cheminformatics*, *3*(10), 33. http://dx.doi.org/10.1186/1758-2946-3-33.

Ostmeyer, J., Chakrapani, S., Pan, A. C., Perozo, E., & Roux, B. (2013). Recovery from slow inactivation in K^+ channels is controlled by water molecules. *Nature*, *501*(7465), 121–124. http://dx.doi.org/10.1038/nature12395.

Perozo, E., Cortes, D. M., & Cuello, L. G. (1998). Three-dimensional architecture and gating mechanism of a K^+ channel studied by EPR spectroscopy. *Nature Structural Biology*, *5*(6), 459–469. http://dx.doi.org/10.1038/Nsb0698-459.

Pless, S. A., & Ahern, C. A. (2013). Unnatural amino acids as probes of ligand–receptor interactions and their conformational consequences. *Annual Review of Pharmacology and Toxicology*, *53*, 211–229.

Pless, S. A., Galpin, J. D., Niciforovic, A. P., & Ahern, C. A. (2011). Contributions of counter-charge in a potassium channel voltage-sensor domain. *Nature Chemical Biology*, 7(9), 617–623.

Qi, Y., Cheng, X., Han, W., Jo, S., Schulten, K., & Im, W. (2014). CHARMM-GUI PACE CG builder for solution, micelle, and bilayer coarse-grained simulations. *Journal of Chemical Information and Modeling*, *54*(3), 1003–1009. http://dx.doi.org/10.1021/ci500007n.

Raghuraman, H., Islam, S. M., Mukherjee, S., Roux, B., & Perozo, E. (2014). Dynamics transitions at the outer vestibule of the KcsA potassium channel during gating. *Proceedings of the National Academy of Sciences of the United States of America, 111*(5), 1831–1836. http://dx.doi.org/10.1073/Pnas.1314875111.

Rappé, A. K., Casewit, C. J., Colwell, K. S., Goddard, W. A., III, & Skiff, W. M. (1992). UFF, a full periodic table force field for molecular mechanics and molecular dynamics simulations. *Journal of the American Chemical Society, 114*(25), 10024–10035. http://dx.doi.org/10.1021/ja00051a040.

Rhee, Y. M., Sorin, E. J., Jayachandran, G., Lindahl, E., & Pande, V. S. (2004). Simulations of the role of water in the protein-folding mechanism. *Proceedings of the National Academy of Sciences of the United States of America, 101*(17), 6456–6461. http://dx.doi.org/10.1073/pnas.0307898101.

Roberts, J. A., Digby, H. R., Kara, M., El Ajouz, S., Sutcliffe, M. J., & Evans, R. J. (2008). Cysteine substitution mutagenesis and the effects of methanethiosulfonate reagents at P2X2 and P2X4 receptors support a core common mode of ATP action at P2X receptors. *The Journal of Biological Chemistry, 283*(29), 20126–20136. http://dx.doi.org/10.1074/jbc.M800294200.

Roberts, D. D., Lewis, S. D., Ballou, D. P., Olson, S. T., & Shafer, J. A. (1986). Reactivity of small thiolate anions and cysteine-25 in papain toward methyl methanethiosulfonate. *Biochemistry, 25*(19), 5595–5601.

Rose, P. W., Beran, B., Bi, C., Bluhm, W. F., Dimitropoulos, D., Goodsell, D. S., et al. (2011). The RCSB Protein Data Bank: Redesigned web site and web services. *Nucleic Acids Research, 39*(Database issue), D392–D401. http://dx.doi.org/10.1093/nar/gkq1021.

Roux, B., & Islam, S. M. (2013). Restrained-ensemble molecular dynamics simulations based on distance histograms from double electron–electron resonance spectroscopy. *Journal of Physical Chemistry B, 117*(17), 4733–4739. http://dx.doi.org/10.1021/Jp3110369.

Roux, B., & MacKinnon, R. (1999). The cavity and pore helices in the KcsA K^+ channel: Electrostatic stabilization of monovalent cations. *Science (New York, NY), 285*(5424), 100–102. http://dx.doi.org/10.1126/science.285.5424.100.

Savelyev, A., & MacKerell, A. D., Jr. (2014). All-atom polarizable force field for DNA based on the classical drude oscillator model. *Journal of Computational Chemistry, 35*(16), 1219–1239. http://dx.doi.org/10.1002/jcc.23611.

Schmidt, T. H., & Kandt, C. (2012). LAMBADA and InflateGRO2: Efficient membrane alignment and insertion of membrane proteins for molecular dynamics simulations. *Journal of Chemical Information and Modeling, 52*(10), 2657–2669. http://dx.doi.org/10.1021/ci3000453.

Schrödinger, LLC. (2010). The PyMOL molecular graphics system, Software Version 1.3r1.

Schüttelkopf, A. W., & van Aalten, D. M. F. (2004). PRODRG: A tool for high-throughput crystallography of protein–ligand complexes. *Acta Crystallographica Section D, Biological Crystallography, 60*(Pt 8), 1355–1363. http://dx.doi.org/10.1107/S0907444904011679.

Sezer, D., Freed, J. H., & Roux, B. (2008). Parametrization, molecular dynamics simulation, and calculation of electron spin resonance spectra of a nitroxide spin label on a polyalanine alpha-helix. *The Journal of Physical Chemistry B, 112*(18), 5755–5767. http://dx.doi.org/10.1021/jp711375x.

Sezer, D., Freed, J. H., & Roux, B. (2009). Multifrequency electron spin resonance spectra of a spin-labeled protein calculated from molecular dynamics simulations. *Journal of the American Chemical Society, 131*(7), 2597–2605. http://dx.doi.org/10.1021/Ja8073819.

Shukla, D., Meng, Y., Roux, B., & Pande, V. S. (2014). Activation pathway of Src kinase reveals intermediate states as targets for drug design. *Nature Communications, 5*, 3397. http://dx.doi.org/10.1038/ncomms4397.

Skjevik, A. A., Madej, B. D., Walker, R. C., & Teigen, K. (2012). LIPID11: A modular framework for lipid simulations using amber. *Journal of Physical Chemistry B, 116*(36), 11124–11136. http://dx.doi.org/10.1021/jp3059992.

Staritzbichler, R., Anselmi, C., Forrest, L. R., & Faraldo-Gómez, J. D. (2011). GRIFFIN: A versatile methodology for optimization of protein–lipid interfaces for membrane protein simulations. *Journal of Chemical Theory and Computation*, 7(4), 1167–1176. http://dx.doi.org/10.1021/ct100576m.

Toledo Warshaviak, D., Cascio, D., Khramtsov, V. V., & Hubbell, W. L. (2010). *Crystal structure of spin labeled T4 lysozyme mutant K65V1/R76V1*. PDB Data Bank.

Tombola, F., Pathak, M. M., & Isacoff, E. Y. (2006). How does voltage open an ion channel? *Annual Review of Cell and Developmental Biology, 22*, 23–52. http://dx.doi.org/10.1146/Annurev.Cellbio.21.020404.145837.

Vanommeslaeghe, K., Hatcher, E. R., Acharya, C., Kundu, S., Zhong, S., Shim, J., et al. (2010). CHARMM general force field: A force field for drug-like molecules compatible with the CHARMM all-atom additive biological force fields. *Journal of Computational Chemistry, 31*(4), 671–690. http://dx.doi.org/10.1002/jcc.21367.

Vanommeslaeghe, K., & MacKerell, A. D., Jr. (2012). Automation of the CHARMM General Force Field (CGenFF) I: Bond perception and atom typing. *Journal of Chemical Information and Modeling, 52*(12), 3144–3154. http://dx.doi.org/10.1021/ci300363c.

Vanommeslaeghe, K., Raman, E. P., & MacKerell, A. D., Jr. (2012). Automation of the CHARMM General Force Field (CGenFF) II: Assignment of bonded parameters and partial atomic charges. *Journal of Chemical Information and Modeling, 52*(12), 3155–3168. http://dx.doi.org/10.1021/ci3003649.

Wang, J., Wang, W., Kollman, P. A., & Case, D. A. (2006). Automatic atom type and bond type perception in molecular mechanical calculations. *Journal of Molecular Graphics & Modelling, 25*(2), 247–260. http://dx.doi.org/10.1016/j.jmgm.2005.12.005.

Wang, J., Wolf, R. M., Caldwell, J. W., Kollman, P. A., & Case, D. A. (2004). Development and testing of a general amber force field. *Journal of Computational Chemistry, 25*(9), 1157–1174. http://dx.doi.org/10.1002/jcc.20035.

Weiner, S. J., Kollman, P. A., Case, D. A., Singh, U. C., Ghio, C., Alagona, G., et al. (1984). A new force field for molecular mechanical simulation of nucleic acids and proteins. *Journal of the American Chemical Society, 106*(3), 765–784. http://dx.doi.org/10.1021/ja00315a051.

Yesselman, J. D., Price, D. J., Knight, J. L., & Brooks, C. L., III (2012). MATCH: An atom-typing toolset for molecular mechanics force fields. *Journal of Computational Chemistry, 33*(2), 189–202. http://dx.doi.org/10.1002/jcc.21963.

Zoete, V., Cuendet, M. A., Grosdidier, A., & Michielin, O. (2011). SwissParam: A fast force field generation tool for small organic molecules. *Journal of Computational Chemistry, 32*(11), 2359–2368. http://dx.doi.org/10.1002/jcc.21816.

CHAPTER NINE

High-Resolution Modeling of Protein Structures Based on Flexible Fitting of Low-Resolution Structural Data

Wenjun Zheng*,[1], Mustafa Tekpinar†
*Department of Physics, University at Buffalo, Buffalo, New York, USA
†Department of Physics, Yuzuncu Yil University, Kampus, Turkey
[1]Corresponding author: e-mail address: wjzheng@buffalo.edu

Contents

Abstract

To circumvent the difficulty of directly solving high-resolution biomolecular structures, low-resolution structural data from Cryo-electron microscopy (EM) and small angle solution X-ray scattering (SAXS) are increasingly used to explore multiple conformational states of biomolecular assemblies. One promising venue to obtain high-resolution structural models from low-resolution data is via data-constrained flexible fitting. To this end, we have developed a new method based on a coarse-grained C_α-only protein representation, and a modified form of the elastic network model (ENM) that allows large-scale conformational changes while maintaining the integrity of local structures including pseudo-bonds and secondary structures. Our method minimizes a pseudo-energy which linearly combines various terms of the modified ENM energy with an EM/SAXS-fitting score and a collision energy that penalizes steric collisions. Unlike some previous flexible fitting efforts using the lowest few normal modes, our method effectively utilizes all normal modes so that both global and local structural changes can be fully modeled

Advances in Protein Chemistry and Structural Biology, Volume 96
ISSN 1876-1623
http://dx.doi.org/10.1016/bs.apcsb.2014.06.004

with accuracy. This method is also highly efficient in computing time. We have demonstrated our method using adenylate kinase as a test case which undergoes a large open-to-close conformational change. The EM-fitting method is available at a web server (http://enm.lobos.nih.gov), and the SAXS-fitting method is available as a precompiled executable upon request.

1. INTRODUCTION

The biological functions of many biomolecules involve multiple conformational states and dynamic transitions in between. Despite rapid progress in structural biology, it remains highly difficult to directly solve all conformational states of biomolecules by high-resolution structural determination protocols such as X-ray crystallography and nuclear magnetic resonance spectroscopy. As attractive alternatives, low-resolution structure-determining techniques, including cryo-electron microscopy (cryo-EM) and small angle solution X-ray scattering (SAXS) are widely used. Cryo-EM constructs three-dimensional electron density maps for large biomolecular complexes at near/subnanometer resolutions based on a large number of two-dimensional images (Chiu, Baker, Jiang, Dougherty, & Schmid, 2005; Saibil, 2000). SAXS measures orientationally averaged X-ray scattering intensity for biomolecules in solution which contains information about the size and shape of the biomolecules (Koch, Vachette, & Svergun, 2003; Mertens & Svergun, 2010; Putnam, Hammel, Hura, & Tainer, 2007). These low-resolution techniques alone cannot generate unique high-resolution structural models with atomistic details. However, they offer highly informative constraints for generating and selecting high-resolution structural models using computational methods (Esquivel-Rodriguez & Kihara, 2013; Fabiola & Chapman, 2005; Lindert, Stewart, & Meiler, 2009). One viable venue of utilizing such constraints is "flexible fitting" (Flores, 2014; Lopez-Blanco & Chacon, 2013; Pandurangan, Shakeel, Butcher, & Topf, 2014; Wriggers & Birmanns, 2001), which flexibly deforms an initial protein structure to fit the given low-resolution data from cryo-EM or SAXS. An alternative venue is "rigid-body fitting" (Arai, Wriggers, Nishikawa, Nagamune, & Fujisawa, 2004; Bernado, Mylonas, Petoukhov, Blackledge, & Svergun, 2007; Bernado et al., 2009; Bernado, Perez, Svergun, & Pons, 2008; Rawat et al., 2003; Shiozawa, Konarev, Neufeld, Wilmanns, & Svergun, 2009; Volkmann et al., 2000; Wendt, Taylor, Trybus, & Taylor, 2001) which divides a protein complex into multiple domains and then fits them separately as rigid bodies. The rigid-body fitting methods depend on a subjective and error-prone partition of a biomolecule into rigid domains and ignore coupled motions between domains which may be functionally important. The success of the flexible fitting methods requires careful validation of the fitted models to avoid overfitting of the low-resolution data (Falkner & Schroder, 2013; Vashisth, Skiniotis, & Brooks, 2013). This is particularly important for SAXS data which contains much less structural information than cryo-EM maps and there is additional contribution from hydration shell near the surface of biomolecules.

The flexibility of biomolecules can be simulated by various computational methods at different levels of details. Molecular dynamics (MD) simulation is, in principle, able to describe the dynamics of biomolecules under physiological conditions (i.e., in the presence of water and ions) with atomic details, which makes it a method of choice for flexible fitting (Li & Frank, 2007). Recently, several MD-based methods have been introduced for cryo-EM fitting with full flexibility. The common strategy of these methods is to bias the MD simulation toward a conformation that optimally fits the cryo-EM data by using a biasing potential function (Caulfield & Harvey, 2007; Chan, Trabuco, Schreiner, & Schulten, 2012; Noda et al., 2006; Orzechowski & Tama, 2008; Trabuco, Villa, Mitra, Frank, & Schulten, 2008; Trabuco, Villa, Schreiner, Harrison, & Schulten, 2009; Wu, Subramaniam, Case, Wu, & Brooks, 2013). The application of these methods, however, has been limited by the high-computational cost of running MD simulations for large biomolecular systems.

As an efficient alternative, the flexibility of biomolecules can be analyzed by coarse-grained models where a group of atoms are represented by a coarse-grained bead (Tozzini, 2005). For example, the elastic network model (ENM) (Atilgan et al., 2001; Hinsen, 1998; Tama & Sanejouand, 2001) represents a protein structure as a network of C_α atoms with neighboring ones connected by springs with a uniform force constant (Tirion, 1996). The ENM has been successfully used to assist the fitting of cryo-EM and X-ray data (Schroder, Brunger, & Levitt, 2007; Schroder, Levitt, & Brunger, 2010; Tan, Devkota, & Harvey, 2008). In particular, ENM-based normal mode analysis (NMA) has been widely utilized to flexibly fit high-resolution structures to low-resolution structural data (Delarue & Dumas, 2004; Falke, Tama, Brooks, Gogol, & Fisher, 2005; Gorba, Miyashita, & Tama, 2008; Hinsen, Reuter, Navaza, Stokes, & Lacapere, 2005; Mitra et al., 2005; Suhre, Navaza, & Sanejouand, 2006;

Tama, Miyashita, & Brooks, 2004a, 2004b; Tama, Ren, Brooks, & Mitra, 2006), or satisfy a few pair wise distance constraints (Zheng & Brooks, 2005, 2006). Despite great success, the ENM/NMA-based flexible fitting methods are limited in accuracy because they usually only use a few low-frequency normal modes solved from ENM, which are less accurate for describing small local conformational changes (like rearrangement of helices inside a densely packed region) than large global ones (like domain motions) (Tama & Sanejouand, 2001). Indeed, it was found that many (>20) normal modes are needed to accurately describe several observed conformational changes in proteins (Petrone & Pande, 2006).

Recently, to achieve both accuracy and efficiency in the flexible fitting of cryo-EM (Zheng, 2011) and SAXS (Zheng & Tekpinar, 2011) data, we have developed a coarse-grained method based on a modified form of the ENM (named mENM) that combines the harmonic interactions for maintaining pseudo-bonds and secondary structures, and the anharmonic interactions between nonbonded residues to allow them to move apart readily. As a result, mENM allows large global structural changes without distorting local structures. Our method is based on minimization of a pseudo-energy which linearly combines various terms of the mENM energy with an EM/SAXS-fitting score and a collision energy that penalizes steric collisions (see Section 2). Our minimization-based implementation has two advantages: (a) unlike some previous flexible fitting efforts using the lowest few normal modes (Tama et al., 2004a, 2004b), our method effectively utilizes all normal modes so that both global and local structural changes can be fully modeled with accuracy. (b) It is efficiently implemented using the Newton–Raphson algorithm based on a sparse linear-equation solver (Chen, Davis, Hager, & Rajamanickam, 2008) which is significantly faster than NMA.

In this chapter, we will describe the methodological details of our flexible fitting method (see Section 2), and demonstrate its usage by applying it to the test case of adenylate kinase using simulated EM/SAXS data (see Section 3). Please refer to our papers (Zheng, 2011; Zheng & Tekpinar, 2011) for more test cases with both simulated and experimental data.

2. METHODS

2.1. Modified elastic network model (mENM)

A C_α-only ENM is constructed from the atomic coordinates of a protein structure (available from the Protein Data Bank). Each residue is represented by a bead located at the C_α atom. The original form of the ENM potential energy (Tirion, 1996) is

$$E_{\mathrm{ENM}} = \frac{1}{2}\sum_{i<j} C_{ij}\theta\left(R_c - d_{ij,0}\right)\left(d_{ij} - d_{ij,0}\right)^2, \quad (1)$$

where d_{ij} is the distance between bead i and j, and $d_{ij,0}$ is the value of d_{ij} given by the native structure, $\theta(x)$ is the Heaviside function, R_c is the cutoff distance chosen to be 10 Å following our previous study (Zheng, 2011), and C_{ij} is the force constant of the spring between bead i and j. C_{ij} can be set to a uniform constant for all residue pairs (Atilgan et al., 2001), or two different values for bonded and nonbonded residues (Tekpinar & Zheng, 2010), or varies as a function of distance $d_{ij,0}$ (Yang, Song, & Jernigan, 2009). The unit of C_{ij} can be arbitrarily chosen without changing the modeling results.

To allow nonbonded beads to move apart readily while maintaining pseudo-bonds and secondary structures, we have adopted the following modified form of the ENM potential energy:

$$\begin{aligned} E_{\mathrm{mENM}} &= E_{\mathrm{b}} + E_{\mathrm{SS}} + E_{\mathrm{nb}} \\ E_{\mathrm{b}} &= \frac{1}{2}\sum_{<ij>\in P_{\mathrm{b}}} C_{\mathrm{b}}\left(d_{ij} - d_{ij,0}\right)^2 \\ E_{\mathrm{SS}} &= \frac{1}{2}\sum_{<ij>\in P_{\mathrm{SS}}} C_{\mathrm{SS}}\left(d_{ij} - d_{ij,0}\right)^2 \\ E_{\mathrm{nb}} &= \frac{1}{2}\sum_{\substack{<ij>\notin P_{\mathrm{b}} \\ \& <ij>\notin P_{\mathrm{SS}}}} C_{\mathrm{nb}}\theta\left(R_c - d_{ij,0}\right) \times \frac{d_{ij,0}^2}{e^2}\left(1 - \frac{d_{ij,0}^e}{d_{ij}^e}\right)^2 \end{aligned} \quad (2)$$

where E_{b} is the pseudo-bonded energy (P_{b} is the set of pseudo-bonded bead pairs, the bonded force constant $C_{\mathrm{b}}=10$); E_{SS} is the nonbonded energy which maintains the local structure of α-helices and β-strands (P_{SS} is the set of C_α atom pairs which are either in a helix with a sequential offset ≤4, or in a β-strand with a sequential offset ≤3, the associated force constant $C_{\mathrm{SS}}=1$); E_{nb} is the remaining nonbonded energy with a new parameter $e=6$, corresponding to the Lennard–Jones potential—it has a minimum at $d_{ij,0}$, saturates as d_{ij} goes to infinity, and diverges as d_{ij} approaches zero (the nonbonded force constant $C_{\mathrm{nb}}=1$). Therefore, unlike the harmonic potential in Eq. (1), the E_{nb} terms in the mENM energy allow two nonbonded beads to move apart at a finite energy cost.

The mENM energy in Eq. (2) can be expanded near a given conformation X_* to the second order as follows:

$$E_{\mathrm{mENM}}(X) \approx E_{\mathrm{mENM}}(X_*) + \delta X^{\mathrm{T}} G + \frac{1}{2}\delta X^{\mathrm{T}} H \delta X, \tag{3}$$

where $\delta X = X - X_*$, $G = \nabla E_{\mathrm{mENM}}|_{X=X_*}$ is the gradient of E_{mENM} at $X = X_*$, and H is the $3N \times 3N$ Hessian matrix comprised of the following 3×3 blocks:

$$H_{ij} = \begin{bmatrix} \left.\frac{\partial^2 E_{\mathrm{mENM}}}{\partial x_i \partial x_j}\right|_{X=X_*} & \left.\frac{\partial^2 E_{\mathrm{mENM}}}{\partial x_i \partial y_j}\right|_{X=X_*} & \left.\frac{\partial^2 E_{\mathrm{mENM}}}{\partial x_i \partial z_j}\right|_{X=X_*} \\ \left.\frac{\partial^2 E_{\mathrm{mENM}}}{\partial y_i \partial x_j}\right|_{X=X_*} & \left.\frac{\partial^2 E_{\mathrm{mENM}}}{\partial y_i \partial y_j}\right|_{X=X_*} & \left.\frac{\partial^2 E_{\mathrm{mENM}}}{\partial y_i \partial z_j}\right|_{X=X_*} \\ \left.\frac{\partial^2 E_{\mathrm{mENM}}}{\partial z_i \partial x_j}\right|_{X=X_*} & \left.\frac{\partial^2 E_{\mathrm{mENM}}}{\partial z_i \partial y_j}\right|_{X=X_*} & \left.\frac{\partial^2 E_{\mathrm{mENM}}}{\partial z_i \partial z_j}\right|_{X=X_*} \end{bmatrix}, \tag{4}$$

where x_i, y_i, z_i (x_j, y_j, z_j) is the x, y, and z component of the coordinates of bead i (j). The calculated gradient and Hessian matrix will be used in the flexible fitting protocol based on Newton–Raphson algorithm (see Eq. 13).

2.2. Calculation of cryo-EM map from a protein structure

Given a set of atomic coordinates, we calculate simulated cryo-EM maps by locating three-dimensional Gaussian functions on each atom (or each coarse-grained bead) and integrating these functions for each voxel to get the following density function (Tama et al., 2004a, 2004b):

$$\rho(i,j,k) = \frac{1}{\left(\sqrt{2\pi/3}\sigma u\right)^3} \sum_{n=1}^{N} \int_{V_{ijk}} \exp\left\{-\frac{3}{2\sigma^2}\left[(x-x_n)^2 + (y-y_n)^2 + (z-z_n)^2\right]\right\} \mathrm{d}x\mathrm{d}y\mathrm{d}z, \tag{5}$$

where σ is half the map resolution (Wriggers & Birmanns, 2001), u is the grid spacing (the edge length of a cubic voxel), N is the number of atoms, $V_{ijk} = u^3$ denotes the volume of a cubic voxel centered at $(x_i, y_j, z_k) = (i, j, k) \times u$, (x_n, y_n, z_n) is the coordinate of atom n. We set the grid spacing $u = \sigma$, which allows significant reduction of computing time without losing modeling accuracy.

Equation (5) can be rewritten as follows:

$$\rho(i,j,k) = \sum_{n=1}^{N} f(x_i - x_n) f\left(y_j - y_n\right) f(z_k - z_n), \tag{6}$$

where $f(w) = \frac{1}{2u}\left[\operatorname{erf}\left(\frac{w+u/2}{\sqrt{2\sigma^2/3}}\right) - \operatorname{erf}\left(\frac{w-u/2}{\sqrt{2\sigma^2/3}}\right)\right]$ and $\operatorname{erf}(w) = \frac{2}{\sqrt{\pi}}\int_0^w e^{-t^2}\,dt$.

Simulated cryo-EM maps are generated at four resolutions (5, 10, 15, and 20 Å) and used as target maps in the flexible fitting. Among these resolution values, 10 Å represents a typical resolution for state-of-the-art cryo-EM maps, so it is used as the default resolution to demonstrate the performance of our method (see Section 3).

2.3. Calculation of SAXS profile for a protein surrounded by a hydration shell

SAXS measures the X-ray scattering intensity of proteins in solution as a function of $s = 4\pi \sin(\theta)/\lambda$, where s is the magnitude of scattering vector ($s \in [0, 0.5$ Å$^{-1}]$), 2θ is the scattering angle, and λ is the X-ray wavelength. The SAXS intensity profile $I(s)$ can be calculated from a protein structure using the atomic coordinates and atomic form factors, after including the contribution from a thin shell of water molecules (termed hydration shell; Svergun, Barberato, & Koch, 1995). To reduce computing cost, a coarse-grained one-bead-per-residue representation was used to define residue form factors for calculating $I(s)$ (Yang, Park, Makowski, & Roux, 2009). The hydration shell is constructed by submerging a protein in a pre-equilibrated water box and keeping those water molecules whose oxygen atom is at a minimal distance of 3.5–6.5 Å from the C_α atoms (Yang, Park, Makowski, & Roux, 2009).

To further reduce system size and computing cost, we have developed an implicit model of the hydration shell, which combines residue i and its nearby water glob (comprised of those water molecules within the hydration shell whose nearest residue is residue i) into a "composite glob." The coarse-grained form factor of a composite glob is calculated as follows (Zheng & Tekpinar, 2011):

$$F_i'(s) = \sqrt{F_i^2(s) + w^2 f_i^2(s) + 2wF_i(s) f_i(s)\frac{\sin(sD_i)}{sD_i}}, \tag{7}$$

where F_i is the coarse-grained form factor of residue i (Yang, Park, Makowski, & Roux, 2009), f_i is the coarse-grained form factor of the water glob near residue i (termed water glob i). D_i is the distance between the center of electron density distribution in residue i and the geometric center of water glob i. $w=\delta\rho_w/\rho_w$ is the relative contrast of hydration shell, where $\rho_w=0.334\ e/\text{Å}^3$ is the electron density of pure water, and $\delta\rho_w$ is the contrast of hydration shell which can be adjusted to optimize the fitting of experimental SAXS data (Svergun et al., 1995). w usually varies from 0% to 10%.

Based on the implicit model of hydration shell, the SAXS profile $I'(s)$ can be calculated using a coarse-grained representation of N composite globs (each composed of a residue and its nearby water glob) as follows (Zheng & Tekpinar, 2011):

$$I'(s)=\sum_{i=1}^{N}\sum_{j=1}^{N}F_i'(s)F_j'(s)\frac{\sin\left(sd'_{ij}\right)}{sd'_{ij}}, \tag{8}$$

where $F_i'(s)$ and $F_j'(s)$ are calculated in Eq. (7), and d'_{ij} is the distance between the centers of electron density distribution in composite glob i and j.

2.4. mENM-based flexible fitting

We start from two given inputs: the initial protein structure and the target data (cryo-EM map or SAXS profile simulated from the target structure). To flexibly fit the initial structure toward the target data, we minimize the following pseudo-energy:

$$E_{\text{total}}=\lambda E_{\text{nb}}+(1-\lambda)E_{\text{fitting}}+E_{\text{b}}+E_{\text{SS}}+E_{\text{col}}, \tag{9}$$

where the various terms are described as follows:

E_{nb} is the nonbonded energy based on the initial structure (see Eq. 2), E_{b} is the pseudo-bonded energy (see Eq. 2), E_{SS} is the nonbonded energy which maintains the secondary structures of α-helices and β-strands (see Eq. 2)

E_{col} is the collision energy between two nonbonded beads defined as follows (Tekpinar & Zheng, 2010):

$$E_{\text{col}}=\frac{1}{2}\sum_{\substack{<ij>\notin P_{\text{b}}\\ \&<ij>\notin P_{\text{SS}}}}C_{\text{col}}\theta\left(R_{\text{col}}-d_{ij}\right)\left(d_{ij}-R_{\text{col}}\right)^2, \tag{10}$$

where the collision force constant $C_{col} = 10$, R_{col} is the minimal distance between nonbonded beads in the initial structure (those bead pairs considered in E_b and E_{SS} are excluded from the summation). The addition of E_{col} penalizes steric collisions between residues whose C_α atoms are within a distance of R_{col}.

$E_{fitting}$ is either the EM-fitting score E_{map} or the SAXS-fitting score E_{SAXS} defined as follows:

$$E_{map} = 100 \times \frac{\sum_{i,j,k} [\rho_m(i,j,k) - \rho_t(i,j,k)]^2}{\sum_{i,j,k} [\rho_t(i,j,k)]^2}, \tag{11}$$

where ρ_m is the model density function calculated from the C_α coordinates of a coarse-grained model (using Eq. 6), ρ_t is the target density function, and (i, j, k) is the voxel index. The denominator in Eq. (11) normalizes E_{map} to a value that only weakly depends on protein size and EM map size.

$$E_{SAXS} = f_{SAXS} \min_c \left\{ \sum_l [cI_m(s_l) - I_t(s_l)]^2 \right\}, \tag{12}$$

where the constant pre-factor is $f_{SAXS} = 3 \times 10^7$ (a large f_{SAXS} is chosen to enable fast fitting), s_l is the scattering vector ranging from 0 to 0.5 Å^{-1}, I_m is the model SAXS profile calculated using Eq. (8), I_t is the target SAXS profile measured experimentally or simulated by CRYSOL (Svergun et al., 1995). Prior to the flexible fitting, I_t is rescaled so that $I_t(0) = 1$. During the flexible fitting, I_m is rescaled by a factor c to minimize the SAXS-fitting score in Eq. (12).

In Eq. (9), the weight parameter $\lambda \in [0, 1]$ controls the degree of data fitting: at $\lambda \sim 1$, the conformational search is restricted near the initial structure with weak fitting to the target data; at $\lambda \sim 0$, the influence of initial structure is very weak with strong fitting to the target data. To achieve optimal fitting of the target data without compromising structural integrity, we minimize the pseudo-energy progressively as λ decreases gradually from 1 to 0 until a termination criterion is met (see below). We employ the Newton–Raphson algorithm to solve $\nabla E_{total}(\lambda, X_{min}) = 0$ as described in the following iterative procedure:

1. *Initialization*: set $n = 0$, $\lambda_0 = 0.5$, and $X_0 = X_{min,0} = X_i$, where X_i represents the bead coordinates of the initial structure.
2. If $n > 0$, decrease λ_n to double the ratio $(1 - \lambda_n)/\lambda_n$.

3. For conformation X_n, calculate the pseudo-energy E_n using Eq. (9), then set $X_{\min,n}=X_n$ if E_n reaches a new low.
4. If E_n fails to be lowered after 20 iterations, stop minimization and go to Step 7.
5. Displace X_n by the following incremental displacement:

$$\begin{aligned}\delta X_n = &-\left(\lambda_n H_{\text{nb}} + (1-\lambda_n) H_{\text{fitting}} + H_{\text{b}} + H_{\text{SS}} + H_{\text{col}}\right)^{-1} \\ &\left(\lambda_n \nabla E_{\text{nb}} + (1-\lambda_n)\nabla E_{\text{fitting}} + \nabla E_{\text{b}} + \nabla E_{\text{SS}} + \nabla E_{\text{col}}\right),\end{aligned} \quad (13)$$

where H_{nb}, H_{fitting}, H_{b}, H_{SS}, and H_{col} are the Hessian matrices calculated from E_{nb}, E_{fitting}, E_{b}, E_{SS}, and E_{col}, respectively.

6. Go to Step 3.
7. Stop if the root mean squared deviation (RMSD) to the initial structure saturates (i.e., it increases by <5% from the previous minimization round).
8. Set $n \leftarrow n+1$ and $X_n = X_{\min,n} = X_{\min,n-1}$, then go to Step 2.

To reduce accumulation of structural distortions, we limit the magnitude of each incremental displacement (see Eq. 13) ≤ 0.2 Å in RMSD. This is attained by adding εI (where I is an identity matrix and ε is an adjustable parameter) to the sum of Hessian matrices such that the linear-equation solution in Eq. (13) satisfies this condition.

3. RESULTS

We demonstrate our flexible fitting method using a test case of adenylate kinase, which consists of two high-resolution X-ray structures corresponding to a closed and open form (PDB code: 1ake and 4ake) at the two ends of a large conformational transition (7.1 Å in RMSD). One of these structures (target structure) is used to simulate the target cryo-EM map or SAXS profile for fitting, the other one is used as the starting model for flexible fitting. The target SAXS profile is simulated using CRYSOL, which is a widely used program that gives fast and accurate simulation of SAXS profiles from known protein structures by taking into account contributions from both the protein and its surrounding hydration shell (Svergun et al., 1995). Please refer to our papers (Zheng, 2011; Zheng & Tekpinar, 2011) for more test cases with both simulated and experimental data.

To illustrate flexible fitting of cryo-EM data for adenylate kinase, we start from the closed form of adenylate kinase and iteratively deform the initial structure to minimize a pseudo-energy that combines several energy terms that maintain the initial tertiary structure, pseudo-bonds, secondary structures, and penalize steric collisions, together with a EM-fitting score that favors fitting of the model into a target cryo-EM map simulated from the open form of adenylate kinase with 10 Å resolution (see Section 2). During the flexible fitting process, the RMSD relative to the target structure (termed $RMSD_t$) decreases gradually from 7.1 to 0.74 Å as a result of improving fitting to the target map (see Fig. 1A). The final model compares very well with the target structure without local structural distortion (see Fig. 1B). For comparison, the same test case was run using a normal mode flexible fitting program (Tama et al., 2004a, 2004b; fitted using the lowest 26 modes solved from a C_α-only ENM) and a Go-model-based biased MD method (Grubisic, Shokhirev, Orzechowski, Miyashita, & Tama, 2010), which yielded models with $RMSD_t$ of 3.4 and 1.8 Å, respectively. To assess the dependence of flexible fitting results on the map resolution, we have also performed flexible fitting with lower-resolution maps, and got models with $RMSD_t$ of 0.86 and 0.99 Å for resolution 15 and 20 Å, respectively. Therefore, lower map resolution only leads to slight degradation of model quality. Next, we evaluate the effect of random noise (mimicking the experimental noise present in real cryo-EM maps) on flexible fitting results. To this end, Gaussian noise with a given signal–noise

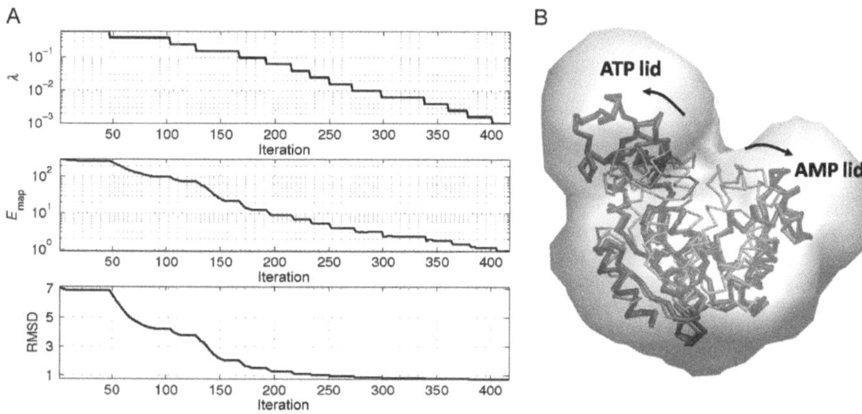

Figure 1 The result of flexible fitting of a closed form structure of adenylate kinase to a 10-Å resolution cryo-EM map simulated from an open form structure of adenylate kinase: (A) Weight parameter λ, EM-fitting score E_{map}, and RMSD (relative to the target structure) as a function of iteration step are shown. (B) The initial structure (green (light gray in the print version) trace), target structure (blue (black in the print version) trace), and final model (red (dark gray in the print version) trace) are shown, and the target cryo-EM map is shown as solid surface. Domain motions from the initial structure to the target structure are indicated by arrows.

ratio (SNR) is added to the simulated maps (see Jolley, Wells, Fromme, & Thorpe, 2008). We have explored SNR of 1, 2, and 5, respectively. As expected, the flexible fitting accuracy worsens as the noise level increases—for adenylate kinase (with map resolution of 10 Å), the final $RMSD_t$ increases to 1.3, 1.7, and 2.3 Å, as SNR decreases from infinity to 5, 2, and 1.

To illustrate flexible fitting of SAXS data for adenylate kinase, we start from the open form of adenylate kinase and iteratively deform the initial structure to minimize a pseudo-energy that combines the mENM energy, a term that penalizes steric collisions, and a SAXS-fitting score favoring fitting of the model's SAXS profile with a target SAXS profile simulated from the closed form of adenylate kinase (see Section 2). During the flexible fitting, the influence of the initial tertiary structure is gradually weakened while the fitting to SAXS data is gradually strengthened (see Section 2). To avoid overfitting, the flexible fitting process is terminated when the RMSD between the fitted model and the initial structure (denoted as $RMSD_i$) has saturated (i.e., $RMSD_i$ does not increase by >5% from previous minimization). As a result of improving fitting to the target SAXS data, $RMSD_t$ decreases from 7.1 to 2.0 Å (see Fig. 2). Notably, to achieve the above result, we have incorporated an implicit hydration shell (see Eq. 7) with a relative contrast $w = \delta\rho_w/\rho_w$ matching the corresponding value (3%) used to simulate the target SAXS profile by CRYSOL. Indeed, without

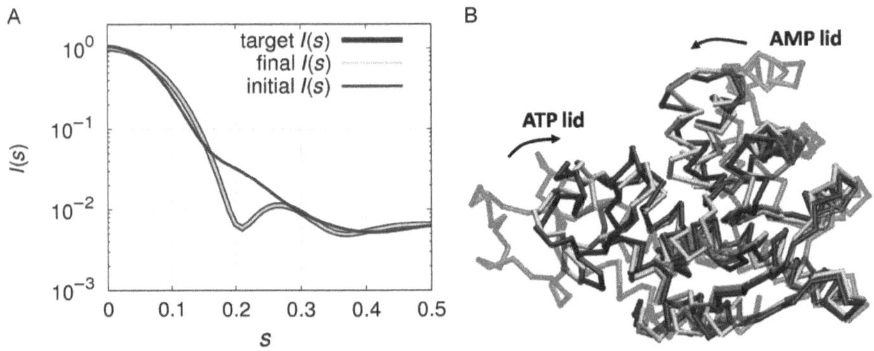

Figure 2 The result of flexible fitting of an open form structure of adenylate kinase to the SAXS profile simulated from a closed form structure of adenylate kinase: (A) The target SAXS profile is colored red (thick black in the print version), the calculated SAXS profiles at the beginning and end of flexible fitting are colored blue (dark gray in the print version) and green (light gray in the print version), respectively. The magnitude of scattering vector s is in unit of $Å^{-1}$. $I(s)$ is rescaled so that $I(0) = 1$. (B) The initial structure (blue (dark gray in the print version) trace), target structure (red (black in the print version) trace), and final model (green (light gray in the print version) trace) are shown. Domain motions from the initial structures to the target structure are indicated by arrows.

including the contribution of hydration shell to SAXS data (i.e., $w=0$), we have got higher $RMSD_t=2.9$ Å. We have also tried flexible fitting with other values of w between 0% and 10%. The result worsens as w deviates further from 3% (see Zheng & Tekpinar, 2011). Therefore, it is important to accurately account for the hydration shell when flexibly fitting SAXS data. However, w could vary from protein to protein, and it is unknown for experimental SAXS data. How to perform flexible fitting of SAXS data without knowing w? To meet this challenge, we have repeated the flexible fitting for a range of w from 0% to 10% (with increment of 1%), and collected 11 final models together with their final SAXS-fitting scores (see Zheng & Tekpinar, 2011). Because a mismatch in w would cause poorer fitting of the target SAXS data, one can select the model with the lowest SAXS-fitting score, which should most likely correspond to a roughly correct w value.

4. DISCUSSION

We have presented a new computational method to flexibly fit a given protein structure to a cryo-EM map or a SAXS profile using a coarse-grained C_α-only model. Our goal is to build a high-resolution structural model compatible with the given low-resolution data while maintaining its local structural integrity. Our method uses a modified form of ENM that allows large-scale conformational changes while maintaining pseudo-bonds and secondary structures and avoiding residue collisions. Our method effectively utilizes all normal modes so that both global and local structural changes can be adequately modeled. Our method is also highly efficient with computing time of several minutes on an Intel-Xeon-based workstation.

To fit cryo-EM data, our flexible fitting method should be combined with and preceded by the use of a rigid-body fitting program like SITUS (Wriggers, Milligan, & McCammon, 1999) which is to find an approximately correct orientation of an initial structure relative to a target map. In our test, the best model produced by the qdock command of SITUS gives a good initial model for subsequent flexible fitting. In practice, flexible fitting should be done starting from several initial models, and convergence toward a common conformation should be verified to ensure the reliability of fitted models. As shown by the test case of adenylate kinase (see Section 3), our method is highly robust for fitting cryo-EM maps with various resolutions (5–20 Å) in the presence of moderate random noise (with $SNR \geq 2$).

To fit SAXS data, it is important to properly account for the contribution of hydration shell to SAXS data. To this end, we have modeled the hydration shell implicitly by combining each residue and its nearby "water glob" into a composite glob, which allows accurate and efficient calculation of SAXS profiles and convenient modeling of changing shape of hydration shell as a protein undergoes large conformational changes. The quality of SAXS-based flexible fitting ultimately depends on the level of degeneracy of SAXS data (i.e., the abundance of alternative conformations with low SAXS-fitting score and their accessibility from the initial structure). The higher the degeneracy is, the harder it is to construct the correct model from SAXS data. To improve the sampling of conformations compatible with the given SAXS profile, one can start flexible fitting from multiple initial structures (for example, different X-ray or NMR structures or snapshots from extensive MD simulations). The selection of correct model from the ensemble of fitted models may require additional information and processing. Given the limitation of SAXS data, it is highly desirable to integrate it with other structural information and modeling techniques to achieve optimal performance (Alber, Forster, Korkin, Topf, & Sali, 2008).

In principle, the flexible fitting of a high-resolution structure to low-resolution data is prone to overfitting, because many degrees of freedom involved in fitting are generally insufficiently constrained by the given data. In our method, we take the following measures to control overfitting: (1) The total number of degrees of freedom is reduced using a coarse-grained model and maintaining the pseudo-bonds and the secondary structures. (2) The flexible fitting is terminated when the RMSD relative to the initial structure starts to saturate. We also visually inspect the fitted models to ensure there is no serious structural distortion indicative of overfitting.

Our flexible fitting method complements alternative methods that optimize the fitting of structural data (SAXS or cryo-EM) using Monte Carlo sampling (Forster et al., 2008; Petoukhov & Svergun, 2005), MD simulation (Grubisic et al., 2010; Orzechowski & Tama, 2008; Trabuco et al., 2008, 2009), and coarse-grained simulation (Yang, Blachowicz, Makowski, & Roux, 2010). Our method is based on the minimization of a pseudo-energy which is computationally fast but susceptible to trapping at local minima, while the alternative methods like Monte Carlo and MD are computationally more expensive but capable of more extensive sampling.

Despite its success, our method has its limitations—it is most suited for proteins undergoing en-block motions of domains or secondary structural elements. A more detailed simulation using MD and all-atom force field (Trabuco et al., 2008) will be needed to model local changes involving folding/unfolding of secondary structures or restructuring of surface loops or sidechains.

Beyond the flexible fitting of cryo-EM and SAXS data, one can use mENM to fit other structural data such as the distances between given pairs of residues as measured by NMR or Förster resonance energy transfer experiments. Such distances-based flexible fitting allows the construction of new structural models satisfying given distance constraints (Zheng & Brooks, 2005, 2006), and it can be adapted for coarse-grained simulation of constant-speed pulling of a protein (Zheng, 2014). The idea of linearly interpolating between two distinct energy/score functions (see Eq. 9) was also found useful in modeling a transition pathway that links two given protein structures (Tekpinar & Zheng, 2010). Interested readers are referred to our web server at http://enm.lobos.nih.gov.

ACKNOWLEDGMENT

We thank the funding support from NSF (grant #0952736).

REFERENCES

Alber, F., Forster, F., Korkin, D., Topf, M., & Sali, A. (2008). Integrating diverse data for structure determination of macromolecular assemblies. *Annual Review of Biochemistry*, *77*, 443–477.

Arai, R., Wriggers, W., Nishikawa, Y., Nagamune, T., & Fujisawa, T. (2004). Conformations of variably linked chimeric proteins evaluated by synchrotron X-ray small-angle scattering. *Proteins—Structure Function and Bioinformatics*, *57*(4), 829–838.

Atilgan, A. R., Durell, S. R., Jernigan, R. L., Demirel, M. C., Keskin, O., & Bahar, I. (2001). Anisotropy of fluctuation dynamics of proteins with an elastic network model. *Biophysical Journal*, *80*(1), 505–515.

Bernado, P., Mylonas, E., Petoukhov, M. V., Blackledge, M., & Svergun, D. I. (2007). Structural characterization of flexible proteins using small-angle X-ray scattering. *Journal of the American Chemical Society*, *129*(17), 5656–5664.

Bernado, P., Perez, Y., Blobel, J., Fernandez-Recio, J., Svergun, D. I., & Pons, M. (2009). Structural characterization of unphosphorylated STAT5a oligomerization equilibrium in solution by small-angle X-ray scattering. *Protein Sciences*, *18*(4), 716–726.

Bernado, P., Perez, Y., Svergun, D. I., & Pons, M. (2008). Structural characterization of the active and inactive states of Src kinase in solution by small-angle X-ray scattering. *Journal of Molecular Biology*, *376*(2), 492–505.

Caulfield, T. R., & Harvey, S. C. (2007). Conformational fitting of atomic models to cryogenic-electron microscopy maps using Maxwell's demon molecular dynamics. *Biophysical Journal, Biophysical Society Meeting Abstracts*, *368a*.

Chan, K. Y., Trabuco, L. G., Schreiner, E., & Schulten, K. (2012). Cryo-electron microscopy modeling by the molecular dynamics flexible fitting method. *Biopolymers*, *97*(9), 678–686.

Chen, Y., Davis, T. A., Hager, W. W., & Rajamanickam, S. (2008). Algorithm 887: CHOLMOD, supernodal sparse cholesky factorization and update/downdate. *ACM Transactions on Mathematical Software*, *35*(3), 1–14.

Chiu, W., Baker, M. L., Jiang, W., Dougherty, M., & Schmid, M. F. (2005). Electron cryomicroscopy of biological machines at subnanometer resolution. *Structure*, *13*(3), 363–372.

Delarue, M., & Dumas, P. (2004). On the use of low-frequency normal modes to enforce collective movements in refining macromolecular structural models. *Proceedings of the National Academy of Sciences of the United States of America, 101*(18), 6957–6962.

Esquivel-Rodriguez, J., & Kihara, D. (2013). Computational methods for constructing protein structure models from 3D electron microscopy maps. *Journal of Structural Biology, 184*(1), 93–102.

Fabiola, F., & Chapman, M. S. (2005). Fitting of high-resolution structures into electron microscopy reconstruction images. *Structure, 13*(3), 389–400.

Falke, S., Tama, F., Brooks, C. L., 3rd., Gogol, E. P., & Fisher, M. T. (2005). The 13 angstroms structure of a chaperonin GroEL-protein substrate complex by cryo-electron microscopy. *Journal of Molecular Biology, 348*(1), 219–230.

Falkner, B., & Schroder, G. F. (2013). Cross-validation in cryo-EM-based structural modeling. *Proceedings of the National Academy of Sciences of the United States of America, 110*(22), 8930–8935.

Flores, S. C. (2014). Fast fitting to low resolution density maps: Elucidating large-scale motions of the ribosome. *Nucleic Acids Research, 42*(2), e9.

Forster, F., Webb, B., Krukenberg, K. A., Tsuruta, H., Agard, D. A., & Sali, A. (2008). Integration of small-angle X-ray scattering data into structural modeling of proteins and their assemblies. *Journal of Molecular Biology, 382*(4), 1089–1106.

Gorba, C., Miyashita, O., & Tama, F. (2008). Normal-mode flexible fitting of high-resolution structure of biological molecules toward one-dimensional low-resolution data. *Biophysical Journal, 94*(5), 1589–1599.

Grubisic, I., Shokhirev, M. N., Orzechowski, M., Miyashita, O., & Tama, F. (2010). Biased coarse-grained molecular dynamics simulation approach for flexible fitting of X-ray structure into cryo electron microscopy maps. *Journal of Structural Biology, 169*(1), 95–105.

Hinsen, K. (1998). Analysis of domain motions by approximate normal mode calculations. *Proteins, 33*(3), 417–429.

Hinsen, K., Reuter, N., Navaza, J., Stokes, D. L., & Lacapere, J. J. (2005). Normal mode-based fitting of atomic structure into electron density maps: Application to sarcoplasmic reticulum Ca-ATPase. *Biophysical Journal, 88*(2), 818–827.

Jolley, C. C., Wells, S. A., Fromme, P., & Thorpe, M. F. (2008). Fitting low-resolution cryo-EM maps of proteins using constrained geometric simulations. *Biophysical Journal, 94*(5), 1613–1621.

Koch, M. H., Vachette, P., & Svergun, D. I. (2003). Small-angle scattering: A view on the properties, structures and structural changes of biological macromolecules in solution. *Quarterly Reviews of Biophysics, 36*(2), 147–227.

Li, W., & Frank, J. (2007). Transfer RNA in the hybrid P/E state: Correlating molecular dynamics simulations with cryo-EM data. *Proceedings of the National Academy of Sciences of the United States of America, 104*(42), 16540–16545.

Lindert, S., Stewart, P. L., & Meiler, J. (2009). Hybrid approaches: Applying computational methods in cryo-electron microscopy. *Current Opinion in Structural Biology, 19*(2), 218–225.

Lopez-Blanco, J. R., & Chacon, P. (2013). iMODFIT: Efficient and robust flexible fitting based on vibrational analysis in internal coordinates. *Journal of Structural Biology, 184*(2), 261–270.

Mertens, H. D. T., & Svergun, D. I. (2010). Structural characterization of proteins and complexes using small-angle X-ray solution scattering. *Journal of Structural Biology, 172*(1), 128–141.

Mitra, K., Schaffitzel, C., Shaikh, T., Tama, F., Jenni, S., Brooks, C. L., 3rd., et al. (2005). Structure of the E. coli protein-conducting channel bound to a translating ribosome. *Nature, 438*(7066), 318–324.

Noda, K., Nakamura, M., Nishida, R., Yoneda, Y., Yamaguchi, Y., Tamura, Y., et al. (2006). Atomic model construction of protein complexes from electron micrographs and visualization of their 3D structure using a virtual reality system. *Journal of Plasma Physics*, *72*(6), 1037–1040.

Orzechowski, M., & Tama, F. (2008). Flexible fitting of high-resolution X-ray structures into cryoelectron microscopy maps using biased molecular dynamics simulations. *Biophysical Journal*, *95*(12), 5692–5705.

Pandurangan, A. P., Shakeel, S., Butcher, S. J., & Topf, M. (2014). Combined approaches to flexible fitting and assessment in virus capsids undergoing conformational change. *Journal of Structural Biology*, *185*(3), 427–439.

Petoukhov, M. V., & Svergun, D. I. (2005). Global rigid body modeling of macromolecular complexes against small-angle scattering data. *Biophysical Journal*, *89*(2), 1237–1250.

Petrone, P., & Pande, V. S. (2006). Can conformational change be described by only a few normal modes? *Biophysical Journal*, *90*(5), 1583–1593.

Putnam, C. D., Hammel, M., Hura, G. L., & Tainer, J. A. (2007). X-ray solution scattering (SAXS) combined with crystallography and computation: Defining accurate macromolecular structures, conformations and assemblies in solution. *Quarterly Reviews of Biophysics*, *40*(3), 191–285.

Rawat, U. B. S., Zavialov, A. V., Sengupta, J., Valle, M., Grassucci, R. A., Linde, J., et al. (2003). A cryo-electron microscopic study of ribosome-bound termination factor RF2. *Nature*, *421*(6918), 87–90.

Saibil, H. R. (2000). Conformational changes studied by cryo-electron microscopy. *Nature Structural Biology*, *7*(9), 711–714.

Schroder, G. F., Brunger, A. T., & Levitt, M. (2007). Combining efficient conformational sampling with a deformable elastic network model facilitates structure refinement at low resolution. *Structure*, *15*(12), 1630–1641.

Schroder, G. F., Levitt, M., & Brunger, A. T. (2010). Super-resolution biomolecular crystallography with low-resolution data. *Nature*, *464*(7292), 1218–1222.

Shiozawa, K., Konarev, P. V., Neufeld, C., Wilmanns, M., & Svergun, D. I. (2009). Solution structure of human Pex5.Pex14.PTS1 protein complexes obtained by small angle X-ray scattering. *Journal of Biological Chemistry*, *284*(37), 25334–25342.

Suhre, K., Navaza, J., & Sanejouand, Y. H. (2006). NORMA: A tool for flexible fitting of high-resolution protein structures into low-resolution electron-microscopy-derived density maps. *Acta Crystallographica. Section D, Biological Crystallography*, *62*(Pt 9), 1098–1100.

Svergun, D., Barberato, C., & Koch, M. H. J. (1995). CRYSOL—A program to evaluate x-ray solution scattering of biological macromolecules from atomic coordinates. *Journal of Applied Crystallography*, *28*, 768–773.

Tama, F., Miyashita, O., & Brooks, C. L., 3rd. (2004a). Flexible multi-scale fitting of atomic structures into low-resolution electron density maps with elastic network normal mode analysis. *Journal of Molecular Biology*, *337*(4), 985–999.

Tama, F., Miyashita, O., & Brooks, C. L., 3rd. (2004b). NMFF: Flexible high-resolution annotation of low-resolution experimental data from cryo-EM maps using normal mode analysis. *Journal of Structural Biology*, *147*, 315–326.

Tama, F., Ren, G., Brooks, C. L., 3rd, & Mitra, A. K. (2006). Model of the toxic complex of anthrax: Responsive conformational changes in both the lethal factor and the protective antigen heptamer. *Protein Sciences*, *15*(9), 2190–2200.

Tama, F., & Sanejouand, Y. H. (2001). Conformational change of proteins arising from normal mode calculations. *Protein Engineering*, *14*(1), 1–6.

Tan, R. K., Devkota, B., & Harvey, S. C. (2008). YUP.SCX: Coaxing atomic models into medium resolution electron density maps. *Journal of Structural Biology*, *163*(2), 163–174.

Tekpinar, M., & Zheng, W. (2010). Predicting order of conformational changes during protein conformational transitions using an interpolated elastic network model. *Proteins—Structure Function and Bioinformatics*, *78*(11), 2469–2481.

Tirion, M. M. (1996). Large amplitude elastic motions in proteins from a single-parameter, atomic analysis. *Physical Review Letters*, 77(9), 1905–1908.

Tozzini, V. (2005). Coarse-grained models for proteins. *Current Opinion in Structural Biology*, *15*(2), 144–150.

Trabuco, L. G., Villa, E., Mitra, K., Frank, J., & Schulten, K. (2008). Flexible fitting of atomic structures into electron microscopy maps using molecular dynamics. *Structure*, *16*(5), 673–683.

Trabuco, L. G., Villa, E., Schreiner, E., Harrison, C. B., & Schulten, K. (2009). Molecular dynamics flexible fitting: A practical guide to combine cryo-electron microscopy and X-ray crystallography. *Methods*, *49*(2), 174–180.

Vashisth, H., Skiniotis, G., & Brooks, C. L., 3rd. (2013). Enhanced sampling and overfitting analyses in structural refinement of nucleic acids into electron microscopy maps. *The Journal of Physical Chemistry. B*, *117*(14), 3738–3746.

Volkmann, N., Hanein, D., Ouyang, G., Trybus, K. M., DeRosier, D. J., & Lowey, S. (2000). Evidence for cleft closure in actomyosin upon ADP release. *Nature Structural Biology*, 7(12), 1147–1155.

Wendt, T., Taylor, D., Trybus, K. M., & Taylor, K. (2001). Three-dimensional image reconstruction of dephosphorylated smooth muscle heavy meromyosin reveals asymmetry in the interaction between myosin heads and placement of subfragment 2. *Proceedings of the National Academy of Sciences of the United States of America*, *98*(8), 4361–4366.

Wriggers, W., & Birmanns, S. (2001). Using situs for flexible and rigid-body fitting of multiresolution single-molecule data. *Journal of Structural Biology*, *133*(2–3), 193–202.

Wriggers, W., Milligan, R. A., & McCammon, J. A. (1999). Situs: A package for docking crystal structures into low-resolution maps from electron microscopy. *Journal of Structural Biology*, *125*(2–3), 185–195.

Wu, X., Subramaniam, S., Case, D. A., Wu, K. W., & Brooks, B. R. (2013). Targeted conformational search with map-restrained self-guided Langevin dynamics: Application to flexible fitting into electron microscopic density maps. *Journal of Structural Biology*, *183*(3), 429–440.

Yang, S. C., Blachowicz, L., Makowski, L., & Roux, B. (2010). Multidomain assembled states of Hck tyrosine kinase in solution. *Proceedings of the National Academy of Sciences of the United States of America*, *107*(36), 15757–15762.

Yang, S., Park, S., Makowski, L., & Roux, B. (2009). A rapid coarse residue-based computational method for X-ray solution scattering characterization of protein folds and multiple conformational states of large protein complexes. *Biophysical Journal*, *96*(11), 4449–4463.

Yang, L., Song, G., & Jernigan, R. L. (2009). Protein elastic network models and the ranges of cooperativity. *Proceedings of the National Academy of Sciences of the United States of America*, *106*(30), 12347–12352.

Zheng, W. (2011). Accurate flexible fitting of high-resolution protein structures into cryo-electron microscopy maps using coarse-grained pseudo-energy minimization. *Biophysical Journal*, *100*(2), 478–488.

Zheng, W. (2014). All-atom and coarse-grained simulations of the forced unfolding pathways of the SNARE complex. *Proteins*, *82*, 1376–1386.

Zheng, W., & Brooks, B. R. (2005). Normal-modes-based prediction of protein conformational changes guided by distance constraints. *Biophysical Journal*, *88*(5), 3109–3117.

Zheng, W., & Brooks, B. R. (2006). Modeling protein conformational changes by iterative fitting of distance constraints using reoriented normal modes. *Biophysical Journal*, *90*(12), 4327–4336.

Zheng, W., & Tekpinar, M. (2011). Accurate flexible fitting of high-resolution protein structures to small-angle X-ray scattering data using a coarse-grained model with implicit hydration shell. *Biophysical Journal*, *101*(12), 2981–2991.

AUTHOR INDEX

Note: Page numbers followed by "*f*" indicate figures, "*t*" indicate tables and "*np*" indicate footnotes.

C

D

E

H

I

L

M

N

O

P

Q

R

S

X

Y

Z

SUBJECT INDEX

Note: Page numbers followed by "*f*" indicate figures and "*t*" indicate tables.

D

E

F

G

H

I

Q

R

S

T

U

V

W

X

Y

Z

FSC
www.fsc.org
MIX
Papier | Fördert
gute Waldnutzung
FSC® C083411

Kolibri 360 GmbH
Ferdinand-Jühlke-Straße 7
99095 Erfurt, Deutschland
produktsicherheit@kolibri360.de